GRADUATE STUDIES IN MATHEMATICS 190

Lectures on Finite Fields

Xiang-dong Hou

AMERICAN MATHEMATICAL SOCIETY
Providence, Rhode Island

2010 *Mathematics Subject Classification.* Primary 11-01, 11Exx, 11Rxx, 11Txx.

For additional information and updates on this book, visit
www.ams.org/bookpages/gsm-190

Library of Congress Cataloging-in-Publication Data

Names: Hou, Xiang-dong, 1962- author.
Title: Lectures on finite fields / Xiang-dong Hou.
Description: Providence, Rhode Island : American Mathematical Society, [2018] | Series: Graduate studies in mathematics ; volume 190 | Includes bibliographical references and index.
Identifiers: LCCN 2017049952 | ISBN 9781470442897 (alk. paper)
Subjects: LCSH: Finite fields (Algebra) | AMS: Number theory – Instructional exposition (textbooks, tutorial papers, etc.). msc | Number theory – Forms and linear algebraic groups – Forms and linear algebraic groups. msc | Number theory – Algebraic number theory: global fields – Algebraic number theory: global fields. msc | Number theory – Finite fields and commutative rings (number-theoretic aspects) – Finite fields and commutative rings (number-theoretic aspects). msc
Classification: LCC QA247.3 .H68 2018 | DDC 512/.3–dc23
LC record available at https://lccn.loc.gov/2017049952

∞ The paper used in this book is acid-free and falls within the guidelines
established to ensure permanence and durability.
Visit the AMS home page at `http://www.ams.org/`

10 9 8 7 6 5 4 3 2 1 23 22 21 20 19 18

To Dong-lin, Wendy, and Elaine

Contents

Preface

This book is partially based on the lecture notes of several graduate courses that I taught at the University of South Florida since 2005. The first draft was written in 2006. The manuscript went through a thorough revision between 2015 and 2016 and finally evolved into the present form.

The subject of finite fields is at the intersection of algebra, combinatorics, and number theory, and is a source of widespread applications in information theory and computer science; as such, its boundary is not always easy to define. The following is a partial list of some areas that are traditionally considered important in the subject: (i) algebraic structures of and related to finite fields; (ii) number theory of finite fields and function fields over finite fields; (iii) finite geometry and combinatorics of finite fields; (iv) applications of finite fields in coding theory and cryptography. The standard references for finite fields are *Finite Fields* [**27**] by R. Lidl and H. Niederreiter and *Handbook of Finite Fields* [**28**] edited by G. Mullen and D. Panario. The former is a treatise on the theory and applications of finite fields with a comprehensive bibliography up to the early 1980s. The latter is the first handbook of finite fields and contains significant results from all areas of finite fields up to the early 2010s.

The present book is intended to be an exposition of selected topics in the theory of finite fields that can be used as a textbook for a graduate course. More precisely, my expectation of the finished work is a volume with a limited scope that covers the fundamentals of finite fields and explores additional selected topics without excessive overlap with other existing books on finite fields. Material gathering for the book was guided by these objectives. Inevitably, the topics selected reflect my own perspectives on the subject. To limit the scope of the book, I have resisted the temptation to include other topics that are arguably both important and interesting, and the temptation to expand on some topics that are already in the book. In particular, applications of finite fields are not explored except for the Reed-Muller codes, which are treated in Chapters 2 and 5 under the guise of polynomials over finite fields. I hope this shortcoming is remedied by the fact that there are many excellent books devoted to applications of finite fields. I wish to mention a few unique features of the book. It contains some nontrivial results that are not so well known but are quite useful (e.g., the formula for the cardinalities of the conjugacy classes of the affine linear group $\mathrm{AGL}(n, \mathbb{F}_q)$); it also contains simplified proofs of several important theorems (e.g., the author's proof of the Katz theorem and Leducq's proof of the Delsarte-Goethals-MacWilliams theorem).

Here are the outlines of the chapters:

Chapter 1: The first section provides the preliminaries for the rest of the book. All basic facts about finite fields are proved there. Section 1.2 is devoted to partially ordered sets and the Möbius function, which are used later to count the number of irreducible polynomials over finite fields.

Chapter 2: We address a number of issues related to the algebra and combinatorics of polynomials over finite fields, except for questions concerning zeros of polynomials over finite fields, which are discussed later in Chapter 5. The topics include Berlekamp's factorization algorithm, counting for irreducible polynomials and irreducible factors, polynomial representation of functions, permutation polynomials, Dickson polynomials, linearized polynomials, and a generalization of a theorem by S. Payne on linearized polynomials. I have resisted the temptation to expand the coverage of permutation polynomials, which constitute an active research area of finite fields; interested readers are referred to a recent survey [**17**] on permutation polynomials. The last section on Payne's theorem is rather technical; the reader may choose to skip it at first reading.

Chapter 3: After a discussion of characters of finite abelian groups, Gauss sums are introduced. The highlights of the chapter are the Davenport-Hasse theorem on the Gauss sum of a lifted character and the calculation of the Gauss quadratic sum.

Chapter 4: This chapter is essentially a tailored introduction to algebraic number theory. No prerequisites other than graduate algebra and elementary number theory are required. Basic properties of number fields are proved and prime factorization in an arbitrary number field is discussed. In section 4.5, we focus on cyclotomic fields and determine how primes factor in such fields. In the last section, the results on cyclotomic fields are used to prove the Stickelberger congruence for Gauss sums.

Chapter 5: Zeros of polynomials over finite fields are an area where sophisticated methods are developed and profound results are proved. In this chapter, we introduce several theorems on zeros of polynomials over finite fields that are of fundamental importance. The theorems of Ax and Katz give sharp lower bounds for the p-adic order of the number of zeros of one or several polynomials over a finite field of characteristic p. The proof of Ax's theorem relies on Stickelberger's congruence for Gauss sums. The proof of Katz's theorem adopted here, found by the author, is much simpler than the original. Theorem 5.9 is a sharp lower bound for the number of common zeros of several polynomials, and Theorem 5.11 is a sharp upper bound for the number of zeros of one polynomial. The Delsarte-Goethals-MacWilliams theorem completely determines the polynomials meeting the upper bound in Theorem 5.11. The Delsarte-Goethals-MacWilliams theorem originally appeared as a characterization of minimal-weight codewords in the q-ary Reed-Muller code [**9**]; unfortunately, this strong result does not seem to be well known outside the coding theory community. The proof of the Delsarte-Goethals-MacWilliams theorem included here, recently discovered by Leducq, is also much simpler than the original. The last major theorem of the chapter is the Hasse-Weil bound on the number of zeros of an absolutely irreducible polynomial over a finite field. The result is easily stated, but its proof is beyond the scope of the present book. We attempt to alleviate the predicament by including a sketchy and informal introduction to function fields; section 5.4 is devoted to outlining the components of function fields that lead to the Hasse-Weil bound. Along the theme-line "places – the Riemann-Roch theorem – extensions – the zeta function – Riemann's hypothesis for function fields – the Hasse-Weil bound", notions and concepts are defined and theorems are stated without proof. For readers with some knowledge of function fields, section 5.4 serves as a review; for those without such knowledge, the section serves as a preview.

Chapter 6: The last chapter is an introduction to classical groups over finite fields. For a considerable part of this chapter, the field F is assumed to be more general than finite. We prove the simplicity of $\mathrm{PSL}(n, F)$ and derive formulas for the cardinalities of the conjugacy classes of the general linear group $\mathrm{GL}(n, \mathbb{F}_q)$ and the affine linear group $\mathrm{AGL}(n, \mathbb{F}_q)$. The formula for $\mathrm{AGL}(n, \mathbb{F}_q)$, which is useful for studying $\mathrm{AGL}(n, \mathbb{F}_q)$-actions on sets, does not seem to have appeared in any book. The last two sections are devoted to bilinear forms, unitary forms, quadratic forms, and the classical groups associated to such forms. When the field is finite, the forms are classified and the orders of the associated classical groups are determined.

Each chapter contains a set of exercises ranging from easy to challenging. The book is mostly self-contained. Except for section 5.4, almost all results in the book are proved in detail. The reader is assumed to have a basic knowledge of graduate algebra. Throughout the book, all rings are with identity, all modules are unitary, a subring has the same identity as the ambient ring, and a ring homomorphism maps identity to identity.

Clarity through conciseness is a mantra that I aspired to throughout the preparation of this book. I would be gratified if a fraction of this goal is achieved.

I owe my special thanks to Professor Gary Mullen; without his encouragement and mentorship, this project would not have come to fruition. I am grateful to the anonymous referees for their careful reading of the manuscript and for their insightful comments and valuable suggestions. I also wish to express my gratitude to the AMS editors and staff members for their patience during my preparation and revision of the manuscript and for their assistance at various stages of the project. Finally, I would like to thank my students for their stimulating input and supportive feedback.

XDH

Tampa, FL 2017

Chapter 1

Preliminaries

1.1. Basic Properties of Finite Fields

Existence and uniqueness. Let F be a field with $|F| < \infty$. Define a ring homomorphism

$$\begin{aligned} f: \mathbb{Z} &\longrightarrow F \\ n &\longmapsto n1_F, \end{aligned}$$

where 1_F is the identity of F. By the first isomorphism theorem we have an embedding $\mathbb{Z}/\ker f \hookrightarrow F$. Thus $\mathbb{Z}/\ker f$ is an integral domain. Therefore, $\ker f$ is a prime ideal of $\mathbb{Z}$, i.e., $\ker f = p\mathbb{Z}$ for some prime p. The prime p is the additive order of 1_F; it is called the *characteristic* of F and is denoted by $\operatorname{char} F$. Since the field $\mathbb{Z}/p\mathbb{Z}$ is embedded in F, we may assume that F contains $\mathbb{Z}/p\mathbb{Z}$ as a subfield. Clearly, F is a vector space over $\mathbb{Z}/p\mathbb{Z}$. Since F is finite, $[F : \mathbb{Z}/p\mathbb{Z}] = \dim_{\mathbb{Z}/p\mathbb{Z}} F < \infty$. Let $n = [F : \mathbb{Z}/p\mathbb{Z}]$. Then $F \cong (\mathbb{Z}/p\mathbb{Z})^n$ as a $(\mathbb{Z}/p\mathbb{Z})$-vector space. In particular, $|F| = p^n$.

To sum up, if F is a finite field, then $|F| = p^n$ for some prime p and some integer $n > 0$.

An immediate question is this: Given a prime p and an integer $n > 0$, does there exist a field F with $|F| = p^n$? The answer is positive. Recall that the splitting field of a polynomial f over a field F is the extension of F generated by the roots of f in the algebraic closure of F.

Theorem 1.1. *Let p be a prime and n a positive integer. The splitting field of $\mathtt{X}^{p^n} - \mathtt{X} \in (\mathbb{Z}/p\mathbb{Z})[\mathtt{X}]$ has precisely p^n elements.*

Proof. Let $f = \mathtt{X}^{p^n} - \mathtt{X}$ and let F be the splitting field of f over $\mathbb{Z}/p\mathbb{Z}$. Since $\gcd(f', f) = \gcd(-1, f) = 1$, f has p^n distinct roots in F. Let $E = \{a \in F : f(a) = 0\}$. We claim that $F = E$. To verify this, it suffices to show that E is a field. (Then, since f splits in E and F is the smallest field in which f splits, we must have $F = E$.)

We claim that

$$\begin{aligned} \phi: \ F &\longrightarrow F \\ a &\longmapsto a^{p^n} \end{aligned}$$

is an automorphism of F. Clearly, $\phi(1) = 1$. Let $a, b \in F$. We have

$$\phi(ab) = (ab)^{p^n} = a^{p^n} b^{p^n} = \phi(a)\phi(b).$$

Since $p = 0$ in F, we also have

$$\phi(a + b) = (a + b)^{p^n} = a^{p^n} + b^{p^n} = \phi(a) + \phi(b).$$

Hence $\phi : F \to F$ is a ring homomorphism. Clearly, $\ker \phi = \{0\}$, so ϕ is one-to-one. Since F is a finite extension over $\mathbb{Z}/p\mathbb{Z}$, $|F| < \infty$. Therefore, ϕ must be onto, making it an automorphism of F.

Now, E is the fixed field of ϕ in F. In particular, E is a field. □

A finite field of a given order (number of elements) is unique up to isomorphism.

Theorem 1.2. *Given a prime p and an integer $n > 0$, all finite fields of order p^n are isomorphic.*

Proof. Let F be a finite field with $|F| = p^n$. As seen at the beginning of this subsection, $\mathbb{Z}/p\mathbb{Z} \subset F$. Since $F \setminus \{0\}$ is a multiplicative group of order $p^n - 1$, we have $a^{p^n - 1} = 1$ for all $a \in F \setminus \{0\}$. Thus $a^{p^n} = a$ for all $a \in F$. In other words, all elements of F are roots of $f = \mathtt{X}^{p^n} - \mathtt{X} \in (\mathbb{Z}/p\mathbb{Z})[\mathtt{X}]$. Since $\deg f = |F|$, F is the set of all roots of f. Therefore, F is a splitting field of f over $\mathbb{Z}/p\mathbb{Z}$.

Since all splitting fields of f over $\mathbb{Z}/p\mathbb{Z}$ are isomorphic, the conclusion of the theorem follows. □

We denote the finite field with p^n elements by $\mathbb{F}_{p^n}$. Thus $\mathbb{F}_p = \mathbb{Z}/p\mathbb{Z}$. We have an $\mathbb{F}_p$-vector space isomorphism (not a ring isomorphism) $\mathbb{F}_{p^n} \cong \mathbb{F}_p^n$.

The multiplicative group of $\mathbb{F}_{p^n}$. The multiplicative group of $\mathbb{F}_{p^n}$ is denoted by $\mathbb{F}_{p^n}^*$.

Theorem 1.3. *$\mathbb{F}_{p^n}^*$ is cyclic. A generator of $\mathbb{F}_{p^n}^*$ is called a* primitive element *of $\mathbb{F}_{p^n}$.*

Proof. Assume to the contrary that $\mathbb{F}_{p^n}^*$ is not cyclic. By the fundamental theorem of finite abelian groups, we have

$$\mathbb{F}_{p^n}^* \cong A \times B, \tag{1.1}$$

where $|A| = a$, $|B| = b$ and $\gcd(a,b) \neq 1$. (The fundamental theorem of finite abelian groups states that every finite abelian group G is isomorphic to

$$(\mathbb{Z}/p_1^{e_1}\mathbb{Z}) \times \cdots \times (\mathbb{Z}/p_k^{e_k}\mathbb{Z})$$

for some primes $p_1, \dots, p_k$ and integers $e_1, \dots, e_k > 0$; G is cyclic if and only if $p_1, \dots, p_k$ are all distinct.) It follows that $p^n - 1 = |\mathbb{F}_{p^n}^*| = ab > \operatorname{lcm}(a,b)$. By (1.1), we have

$$x^{\operatorname{lcm}(a,b)} = 1 \quad \text{for all } x \in \mathbb{F}_{p^n}^*. \tag{1.2}$$

However, the polynomial $\mathtt{X}^{\operatorname{lcm}(a,b)} - 1$ can have at most $\operatorname{lcm}(a,b)$ roots in $\mathbb{F}_{p^n}$, which is a contradiction to (1.2). □

Representation of elements. Let α be a primitive element of $\mathbb{F}_{p^n}$. Then $\mathbb{F}_{p^n} = \{0, 1, \alpha, \dots, \alpha^{p^n-2}\}$. Multiplications in $\mathbb{F}_{p^n}$ are easily performed with this representation of the elements of $\mathbb{F}_{p^n}$. However, to perform additions in $\mathbb{F}_{p^n}$, we need to treat $\mathbb{F}_{p^n}$ as an extension of $\mathbb{F}_p$ by an irreducible polynomial of degree n.

Lemma 1.4. *Let p be a prime and $n > 0$ an integer. Then there exists an irreducible polynomial $f \in \mathbb{F}_p[\mathtt{X}]$ of degree n.*

Proof. Let $\alpha \in \mathbb{F}_{p^n}$ be a primitive element. Clearly, $\mathbb{F}_{p^n} = \mathbb{F}_p(\alpha)$. (Here $\mathbb{F}_p(\alpha)$ is the extension of $\mathbb{F}_p$ obtained by adjoining α to $\mathbb{F}_p$.) Let $f \in \mathbb{F}_p[\mathtt{X}]$ be the minimal polynomial of α over $\mathbb{F}_p$. Then f is irreducible and $\deg f = [\mathbb{F}_p(\alpha) : \mathbb{F}_p] = [\mathbb{F}_{p^n} : \mathbb{F}_p] = n$. □

Lemma 1.4 is an existence result. In Chapter 2, we will determine the exact number of irreducible polynomials of degree n over a finite field. However, finding irreducible polynomials of large degree over a finite field is not easy.

Let $f = \mathtt{X}^n + a_{n-1}\mathtt{X}^{n-1} + \cdots + a_0 \in \mathbb{F}_p[\mathtt{X}]$ be a monic irreducible polynomial of degree n. Then $\mathbb{F}_p[\mathtt{X}]/(f)$ is a field and every element of $\mathbb{F}_p[\mathtt{X}]/(f)$ is uniquely of the form

$$c_0 + c_1\bar{\mathtt{X}} + \cdots + c_{n-1}\bar{\mathtt{X}}^{n-1},$$

where $\bar{\mathtt{X}} = \mathtt{X} + (f) \in \mathbb{F}_p[\mathtt{X}]/(f)$ and $c_0, \dots, c_{n-1} \in \mathbb{F}_p$. Since $|\mathbb{F}_p[\mathtt{X}]/(f)| = p^n$, by Theorem 1.2, $\mathbb{F}_p[\mathtt{X}]/(f) = \mathbb{F}_{p^n}$. An element $g + (f) \in \mathbb{F}_p[\mathtt{X}]/(f)$, where $g \in \mathbb{F}_p[\mathtt{X}]$, is also denoted by g when the meaning is clear from the context. Thus the elements of $\mathbb{F}_p[\mathtt{X}]/(f)$ are polynomials of degree $< n$ in $\mathbb{F}_p[\mathtt{X}]$; the addition of two such elements is simply the polynomial addition; the multiplication of two such elements is the polynomial multiplication followed by a reduction modulo f.

Table 1.1. Multiplication table of $\mathbb{F}_{2^3} = \mathbb{F}_2[\mathtt{X}]/(\mathtt{X}^3 + \mathtt{X} + 1)$

·	000	001	010	011	100	101	110	111
000	000	000	000	000	000	000	000	000
001	000	001	010	011	100	101	110	111
010	000	010	100	110	011	001	111	101
011	000	011	110	101	111	100	001	010
100	000	100	011	111	110	010	101	001
101	000	101	001	100	010	111	011	110
110	000	110	111	001	101	011	010	100
111	000	111	101	010	001	110	100	011

Example 1.5. $f = \mathtt{X}^3 + \mathtt{X} + 1 \in \mathbb{F}_2[\mathtt{X}]$ is irreducible. (A polynomial of degree ≤ 3 over a field F having no root in F is irreducible over F.) Hence $\mathbb{F}_{2^3} = \mathbb{F}_2[\mathtt{X}]/(f)$. Let $g = \mathtt{X}^2 + \mathtt{X} + 1, h = \mathtt{X}^2 + 1 \in \mathbb{F}_2[\mathtt{X}]/(f)$. We have

$$\begin{aligned} fg &= (\mathtt{X}^2 + \mathtt{X} + 1)(\mathtt{X}^2 + 1) = \mathtt{X}^4 + \mathtt{X}^3 + \mathtt{X} + 1 \\ &= \mathtt{X}(\mathtt{X} + 1) \qquad\qquad (\text{since } \mathtt{X}^3 + \mathtt{X} + 1 = 0) \\ &= \mathtt{X}^2 + \mathtt{X}. \end{aligned}$$

The multiplication table of $\mathbb{F}_{2^3} = \mathbb{F}_2[\mathtt{X}]/(f)$ is given in Table 1.1, where $c_2\mathtt{X}^2 + c_1\mathtt{X} + c_0$ is abbreviated as $c_2c_1c_0$.

Lattice of finite fields. Let p be a fixed prime. To describe the relations among all finite fields of characteristic p, we put all such fields in one ambient field. Let $\overline{\mathbb{F}}_p$ be the algebraic closure of $\mathbb{F}_p$. For each integer $n > 0$, since $\overline{\mathbb{F}}_p$ contains a splitting field of $\mathtt{X}^{p^n} - \mathtt{X}$ over $\mathbb{F}_p$, $\mathbb{F}_{p^n}$ is a subfield of $\overline{\mathbb{F}}_p$.

Theorem 1.6. *Let p be a prime.*

(i) *For each integer $n > 0$, $\overline{\mathbb{F}}_p$ has a unique subfield of order p^n.*

(ii) *Let $\mathbb{F}_{p^m} \subset \overline{\mathbb{F}}_p$ and $\mathbb{F}_{p^n} \subset \overline{\mathbb{F}}_p$. Then $\mathbb{F}_{p^m} \subset \mathbb{F}_{p^n}$ if and only if $m \mid n$. In general,*

$$\mathbb{F}_{p^m} \cap \mathbb{F}_{p^n} = \mathbb{F}_{p^{\gcd(m,n)}}, \tag{1.3}$$

$$\mathbb{F}_{p^m}\mathbb{F}_{p^n} = \mathbb{F}_{p^{\mathrm{lcm}[m,n]}}, \tag{1.4}$$

where $\mathbb{F}_{p^m}\mathbb{F}_{p^n}$ is the subfield of $\overline{\mathbb{F}}_p$ generated by $\mathbb{F}_{p^m} \cup \mathbb{F}_{p^n}$.

We already know that a finite field of order p^n is unique up to isomorphism. However, Theorem 1.6 (i) states that in a given algebraic closure of $\mathbb{F}_p$, a finite field of order p^n is not only unique up to isomorphism, but also unique as a set.

Proof of Theorem 1.6. (i) By the proof of Theorem 1.2, a subfield of $\overline{\mathbb{F}}_p$ of order p^n must be $\{a \in \overline{\mathbb{F}}_p : a^{p^n} = a\}$.

(ii) If $\mathbb{F}_{p^m} \subset \mathbb{F}_{p^n}$, then $\mathbb{F}_{p^n}$ is an $[\mathbb{F}_{p^n} : \mathbb{F}_{p^m}]$-dimensional vector space over $\mathbb{F}_{p^m}$. Hence,

$$p^n = |\mathbb{F}_{p^n}| = |\mathbb{F}_{p^m}|^{[\mathbb{F}_{p^n}:\mathbb{F}_{p^m}]} = p^{m[\mathbb{F}_{p^n}:\mathbb{F}_{p^m}]}.$$

Thus $n = m[\mathbb{F}_{p^n} : \mathbb{F}_{p^m}]$.

If $m \mid n$, then

$$\begin{aligned} \mathtt{X}^{p^n} - \mathtt{X} &= \mathtt{X}(\mathtt{X}^{p^n-1} - 1) = \mathtt{X}\left(\mathtt{X}^{\frac{p^n-1}{p^m-1}(p^m-1)} - 1\right) \\ &= \mathtt{X}(\mathtt{X}^{p^m-1} - 1) \sum_{i=0}^{\frac{p^n-1}{p^m-1}-1} \mathtt{X}^{(p^m-1)i} = (\mathtt{X}^{p^m} - \mathtt{X}) \sum_{i=0}^{\frac{p^n-1}{p^m-1}-1} \mathtt{X}^{(p^m-1)i}. \end{aligned}$$

Therefore, in $\overline{\mathbb{F}}_p$, the splitting field of $\mathtt{X}^{p^m} - \mathtt{X}$ is contained in the splitting field of $\mathtt{X}^{p^n} - \mathtt{X}$, i.e., $\mathbb{F}_{p^m} \subset \mathbb{F}_{p^n}$.

To prove (1.3), first observe that $\mathbb{F}_{p^{\gcd(m,n)}} \subset \mathbb{F}_{p^m} \cap \mathbb{F}_{p^n}$. Let $\mathbb{F}_{p^m} \cap \mathbb{F}_{p^n} = \mathbb{F}_{p^s}$. Since $\mathbb{F}_{p^s} \subset \mathbb{F}_{p^m}$ and $\mathbb{F}_{p^s} \subset \mathbb{F}_{p^n}$, from the above, $s \mid m$ and $s \mid n$, and hence $s \mid \gcd(m,n)$. Therefore, $\mathbb{F}_{p^m} \cap \mathbb{F}_{p^n} = \mathbb{F}_{p^s} \subset \mathbb{F}_{p^{\gcd(m,n)}}$.

Equation (1.4) is proved in the same way. □

Proposition 1.7. *Let $\mathbb{F}_{p^m} \subset \mathbb{F}_{p^n}$, where $m \mid n$. If α is a primitive element of $\mathbb{F}_{p^n}$, then $\alpha^{(p^n-1)/(p^m-1)}$ is a primitive element of $\mathbb{F}_{p^m}$.*

Proof. Since $o(\alpha) = p^n - 1$, $o(\alpha^{(p^n-1)/(p^m-1)}) = p^m - 1$. Since $\mathbb{F}_{p^n}^*$ is cyclic, $\mathbb{F}_{p^m}^*$ is the only subgroup of $\mathbb{F}_{p^n}^*$ of order $p^m - 1$. Thus $\mathbb{F}_{p^m}^* = \langle \alpha^{(p^n-1)/(p^m-1)} \rangle$. □

The automorphism group. In general, given fields $F \subset K$, the Galois group of the extension K/F is $\mathrm{Aut}(K/F) = \{f \in \mathrm{Aut}(K) : f(a) = a \text{ for all } a \in F\}$. The extension K/F is said to be Galois if $\{a \in K : f(a) = a \text{ for all } f \in \mathrm{Aut}(K/F)\} = F$. When $[K : F] < \infty$, the following statements are equivalent: (i) K/F is Galois; (ii) K is the splitting field of some separable polynomial over F; (iii) $|\mathrm{Aut}(K/F)| = [K : F]$.

Define a map

$$\begin{aligned} \sigma : \mathbb{F}_{p^n} &\longrightarrow \mathbb{F}_{p^n} \\ a &\longmapsto a^p. \end{aligned}$$

It is clear that $\sigma \in \mathrm{Aut}(\mathbb{F}_{p^n}/\mathbb{F}_p)$; σ is called the *Frobenius map* of $\mathbb{F}_{p^n}$ over $\mathbb{F}_p$.

Theorem 1.8. *The extension* $\mathbb{F}_{p^n}/\mathbb{F}_p$ *is Galois and* $\mathrm{Aut}(\mathbb{F}_{p^n}/\mathbb{F}_p) = \langle\sigma\rangle$. *More generally, if* $m \mid n$, *then the extension* $\mathbb{F}_{p^n}/\mathbb{F}_{p^m}$ *is Galois and* $\mathrm{Aut}(\mathbb{F}_{p^n}/\mathbb{F}_{p^m}) = \langle\sigma^m\rangle$.

Proof. Since $\mathtt{X}^{p^n} - \mathtt{X}$ is a separable polynomial in $\mathbb{F}_p[\mathtt{X}]$ and since $\mathbb{F}_{p^n}$ is the splitting field of $\mathtt{X}^{p^n} - \mathtt{X}$ over $\mathbb{F}_p$, $\mathbb{F}_{p^n}$ is Galois over $\mathbb{F}_p$. Thus $|\mathrm{Aut}(\mathbb{F}_{p^n}/\mathbb{F}_p)| = [\mathbb{F}_{p^n} : \mathbb{F}_p] = n$. Since $\sigma \in \mathrm{Aut}(\mathbb{F}_{p^n}/\mathbb{F}_p)$, to prove that $\mathrm{Aut}(\mathbb{F}_{p^n}/\mathbb{F}_p) = \langle\sigma\rangle$, it suffices to show that $o(\sigma) = n$ or, equivalently, $o(\sigma) \geq n$. Since $\sigma^{o(\sigma)} = \mathrm{id}$, we have

$$0 = \sigma^{o(\sigma)}(a) - a = a^{p^{o(\sigma)}} - a \quad \text{for all } a \in \mathbb{F}_{p^n}. \tag{1.5}$$

The polynomial $\mathtt{X}^{p^{o(\sigma)}} - \mathtt{X}$, being of degree $p^{o(\sigma)}$, has at most $p^{o(\sigma)}$ roots in $\mathbb{F}_{p^n}$. Thus (1.5) implies that $p^n \leq p^{o(\sigma)}$, i.e., $n \leq o(\sigma)$.

If $m \mid n$, then $\mathbb{F}_p \subset \mathbb{F}_{p^m} \subset \mathbb{F}_{p^n}$. Since $\mathbb{F}_{p^n}/\mathbb{F}_p$ is Galois, so is $\mathbb{F}_{p^n}/\mathbb{F}_{p^m}$. Moreover, $\mathrm{Aut}(\mathbb{F}_{p^n}/\mathbb{F}_{p^m})$ is a subgroup of $\mathrm{Aut}(\mathbb{F}_{p^n}/\mathbb{F}_p)$ of order n/m. Since $\mathrm{Aut}(\mathbb{F}_{p^n}/\mathbb{F}_p) = \langle\sigma\rangle$ is cyclic, its only subgroup of order n/m is $\langle\sigma^m\rangle$. Thus $\mathrm{Aut}(\mathbb{F}_{p^n}/\mathbb{F}_{p^m}) = \langle\sigma^m\rangle$. □

The automorphism $\sigma^m \in \mathrm{Aut}(\mathbb{F}_{p^n}/\mathbb{F}_{p^m})$ is defined by $\sigma^m(a) = a^{p^m}$, $a \in \mathbb{F}_{p^n}$, and is called the *Frobenius map* of $\mathbb{F}_{p^n}$ over $\mathbb{F}_{p^m}$.

Trace and norm. Let q be a prime power and n a positive integer. By Theorem 1.8, $\mathrm{Aut}(\mathbb{F}_{q^n}/\mathbb{F}_q) = \langle\tau\rangle$, where

$$\begin{array}{rccc} \tau: & \mathbb{F}_{q^n} & \longrightarrow & \mathbb{F}_{q^n} \\ & a & \longmapsto & a^q \end{array}$$

is the Frobenius map of $\mathbb{F}_{q^n}$ over $\mathbb{F}_q$. For each $a \in \mathbb{F}_{q^n}$, define

$$\mathrm{Tr}_{\mathbb{F}_{q^n}/\mathbb{F}_q}(a) = \sum_{\phi\in\mathrm{Aut}(\mathbb{F}_{q^n}/\mathbb{F}_q)} \phi(a) = \sum_{i=0}^{n-1} \tau^i(a) = \sum_{i=0}^{n-1} a^{q^i}$$

and

$$\mathrm{N}_{\mathbb{F}_{q^n}/\mathbb{F}_q}(a) = \prod_{\phi\in\mathrm{Aut}(\mathbb{F}_{q^n}/\mathbb{F}_q)} \phi(a) = \prod_{i=0}^{n-1} \tau^i(a) = a^{q^0+\cdots+q^{n-1}} = a^{(q^n-1)/(q-1)}.$$

For each $\psi \in \mathrm{Aut}(\mathbb{F}_{q^n}/\mathbb{F}_q)$, we have

$$\begin{aligned} \psi\big(\mathrm{Tr}_{\mathbb{F}_{q^n}/\mathbb{F}_q}(a)\big) &= \psi\Big(\sum_{\phi\in\mathrm{Aut}(\mathbb{F}_{q^n}/\mathbb{F}_q)} \phi(a)\Big) = \sum_{\phi\in\mathrm{Aut}(\mathbb{F}_{q^n}/\mathbb{F}_q)} (\psi\phi)(a) \\ &= \sum_{\phi\in\mathrm{Aut}(\mathbb{F}_{q^n}/\mathbb{F}_q)} \phi(a) = \mathrm{Tr}_{\mathbb{F}_{q^n}/\mathbb{F}_q}(a). \end{aligned}$$

Since $\mathbb{F}_{q^n}/\mathbb{F}_q$ is Galois, we have $\mathrm{Tr}_{\mathbb{F}_{q^n}/\mathbb{F}_q}(a) \in \mathbb{F}_q$. By the same argument, $\mathrm{N}_{\mathbb{F}_{q^n}/\mathbb{F}_q}(a) \in \mathbb{F}_q$. $\mathrm{Tr}_{\mathbb{F}_{q^n}/\mathbb{F}_q}(a)$ and $\mathrm{N}_{\mathbb{F}_{q^n}/\mathbb{F}_q}(a)$ are called the *trace* and the *norm*, respectively, of a from $\mathbb{F}_{q^n}$ to $\mathbb{F}_q$. We usually write $\mathrm{Tr}_{q^n/q}$ for $\mathrm{Tr}_{\mathbb{F}_{q^n}/\mathbb{F}_q}$ and $\mathrm{N}_{q^n/q}$ for $\mathrm{N}_{\mathbb{F}_{q^n}/\mathbb{F}_q}$.

Theorem 1.9.

(i) $\mathrm{Tr}_{q^n/q} : \mathbb{F}_{q^n} \to \mathbb{F}_q$ *is an onto* $\mathbb{F}_q$*-map.*

(ii) *If* $a \in \mathbb{F}_q$, *then* $\mathrm{Tr}_{q^n/q}(a) = na$.

(iii) *For all* $a \in \mathbb{F}_{q^n}$ *and* $\phi \in \mathrm{Aut}(\mathbb{F}_{q^n}/\mathbb{F}_q)$, $\mathrm{Tr}_{q^n/q}(\phi(a)) = \mathrm{Tr}_{q^n/q}(a)$. *In particular,* $\mathrm{Tr}_{q^n/q}(a^q) = \mathrm{Tr}_{q^n/q}(a)$.

Proof. (i) Since $\phi \in \mathrm{Hom}_{\mathbb{F}_q}(\mathbb{F}_{q^n}, \mathbb{F}_q)$ for all $\phi \in \mathrm{Aut}(\mathbb{F}_{q^n}/\mathbb{F}_q)$, we have $\mathrm{Tr}_{q^n/q} = \sum_{\phi\in\mathrm{Aut}(\mathbb{F}_{q^n}/\mathbb{F}_q)} \phi \in \mathrm{Hom}_{\mathbb{F}_q}(\mathbb{F}_{q^n}, \mathbb{F}_q)$. We claim that $\mathrm{Tr}_{\mathbb{F}_{q^n}/\mathbb{F}_q} \neq 0$. As a runs through $\mathbb{F}_{q^n}$, $\mathrm{Tr}_{q^n/q}(a) = \sum_{i=0}^{n-1} a^{q^i}$ is a polynomial of degree q^{n-1} in a and thus cannot be all 0. Therefore, $\mathrm{Tr}_{q^n/q} : \mathbb{F}_{q^n} \to \mathbb{F}_q$ is onto since the target $\mathbb{F}_q$ is of dimension 1 over $\mathbb{F}_q$.

(ii) We have

$$\mathrm{Tr}_{q^n/q}(a) = \sum_{\psi\in\mathrm{Aut}(\mathbb{F}_{q^n}/\mathbb{F}_q)} \psi(a) = \sum_{\psi\in\mathrm{Aut}(\mathbb{F}_{q^n}/\mathbb{F}_q)} a = na.$$

(iii) We have

$$\begin{aligned}\mathrm{Tr}_{q^n/q}(\phi(a)) &= \sum_{\psi\in\mathrm{Aut}(\mathbb{F}_{q^n}/\mathbb{F}_q)} \psi(\phi(a)) = \sum_{\psi\in\mathrm{Aut}(\mathbb{F}_{q^n}/\mathbb{F}_q)} (\psi\phi)(a)\\ &= \sum_{\psi\in\mathrm{Aut}(\mathbb{F}_{q^n}/\mathbb{F}_q)} \psi(a) = \mathrm{Tr}_{q^n/q}(a).\end{aligned}$$

□

Theorem 1.10.

(i) $\mathrm{N}_{q^n/q}(0) = 0$, *and the map* $\mathrm{N}_{q^n/q} : \mathbb{F}_{q^n}^* \to \mathbb{F}_q^*$ *is an onto group homomorphism.*

(ii) *If* $a \in \mathbb{F}_q$, *then* $\mathrm{N}_{q^n/q}(a) = a^n$.

(iii) *For all* $a \in \mathbb{F}_{q^n}$ *and* $\phi \in \mathrm{Aut}(\mathbb{F}_{q^n}/\mathbb{F}_q)$, $\mathrm{N}_{q^n/q}(\phi(a)) = \mathrm{N}_{q^n/q}(a)$. *In particular,* $\mathrm{N}_{q^n/q}(a^q) = \mathrm{N}_{q^n/q}(a)$.

Proof. (i) Clearly, $\mathrm{N}_{q^n/q}(0) = 0$. Since

$$\mathrm{N}_{q^n/q}(a) = a^{(q^n-1)/(q-1)}, \quad a \in \mathbb{F}_{q^n}^*,$$

$\mathrm{N}_{q^n/q} : \mathbb{F}_{q^n}^* \to \mathbb{F}_q^*$ is a group homomorphism. By Proposition 1.7, $\mathrm{N}_{q^n/q}$ maps a generator of $\mathbb{F}_{q^n}^*$ to a generator of $\mathbb{F}_q^*$. Thus $\mathrm{N}_{q^n/q} : \mathbb{F}_{q^n}^* \to \mathbb{F}_q^*$ is onto.

Statements (ii) and (iii) are proved in the same way as (ii) and (iii) of Theorem 1.9. □

Theorem 1.11 (Transitivity of trace and norm). *Let $F \subset K \subset L$ be finite fields and let $a \in L$. Then*

$$\mathrm{Tr}_{K/F}(\mathrm{Tr}_{L/K}(a)) = \mathrm{Tr}_{L/F}(a), \tag{1.6}$$

$$\mathrm{N}_{K/F}(\mathrm{N}_{L/K}(a)) = \mathrm{N}_{L/F}(a). \tag{1.7}$$

Proof. Let $F = \mathbb{F}_q$, $K = \mathbb{F}_{q^s}$, and $L = \mathbb{F}_{q^{st}}$. Let τ be the Frobenius map of L over F. Then τ^s is the Frobenius map of L over K and $\tau|_K$ is the Frobenius map of K over F. Thus,

$$\begin{aligned}
\mathrm{Tr}_{K/F}(\mathrm{Tr}_{L/K}(a)) &= \mathrm{Tr}_{K/F}\Big(\sum_{i=0}^{t-1} \tau^{si}(a)\Big) = \sum_{j=0}^{s-1} \tau^j\Big(\sum_{i=0}^{t-1} \tau^{si}(a)\Big) \\
&= \sum_{i=0}^{t-1}\sum_{j=0}^{s-1} \tau^{si+j}(a) = \sum_{k=0}^{st-1} \tau^k(a) \qquad (k = si + j) \\
&= \mathrm{Tr}_{L/F}(a).
\end{aligned}$$

The proof of (1.7) is the same. □

The next two theorems describe the kernels of $\mathrm{Tr}_{q^n/q}$ and $\mathrm{N}_{q^n/q}$.

Theorem 1.12. *Let ϕ be any generator of* $\mathrm{Aut}(\mathbb{F}_{q^n}/\mathbb{F}_q)$. *Then*

$$\ker(\mathrm{Tr}_{q^n/q}) = \{\phi(x) - x : x \in \mathbb{F}_{q^n}\}. \tag{1.8}$$

Proof. Let $f = \phi - \mathrm{id} \in \mathrm{Hom}_{\mathbb{F}_q}(\mathbb{F}_{q^n}, \mathbb{F}_{q^n})$. Then the right side of (1.8) is $f(\mathbb{F}_{q^n})$. By Theorem 1.9 (iii), $\mathrm{Tr}_{q^n/q} \circ f = \mathrm{Tr}_{q^n/q} \circ \phi - \mathrm{Tr}_{q^n/q} = 0$. Hence, $f(\mathbb{F}_{q^n}) \subset \ker(\mathrm{Tr}_{q^n/q})$. By Theorem 1.9 (i), $\dim_{\mathbb{F}_q} \ker(\mathrm{Tr}_{q^n/q}) = n-1$. Thus, to prove (1.8), it suffices to show that $\dim_{\mathbb{F}_q} f(\mathbb{F}_{q^n}) = n - 1$. Note that $\ker f = \{x \in \mathbb{F}_{q^n} : \phi(x) = x\} = \mathbb{F}_q$ since $\mathbb{F}_{q^n}/\mathbb{F}_q$ is Galois with $\mathrm{Aut}(\mathbb{F}_{q^n}/\mathbb{F}_q) = \langle \phi \rangle$. Thus

$$\dim_{\mathbb{F}_q} f(\mathbb{F}_{q^n}) = n - \dim_{\mathbb{F}_q} \ker f = n - 1.$$

□

Theorem 1.13 (Hilbert's Theorem 90 for finite fields). *Let ϕ be any generator of* $\mathrm{Aut}(\mathbb{F}_{q^n}/\mathbb{F}_q)$. *Then*

$$\ker(\mathrm{N}_{q^n/q} : \mathbb{F}_{q^n}^* \to \mathbb{F}_q^*) = \Big\{\frac{\phi(x)}{x} : x \in \mathbb{F}_{q^n}^*\Big\}. \tag{1.9}$$

Proof. The proof is similar to that of Theorem 1.12. Define a group homomorphism

$$\begin{aligned}
f:\ \mathbb{F}_{q^n}^* &\longrightarrow \mathbb{F}_{q^n}^* \\
x &\longmapsto \phi(x)/x.
\end{aligned}$$

Then the right side of (1.9) is $f(\mathbb{F}_{q^n}^*)$. It is easy to see that

$$f(\mathbb{F}_{q^n}^*) \subset \ker(\mathrm{N}_{q^n/q} : \mathbb{F}_{q^n}^* \to \mathbb{F}_q^*).$$

By Theorem 1.10 (i), $|\ker(\mathrm{N}_{q^n/q} : \mathbb{F}_{q^n}^* \to \mathbb{F}_q^*)| = |\mathbb{F}_{q^n}^*|/|\mathbb{F}_q^*|$. Thus, to prove (1.9), it suffices to show that $|f(\mathbb{F}_{q^n}^*)| = |\mathbb{F}_{q^n}^*|/|\mathbb{F}_q^*|$. We have $\ker f = \{x \in \mathbb{F}_{q^n}^* : \phi(x) = x\} = \mathbb{F}_q^*$ since $\mathbb{F}_{q^n}/\mathbb{F}_q$ is Galois with $\mathrm{Aut}(\mathbb{F}_{q^n}/\mathbb{F}_q) = \langle\phi\rangle$. Thus,

$$|f(\mathbb{F}_{q^n}^*)| = \frac{|\mathbb{F}_{q^n}^*|}{|\ker f|} = \frac{|\mathbb{F}_{q^n}^*|}{|\mathbb{F}_q^*|}.$$

□

By Theorem 1.9 (i), Theorem 1.10 (i), and Theorems 1.12 and 1.13, for any generator ϕ of $\mathrm{Aut}(\mathbb{F}_{q^n}/\mathbb{F}_q)$, we have exact sequences

$$\{0\} \longrightarrow \mathbb{F}_q \hookrightarrow \mathbb{F}_{q^n} \xrightarrow{\phi-\mathrm{id}} \mathbb{F}_{q^n} \xrightarrow{\mathrm{Tr}_{q^n/q}} \mathbb{F}_q \longrightarrow \{0\}$$

and

$$\{1\} \longrightarrow \mathbb{F}_q^* \hookrightarrow \mathbb{F}_{q^n}^* \xrightarrow{\phi/\mathrm{id}} \mathbb{F}_{q^n}^* \xrightarrow{\mathrm{N}_{q^n/q}} \mathbb{F}_q^* \longrightarrow \{1\}.$$

The trace and norm can also be characterized in terms of a linear transformation.

Theorem 1.14. *Let $a \in \mathbb{F}_{q^n}$ and define an $\mathbb{F}_q$-linear map*

$$\begin{aligned} T_a : \mathbb{F}_{q^n} &\longrightarrow \mathbb{F}_{q^n} \\ x &\longmapsto ax. \end{aligned}$$

Then $\mathrm{Tr}_{q^n/q}(a) = \mathrm{Tr}(T_a)$ and $\mathrm{N}_{q^n/q}(a) = \det(T_a)$. (The trace and determinant of a linear transformation T of a finite-dimensional vector space V are defined to be the trace and determinant of the matrix of T with respect to any basis of V.)

Proof. Consider the tower $\mathbb{F}_q \subset \mathbb{F}_q(a) \subset \mathbb{F}_{q^n}$, and let $[\mathbb{F}_q(a) : \mathbb{F}_q] = s$ and $[\mathbb{F}_{q^n} : \mathbb{F}_q(a)] = t$. Then $1, a, \dots, a^{s-1}$ is a basis of $\mathbb{F}_q(a)$ over $\mathbb{F}_q$. Let

$$f = \mathtt{X}^s + b_{s-1}\mathtt{X}^{s-1} + \cdots + b_0 \in \mathbb{F}_q[\mathtt{X}] \tag{1.10}$$

be the minimal polynomial of a over $\mathbb{F}_q$. Then

$$T_a \begin{bmatrix} 1 \\ a \\ \vdots \\ a^{s-1} \end{bmatrix} = A \begin{bmatrix} 1 \\ a \\ \vdots \\ a^{s-1} \end{bmatrix},$$

where

$$A = \begin{bmatrix} 0 & 1 & & & & \\ & 0 & 1 & & & \\ & & \cdot & \cdot & & \\ & & & \cdot & \cdot & \\ & & & & 0 & 1 \\ -b_0 & -b_1 & \cdot & \cdot & \cdot & -b_{s-1} \end{bmatrix}.$$

Let $\epsilon_1, \dots, \epsilon_t$ be a basis of $\mathbb{F}_{q^n}$ over $\mathbb{F}_q(a)$. Then $\epsilon_i a^j$, $1 \le i \le t$, $0 \le j \le s-1$, is a basis of $\mathbb{F}_{q^n}$ over $\mathbb{F}_q$. With respect to this basis, we have

$$T_a \begin{bmatrix} \epsilon_1 a^0 \\ \vdots \\ \epsilon_1 a^{s-1} \\ \vdots \\ \epsilon_t a^0 \\ \vdots \\ \epsilon_t a^{s-1} \end{bmatrix} = \underbrace{\begin{bmatrix} A & & \\ & \ddots & \\ & & A \end{bmatrix}}_{t \text{ blocks}} \begin{bmatrix} \epsilon_1 a^0 \\ \vdots \\ \epsilon_1 a^{s-1} \\ \vdots \\ \epsilon_t a^0 \\ \vdots \\ \epsilon_t a^{s-1} \end{bmatrix}.$$

Therefore,

(1.11)

$$\operatorname{Tr}(T_a) = t \operatorname{Tr}(A) = t(-b_{s-1}) \quad \text{and} \quad \det(T_a) = (\det A)^t = \big[(-1)^s b_0\big]^t.$$

Let τ be the Frobenius map of $\mathbb{F}_q(a)$ over $\mathbb{F}_q$. Then $\tau^0(a), \dots, \tau^{s-1}(a)$ are all roots of f and are all distinct. (If, to the contrary, $\tau^i(a) = \tau^j(a)$ for some $0 \le i < j \le s-1$, then $\tau^{j-i}(a) = a$. Since $\tau^{j-i} \in \operatorname{Aut}(\mathbb{F}_q(a)/\mathbb{F}_q)$ fixes $\mathbb{F}_q$ and a, we must have $\tau^{j-i} = \mathrm{id}$, which is a contradiction since $o(\tau) = s$.) Therefore,

$$\begin{aligned} f(\mathtt{X}) &= \prod_{i=0}^{s-1} (\mathtt{X} - \tau^i(a)) \\ &= \mathtt{X}^s - \Big(\sum_{i=0}^{s-1} \tau^i(a)\Big) \mathtt{X}^{s-1} + \cdots + (-1)^s \prod_{i=0}^{s-1} \tau^i(a) \\ &= \mathtt{X}^s - \operatorname{Tr}_{\mathbb{F}_q(a)/\mathbb{F}_q}(a)\, \mathtt{X}^{s-1} + \cdots + (-1)^s \mathrm{N}_{\mathbb{F}_q(a)/\mathbb{F}_q}(a). \end{aligned} \tag{1.12}$$

A comparison of (1.10) and (1.12) yields

$$-b_{s-1} = \operatorname{Tr}_{\mathbb{F}_q(a)/\mathbb{F}_q}(a) \quad \text{and} \quad (-1)^s b_0 = \mathrm{N}_{\mathbb{F}_q(a)/\mathbb{F}_q}(a). \tag{1.13}$$

Thus, from (1.11) and (1.13), we have

$$\begin{aligned} \operatorname{Tr}(T_a) &= t \operatorname{Tr}_{\mathbb{F}_q(a)/\mathbb{F}_q}(a) = \operatorname{Tr}_{\mathbb{F}_q(a)/\mathbb{F}_q}(ta) \\ &= \operatorname{Tr}_{\mathbb{F}_q(a)/\mathbb{F}_q}\big(\operatorname{Tr}_{\mathbb{F}_{q^n}/\mathbb{F}_q(a)}(a)\big) = \operatorname{Tr}_{\mathbb{F}_{q^n}/\mathbb{F}_q}(a) \end{aligned}$$

and

$$\begin{aligned}\det(T_a) &= \left[\mathrm{N}_{\mathbb{F}_q(a)/\mathbb{F}_q}(a)\right]^t = \mathrm{N}_{\mathbb{F}_q(a)/\mathbb{F}_q}(a^t)\\ &= \mathrm{N}_{\mathbb{F}_q(a)/\mathbb{F}_q}\big(\mathrm{N}_{\mathbb{F}_{q^n}/\mathbb{F}_q(a)}(a)\big) = \mathrm{N}_{\mathbb{F}_{q^n}/\mathbb{F}_q}(a).\end{aligned}$$

□

Normal bases. Let τ be the Frobenius map of $\mathbb{F}_{q^n}$ over $\mathbb{F}_q$ and let $a \in \mathbb{F}_{q^n}$. In general, $\tau^0(a), \tau^1(a), \dots, \tau^{n-1}(a)$ do not necessarily form a basis of $\mathbb{F}_{q^n}$ over $\mathbb{F}_q$; if they do, the basis is called a *normal basis* of $\mathbb{F}_{q^n}$ over $\mathbb{F}_q$.

Theorem 1.15 (Existence of a normal basis). *There exists a normal basis of $\mathbb{F}_{q^n}$ over $\mathbb{F}_q$.*

Proof. Let τ be the Frobenius map of $\mathbb{F}_{q^n}$ over $\mathbb{F}_q$ and view τ as an $\mathbb{F}_q$-linear transformation of $\mathbb{F}_{q^n}$. Since $\tau^n = \mathrm{id}$, the polynomial $\mathtt{X}^n - 1$ annihilates τ. We claim that $\mathtt{X}^n - 1$ is the minimal polynomial of τ. Assume to the contrary that the minimal polynomial of τ is $f = \mathtt{X}^m + a_{m-1}\mathtt{X}^{m-1} + \cdots + a_0 \in \mathbb{F}_q[\mathtt{X}]$, where $0 < m < n$. Then for all $y \in \mathbb{F}_{q^n}$,

(1.14)

$$0 = f(\tau)(y) = (\tau^m + a_{m-1}\tau^{m-1} + \cdots + a_0\tau^0)(y) = y^{q^m} + a_{m-1}y^{q^{m-1}} + \cdots + a_0 y.$$

However, this is impossible since the right side of (1.14) is a polynomial of degree q^m in y and thus has at most q^m roots in $\mathbb{F}_{q^n}$.

Let A be the matrix of τ with respect to any basis of $\mathbb{F}_{q^n}$ over $\mathbb{F}_q$. Then the minimal polynomial of A is $\mathtt{X}^n - 1$. It follows that

$$A \sim \begin{bmatrix} 0 & 1 & & & & \\ & 0 & 1 & & & \\ & & \ddots & \ddots & & \\ & & & \ddots & \ddots & \\ & & & & 0 & 1 \\ 1 & 0 & \cdot & \cdot & \cdot & 0 \end{bmatrix}. \tag{1.15}$$

(The symbol $\sim$ means matrix similarity.) Similarity (1.15) holds since the two matrices have the same invariant factor $\mathtt{X}^n - 1$. Therefore, there is a basis $\epsilon_1, \dots, \epsilon_n$ of $\mathbb{F}_{q^n}$ over $\mathbb{F}_q$ with respect to which the matrix of τ is the matrix on the right side of (1.15). Since

$$\tau \begin{bmatrix} \epsilon_1 \\ \vdots \\ \epsilon_n \end{bmatrix} = \begin{bmatrix} 0 & 1 & & & & \\ & 0 & 1 & & & \\ & & \ddots & \ddots & & \\ & & & \ddots & \ddots & \\ & & & & 0 & 1 \\ 1 & 0 & \cdot & \cdot & \cdot & 0 \end{bmatrix} \begin{bmatrix} \epsilon_1 \\ \vdots \\ \epsilon_n \end{bmatrix} = \begin{bmatrix} \epsilon_2 \\ \vdots \\ \epsilon_n \\ \epsilon_1 \end{bmatrix},$$

we have $\epsilon_2 = \tau(\epsilon_1)$, $\epsilon_3 = \tau(\epsilon_2) = \tau^2(\epsilon_1), \ldots, \epsilon_n = \tau^{n-1}(\epsilon_1)$. Thus $\epsilon_1, \tau(\epsilon_1)$, $\ldots$, $\tau^{n-1}(\epsilon_1)$ is a normal basis of $\mathbb{F}_{q^n}$ over $\mathbb{F}_q$. □

1.2. Partially Ordered Sets and the Möbius Function

Definition 1.16. A *partially ordered set* (*poset*) is a nonempty set X with a binary relation $\leq$ satisfying the following conditions:

(i) (reflexivity) $x \leq x$ for all $x \in X$;

(ii) (transitivity) if $x \leq y$ and $y \leq z$, where $x, y, z \in X$, then $x \leq z$;

(iii) (anti-symmetry) if $x \leq y$ and $y \leq x$, where $x, y \in X$, then $x = y$.

The relation $\leq$ is called a *partial order* on X.

Let $(X, \leq)$ be a poset and $x, y \in X$. The notation "$x < y$" means that $x \leq y$ and $x \neq y$. We define $[x, y] = \{z \in X : x \leq z \leq y\}$, $[x, y) = \{z \in X : x \leq z < y\}$, etc., and call them intervals. A poset $(X, \leq)$ is said to be *locally finite* if for all $x, y \in X$, $|[x, y]| < \infty$.

Definition 1.17 (The Möbius function). Let $(X, \leq)$ be a locally finite poset. The *Möbius function* of $(X, \leq)$ is a function

$$\mu : X \times X \longrightarrow \mathbb{Z}$$

such that if $x \not\leq y$, then $\mu(x, y) = 0$, and if $x \leq y$, then

$$\sum_{z \in [x,y]} \mu(x, z) = \delta(x, y),$$

where

$$\delta(x, y) = \begin{cases} 1 & \text{if } x = y, \\ 0 & \text{if } x \neq y \end{cases}$$

is the *Kronecker symbol.*

The Möbius function of a locally finite poset $(X, \leq)$ exists and is unique. In fact, with a fixed $x \in X$, $\mu(x, y)$ where $y \geq x$ is inductively given by

$$\begin{cases} \mu(x, x) = 1, \\ \mu(x, y) = -\displaystyle\sum_{z \in [x,y)} \mu(x, z) \quad \text{if } y > x. \end{cases} \tag{1.16}$$

Since $(X, \leq)$ is locally finite, the sum in (1.16) is finite for all $y > x$. Note that for $x \leq y$, $\mu(x, y)$ depends only on the sub-poset $([x, y], \leq)$.

The usefulness of the Möbius function lies in the so-called *Möbius inversion* formula.

Theorem 1.18 (The Möbius inversion). *Let $(X, \leq)$ be a locally finite poset with Möbius function μ. Let A be an abelian group and $N : X \to A$ a function. Let $l, m \in X$ be fixed, and for $x \in X$ define*

$$N_{\geq}(x) = \sum_{y \in [x,m]} N(y)$$

and

$$N_{\leq}(x) = \sum_{y \in [l,x]} N(y).$$

Then

$$N(x) = \sum_{y \in [x,m]} \mu(x, y) N_{\geq}(y) \quad \text{for all } x \in X \text{ with } x \leq m \tag{1.17}$$

and

$$N(x) = \sum_{y \in [l,x]} \mu(y, x) N_{\leq}(y) \quad \text{for all } x \in X \text{ with } x \geq l. \tag{1.18}$$

Proof. Let $x \in X$ be such that $x \leq m$. We have

$$\begin{aligned}
&\sum_{y \in [x,m]} \mu(x, y) N_{\geq}(y) \\
&= \sum_{x \leq y \leq m} \mu(x, y) \sum_{y \leq z \leq m} N(z) = \sum_{x \leq y \leq z \leq m} \mu(x, y) N(z) \\
&= \sum_{x \leq z \leq m} N(z) \sum_{x \leq y \leq z} \mu(x, y) = \sum_{x \leq z \leq m} N(z) \delta(x, z) = N(x).
\end{aligned}$$

To prove (1.18), we define a partial order $\geq$ on X such that $x \geq y$ if and only if $y \leq x$. It is obvious that the Möbius function of the poset $(X, \geq)$ is $\eta(x, y) = \mu(y, x)$. Thus (1.18) follows from (1.17) applied to $(X, \geq)$. □

Let $(X_1, \leq_1)$ and $(X_1, \leq_2)$ be two posets. A bijection $f : X_1 \to X_2$ is called an *isomorphism* if for $x, y \in X_1$, $x \leq_1 y$ if and only if $f(x) \leq_2 f(y)$. The posets $(X_1, \leq_1)$ and $(X_2, \leq_2)$ are said to be *isomorphic*, denoted by $(X_1, \leq_1) \cong (X_2, \leq_2)$, if there is an isomorphism from $(X_1, \leq_1)$ to $(X_2, \leq_2)$.

Clearly, isomorphic locally finite posets have the "same" Möbius function. More precisely, let $(X_i, \leq_i)$ be a locally finite poset with Möbius function μ_i, $i = 1, 2$, and let $f : (X_1, \leq_1) \to (X_2, \leq_2)$ be an isomorphism. Then for $x, y \in X_1$,

$$\mu_2\big(f(x), f(y)\big) = \mu_1(x, y). \tag{1.19}$$

On the other hand, the partial order of a locally finite poset is completely determined by its Möbius function.

Proposition 1.19. *Let $(X, \leq)$ be a locally finite poset with Möbius function μ. Then for distinct $x, y \in X$, $x < y$ if and only if there is a finite sequence $x = x_1, x_2, \dots, x_n = y$ such that $\mu(x_i, x_{i+1}) = -1$ for all $1 \leq i < n$.*

Proof. ($\Leftarrow$) Since $\mu(x_i, x_{i+1}) \neq 0$, we have $x_i \leq x_{i+1}$, $1 \leq i < n$. By the transitivity of the partial order, $x = x_1 \leq x_n = y$.

($\Rightarrow$) Let $x_1 = x$. Choose $x_2 \in (x_1, y]$ such that $(x_1, x_2) = \emptyset$. (Such an x_2 exists since $|(x_1, y]| < \infty$.) Then by (1.16), $\mu(x_1, x_2) = -1$. In the same way, choose $x_3, x_4, \dots$ such that $x_1 < x_2 < x_3 < \cdots \leq y$ and $\mu(x_i, x_{i+1}) = -1$, $i = 1, 2, \dots$. Since $|[x_1, y]| < \infty$, the sequence $x_1, x_2, \dots$ must stop with $x_n = y$. □

Let $P_i = (X_i, \leq_i)$, $i = 1, 2$, be posets. For $(x_1, x_2), (y_1, y_2) \in X_1 \times X_2$, define $(x_1, x_2) \leq (y_1, y_2)$ if and only if $x_1 \leq_1 y_1$ and $x_2 \leq_2 y_2$. Clearly, $(X_1 \times X_2, \leq)$ is also a poset; it is called the *product* of P_1 and P_2 and is denoted by $P_1 \times P_2$.

Theorem 1.20. *Let $P_i = (X_i, \leq_i)$ be a locally finite poset with Möbius function μ_i, $i = 1, 2$. Then $P_1 \times P_2$ is a locally finite poset with Möbius function $\mu_1 \times \mu_2$, where*

$$(\mu_1 \times \mu_2)\big((x_1, x_2), (y_1, y_2)\big) = \mu_1(x_1, y_1)\mu_2(x_2, y_2), \quad (x_1, x_2), (y_1, y_2) \in X_1 \times X_2. \tag{1.20}$$

Proof. For $(x_1, x_2), (y_1, y_2) \in X_1 \times X_2$, we have $[(x_1, x_2), (y_1, y_2)] = [x_1, y_1] \times [x_2, y_2]$, which is finite. Thus $P_1 \times P_2$ is locally finite.

To prove that $\mu_1 \times \mu_2$ is the Möbius function of $P_1 \times P_2$, first note that if $(x_1, x_2) \not\leq (y_1, y_2)$, then $\mu_1(x_1, y_1)\mu_2(x_2, y_2) = 0$. Now assume that $(x_1, x_2) \leq (y_1, y_2)$. We have

$$\begin{aligned}
&\sum_{(z_1,z_2)\in[(x_1,x_2),(y_1,y_2)]} \mu_1(x_1, z_1)\mu_2(x_2, z_2) \\
&= \Big(\sum_{z_1\in[x_1,y_1]} \mu_1(x_1, z_1)\Big)\Big(\sum_{z_2\in[x_2,y_2]} \mu_2(x_2, z_2)\Big) \\
&= \delta(x_1, y_1)\delta(x_2, y_2) = \delta\big((x_1, x_2), (y_1, y_2)\big).
\end{aligned}$$

Thus $\mu_1 \times \mu_2$ is indeed the Möbius function of $P_1 \times P_2$. □

We end this section with some well-known examples of locally finite posets and their Möbius functions.

Example 1.21. Let $\leq$ be the ordinary order in $\mathbb{Z}$. It follows immediately from (1.16) that the Möbius function of $(\mathbb{Z}, \leq)$ is

$$\mu_{\mathbb{Z}}(x,y) = \begin{cases} 1 & \text{if } y = x, \\ -1 & \text{if } y = x+1, \\ 0 & \text{otherwise} \end{cases}$$
$$= \begin{cases} (-1)^{y-x} & \text{if } y = x \text{ or } y = x+1, \\ 0 & \text{otherwise.} \end{cases}$$

Example 1.22. Let X be a finite set and $\mathcal{P}(X)$ the set of all subsets of X. Then $(\mathcal{P}(X), \subset)$ is a locally finite poset. To determine the Möbius function μ of $(\mathcal{P}(X), \subset)$, write $X = \{x_1, \ldots, x_n\}$ and define

$$\begin{aligned} f: \ \mathcal{P}(X) &\longrightarrow \{0,1\}^n \\ A &\longmapsto (a_1, \ldots, a_n), \end{aligned} \tag{1.21}$$

where

$$a_i = \begin{cases} 1 & \text{if } x_i \in A, \\ 0 & \text{if } x_i \notin A. \end{cases}$$

We make $E = \{0,1\}$ into a poset by defining $0 \leq 1$. The Möbius function of E is

$$\eta(a,b) = \begin{cases} (-1)^{b-a} & \text{if } a \leq b, \\ 0 & \text{otherwise.} \end{cases}$$

It is easy to see that the map f in (1.21) is an isomorphism from $(\mathcal{P}(X), \subset)$ to $\underbrace{E \times \cdots \times E}_{n}$. Let $A, B \in \mathcal{P}(X)$ be such that $A \subset B$, and write $f(A) = (a_1, \ldots, a_n)$ and $f(B) = (b_1, \ldots, b_n)$. By (1.19) and (1.20), we have

$$\begin{aligned} \mu(A,B) &= (\eta \times \cdots \times \eta)\big((a_1, \ldots, a_n), (b_1, \ldots, b_n)\big) \\ &= \prod_{i=1}^{n} \eta(a_i, b_i) = \prod_{i=1}^{n} (-1)^{b_i - a_i} \\ &= (-1)^{\sum_{i=1}^{n} b_i - \sum_{i=1}^{n} a_i} = (-1)^{|B|-|A|}. \end{aligned}$$

Example 1.23. Let $\mathbb{Z}^+$ be the set of all positive integers. Then $(\mathbb{Z}^+, \mid)$ is a locally finite poset where $x \mid y$ $(x, y \in \mathbb{Z}^+)$ means that x divides y. Let $x, y \in \mathbb{Z}^+$ be such that $x \mid y$. To determine the value $\mu(x,y)$ of the Möbius function μ of $(\mathbb{Z}^+, \mid)$, write $x = p_1^{a_1} \cdots p_n^{a_n}$ and $y = p_1^{b_1} \cdots p_n^{b_n}$, where $p_1, \ldots, p_n$ are distinct primes and $0 \leq a_i \leq b_i$, $1 \leq i \leq n$. With $p_1, \ldots, p_n$ fixed, let

$$X = \{p_1^{c_1} \cdots p_n^{c_n} : c_i \geq 0,\ 1 \leq i \leq n\} \subset \mathbb{Z}^+.$$

Then

$$f: \begin{array}{ccc} X & \longrightarrow & \mathbb{N}^n \\ p_1^{c_1}\cdots p_n^{c_n} & \longmapsto & (c_1,\dots,c_n) \end{array}$$

is an isomorphism from the poset $(X, |\,)$ to $(\mathbb{N}, \le)\times\cdots\times(\mathbb{N}, \le)$, where $(\mathbb{N}, \le)$ is a sub-poset of $(\mathbb{Z}, \le)$. Therefore,

$$\begin{aligned} &\mu(p_1^{a_1}\cdots p_n^{a_n}, p_1^{b_1}\cdots p_n^{b_n}) \\ &= (\mu_{\mathbb{Z}}\times\cdots\times\mu_{\mathbb{Z}})\big((a_1,\dots,a_n),(b_1,\dots,b_n)\big) = \mu_{\mathbb{Z}}(a_1,b_1)\cdots\mu_{\mathbb{Z}}(a_n,b_n) \\ &= \begin{cases} (-1)^{\sum_{i=1}^n (b_i-a_i)} & \text{if } b_i-a_i\in\{0,1\} \text{ for all } 1\le i\le n, \\ 0 & \text{if } b_i-a_i\ge 2 \text{ for some } 1\le i\le n. \end{cases} \end{aligned}$$

Equivalently,

$$\mu(x,y) = \begin{cases} (-1)^s & \text{if } y/x \text{ is a product of } s \text{ distinct primes,} \\ 0 & \text{if } y/x \text{ is divisible by the square of a prime.} \end{cases} \tag{1.22}$$

Example 1.24. Let F be a field and $F[\mathtt{X}]_{\mathrm{m}}$ the set of all monic polynomials in $F[\mathtt{X}]$. Then $(F[\mathtt{X}]_{\mathrm{m}}, |\,)$ is a locally finite poset where $f \mid g$ ($f, g\in F[\mathtt{X}]_{\mathrm{m}}$) means that f divides g. For $f, g \in F[\mathtt{X}]_{\mathrm{m}}$ with $f \mid g$, the value of $\mu(f, g)$ for the Möbius function μ of $(F[\mathtt{X}]_{\mathrm{m}}, |\,)$ is given by

$$\begin{aligned} &\mu(f,g) \\ &= \begin{cases} (-1)^s & \text{if } g/f \text{ is a product of } s \text{ distinct irreducibles in } F[\mathtt{X}]_{\mathrm{m}}, \\ 0 & \text{if } g/f \text{ is divisible by the square of an irreducible in } F[\mathtt{X}]_{\mathrm{m}}. \end{cases} \end{aligned}$$

The above formula follows from the same argument as in Example 1.23.

Example 1.25. Let V be an n-dimensional vector space over $\mathbb{F}_q$ and let $\mathcal{L}(V)$ be the set of all subspaces of V. Clearly, $(\mathcal{L}(V), \subset)$ is a locally finite poset. Denote the Möbius function of $(\mathcal{L}(V), \subset)$ by $\mu_{\mathcal{L}(V)}$. First, note that for $U, W \in \mathcal{L}(V)$ with $U \subset W$, $\mu_{\mathcal{L}(V)}(U, W)$ is determined by $\dim_{\mathbb{F}_q} W/U$. In fact, the poset $([U, W], \subset)$ is isomorphic to $(\mathcal{L}(W/U), \subset)$ by the correspondence between the subspaces of W/U and the subspaces between U and W, and $(\mathcal{L}(W/U), \subset)$ is further isomorphic to $(\mathcal{L}(\mathbb{F}_q^m), \subset)$, where $m = \dim_{\mathbb{F}_q} W/U$. Thus $\mu_{\mathcal{L}(V)}(U, W) = \mu_{\mathcal{L}(\mathbb{F}_q^m)}(\{0\}, \mathbb{F}_q^m)$, which is denoted by μ_m.

The method used here to determine μ_n is from [**3**]. Let $k \ge n$ be an integer and let $\mathrm{Hom}_{\mathbb{F}_q}(V, \mathbb{F}_q^k)$ denote the set of all $\mathbb{F}_q$-maps from V to $\mathbb{F}_q^k$. For each $U \in \mathcal{L}(V)$, let

$$N(U) = |\{f \in \mathrm{Hom}_{\mathbb{F}_q}(V, \mathbb{F}_q^k) : \ker f = U\}|$$

and

$$N_{\supset}(U) = \sum_{\substack{W\in\mathcal{L}(V)\\ W\supset U}} N(W) = |\{f \in \mathrm{Hom}_{\mathbb{F}_q}(V, \mathbb{F}_q^k) : \ker f \supset U\}|.$$

Let $\epsilon_1, \dots, \epsilon_s \in V$ be such that their images in V/U form a basis of V/U, where $s = \dim_{\mathbb{F}_q} V/U = n - \dim_{\mathbb{F}_q} U$. Then a map $f \in \mathrm{Hom}_{\mathbb{F}_q}(V, \mathbb{F}_q^k)$ with $\ker f \supset U$ is uniquely determined by $f(\epsilon_1), \dots, f(\epsilon_s) \in \mathbb{F}_q^k$ which can be arbitrarily chosen. Thus

$$N_{\supset}(U) = |\mathbb{F}_q^k|^{n-\dim_{\mathbb{F}_q} U} = (q^k)^{n-\dim_{\mathbb{F}_q} U}.$$

Let $\mathcal{L}_d(V) = \{U \in \mathcal{L}(V) : \dim_{\mathbb{F}_q} U = d\}$. By (1.17), we have

$$\begin{aligned} N(\{0\}) &= \sum_{U\in\mathcal{L}(V)} \mu_{\mathcal{L}(V)}(\{0\}, U)\, N_{\supset}(U) \\ &= \sum_{d=0}^{n} \sum_{U\in\mathcal{L}_d(V)} \mu_{\mathcal{L}(V)}(\{0\}, U)\, N_{\supset}(U) \\ &= \sum_{d=0}^{n} |\mathcal{L}_d(V)|\, \mu_d \cdot (q^k)^{n-d}. \end{aligned} \tag{1.23}$$

Note that $N(\{0\})$ is the number of injections in $\mathrm{Hom}_{\mathbb{F}_q}(V, \mathbb{F}_q^k)$. Let $\delta_1, \dots, \delta_n$ be a basis of V. Then every injection $f \in \mathrm{Hom}_{\mathbb{F}_q}(V, \mathbb{F}_q^k)$ is uniquely determined by a linearly independent list $f(\delta_1), \dots, f(\delta_n) \in \mathbb{F}_q^k$. The number of choices for $f(\delta_1)$ is $q^k - 1$, the number of choices for $f(\delta_2)$ is $q^k - q$, …, and the number of choices for $f(\delta_n)$ is $q^k - q^{n-1}$. Hence

$$N(\{0\}) = (q^k - 1)(q^k - q) \cdots (q^k - q^{n-1}).$$

Thus (1.23) can be written as

$$(x-1)(x-q)\cdots(x-q^{n-1}) = \sum_{d=0}^{n} |\mathcal{L}_d(V)|\, \mu_d\, x^{n-d}, \tag{1.24}$$

where $x = q^k$. As k runs through all integers $\geq n$, x takes infinitely many values. Thus (1.24) holds for all $x \in \mathbb{R}$. Letting $x = 0$, we have

$$\mu_n = (-1)(-q)\cdots(-q^{n-1}) = (-1)^n q^{\binom{n}{2}}.$$

Exercises

1.1. Find all primitive elements of $\mathbb{F}_{13}$.

1.2. Let

$$F = \left\{ \begin{bmatrix} a & b \\ 2b & a \end{bmatrix} : a, b \in \mathbb{F}_3 \right\} \subset \mathrm{M}_{2\times 2}(\mathbb{F}_3).$$

Prove that $F \cong \mathbb{F}_{3^2}$.

1.3. Let n be a positive integer, and let A and B be the (unique) subgroups of $\mathbb{F}_{2^{2n}}^*$ of orders $2^n - 1$ and $2^n + 1$, respectively.
  (i) Prove that every $x \in \mathbb{F}_{2^{2n}}^*$ has a unique representation $x = ab$, where $a \in A$ and $b \in B$.
  (ii) Let α be a primitive element of $\mathbb{F}_{2^{2n}}^*$ and let $k \in \mathbb{Z}$. Prove that $\alpha^k = \alpha^{2^{n-1}(2^n+1)k}\alpha^{2^{n-1}(2^n-1)k}$, where $\alpha^{2^{n-1}(2^n+1)k} \in A$ and $\alpha^{2^{n-1}(2^n-1)k} \in B$.

1.4. (i) Clearly, $f : \mathbb{F}_{q^n} \times \mathbb{F}_{q^n} \to \mathbb{F}_{q^n}$, $(x, y) \mapsto \mathrm{Tr}_{q^n/q}(xy)$, is a symmetric $\mathbb{F}_q$-bilinear map. Prove that f is *nondegenerate*, i.e., $f(x, y) = 0$ for all $y \in \mathbb{F}_{q^n}$ implies $x = 0$.
  (ii) Define

$$\begin{aligned} \alpha : \ \mathbb{F}_{q^n} &\longrightarrow \mathrm{Hom}_{\mathbb{F}_q}(\mathbb{F}_{q^n}, \mathbb{F}_q) \\ x &\longmapsto \mathrm{Tr}_{q^n/q}(x\,\cdot\,), \end{aligned}$$

  where $\mathrm{Tr}_{q^n/q}(x\,\cdot\,)$ maps $y \in \mathbb{F}_{q^n}$ to $\mathrm{Tr}_{q^n/q}(xy)$. Prove that α is an $\mathbb{F}_q$-module isomorphism.
  (iii) If $g \in \mathrm{Hom}_{\mathbb{F}_q}(\mathbb{F}_{q^n}, \mathbb{F}_q)$ is such that $g \circ \tau = g$, where τ is the Frobenius map of $\mathbb{F}_{q^n}$ over $\mathbb{F}_q$, prove that $g = a\,\mathrm{Tr}_{q^n/q}$ for some $a \in \mathbb{F}_q$.

1.5. Prove that every element in $\mathbb{F}_q$ is a sum of two squares.

1.6. Let $f(\mathtt{X}) = \mathtt{X}^n + a_{n-1}\mathtt{X}^{n-1} + \cdots + a_0 \in \mathbb{F}_q[\mathtt{X}]$ be irreducible with $a_0 \neq 0$. Let $\alpha \in \mathbb{F}_{q^n}$ be a root of f. Prove the following statements.
  (i) The minimal polynomial of α^{-1} over $\mathbb{F}_q$ is $f^* = a_0^{-1}\mathtt{X}^n f(\mathtt{X}^{-1}) = a_0^{-1}(a_0\mathtt{X}^n + \cdots + a_{n-1}\mathtt{X} + 1) \in \mathbb{F}_q[\mathtt{X}]$, which is called the *reciprocal polynomial* of f.
  (ii) $f = f^*$ if and only if $\alpha^{q^m+1} = 1$ for some $m \geq 0$.
  (iii) If $f = f^*$, then either $n = 1$ and $f = \mathtt{X} \pm 1$, or n is even and $a_0 = 1$.

1.7. An element $\alpha \in \mathbb{F}_{q^n}$ is said to be *normal* over $\mathbb{F}_q$ if $\{\sigma(\alpha) : \sigma \in \mathrm{Aut}(\mathbb{F}_{q^n}/\mathbb{F}_q)\}$ is a basis of $\mathbb{F}_{q^n}$ over $\mathbb{F}_q$. Assume that $n = n_1 n_2$, where $\gcd(n_1, n_2) = 1$, and assume that $\alpha_i \in \mathbb{F}_{q^{n_i}}$ is normal over $\mathbb{F}_q$, $i = 1, 2$. Prove that $\alpha_1\alpha_2 \in \mathbb{F}_{q^n}$ is normal over $\mathbb{F}_q$.

1.8. (Lucas's theorem) For integers $n, n_1, \ldots, n_k \geq 0$, define

$$\binom{n}{n_1, \ldots, n_k} = \begin{cases} \dfrac{n!}{n_1! \cdots n_k!} & \text{if } n_1 + \cdots + n_k = n, \\ 0 & \text{otherwise.} \end{cases}$$

Let p be a prime and let the base-p representations of n and n_j ($1 \leq j \leq k$) be

$$n = \sum_{l \geq 0} n^{(l)} p^l, \quad n_j = \sum_{l \geq 0} n_j^{(l)} p^l, \quad 0 \leq n^{(l)}, n_j^{(l)} \leq p - 1.$$

Prove that

$$\binom{n}{n_1, \ldots, n_k} \equiv \prod_{l \geq 0} \binom{n^{(l)}}{n_1^{(l)}, \ldots, n_k^{(l)}} \pmod{p}.$$

1.9. (Kós) Let $a > 1$ be an integer. Prove that all prime divisors of $5a^4 - 5a^2 + 1$ are of the form $20k \pm 1$. (Hint: Let p be a prime divisor of $5a^4 - 5a^2 + 1$ and write $a^{-1} = b + b^{-1}$, where $b \in \mathbb{F}_{p^2}$.)

1.10. (Sutner) Let G be a graph. Each vertex of G is equipped with a light and a switch button. Initially, all lights are off. Pressing a button at a vertex will change the lights at the vertex and at the vertices adjacent to it from off to on and vice versa. Prove that one can turn on all lights by pressing a suitable sequence of buttons. Let $A = (a_{ij})_{n \times n}$ be the adjacency matrix of G considered as a matrix over $\mathbb{F}_2$. The claim is equivalent to the statement that $(A + I)x = (1, \ldots, 1)^T$ has a solution $x \in \mathbb{F}_2^n$.

1.11. Prove that every finite integral domain is a field.

1.12. Wedderburn's theorem states that every finite division ring D is a field. Prove this theorem through the following steps.

(i) Let Z be the center of D with $|Z| = q$, and assume to the contrary that $\dim_Z D = n > 1$. Let $[a_1], \ldots, [a_k]$ be the conjugacy classes of $D^\times$ (the multiplicative group of D) not contained in $Z^\times$. Derive from the class equation of $D^\times$ that

$$q^n - 1 = q - 1 + \sum_{i=1}^{k} \frac{q^n - 1}{q^{d(a_i)} - 1},$$

where $d(a_i) = \dim_Z \mathrm{C}_D(a_i)$ with $\mathrm{C}_D(a_i) = \{x \in D : xa_i = a_i x\}$.

(ii) Let $\zeta = e^{2\pi i/n}$ and let $\Phi_n(\mathtt{X}) = \prod_{1 \leq k \leq n, \gcd(k,n)=1} (\mathtt{X} - \zeta^k) \in \mathbb{Z}[\mathtt{X}]$ be the nth cyclotomic polynomial. Prove that $|\Phi_n(q)| > q - 1$.

(iii) Prove that $\Phi_n \mid (\mathtt{X}^n - 1)/(\mathtt{X}^d - 1)$ in $\mathbb{Z}[\mathtt{X}]$ for all $d \mid n$ with $d < n$. Use (i) to conclude that $\Phi_n(q) \mid q - 1$, which is a contradiction to (ii).

1.13. Let p be a prime. Prove that

$$\mathbb{Z}[i]/(p) \cong \begin{cases} \mathbb{F}_2[\mathtt{X}]/(\mathtt{X}^2) & \text{if } p = 2, \\ \mathbb{F}_p \times \mathbb{F}_p & \text{if } p \equiv 1 \pmod 4, \\ \mathbb{F}_{p^2} & \text{if } p \equiv -1 \pmod 4. \end{cases}$$

(Note: $\mathbb{F}_2[\mathtt{X}]/(\mathtt{X}^2) \not\cong \mathbb{F}_2 \times \mathbb{F}_2$ since the former is not semisimple.)

1.14. (The automorphism group of $\overline{\mathbb{F}}_p$) Let p be a prime and $\overline{\mathbb{F}}_p$ the algebraic closure of $\mathbb{F}_p$. All finite fields $\mathbb{F}_{p^i}$ considered in this question are contained in $\overline{\mathbb{F}}_p$. Define $\sigma : \overline{\mathbb{F}}_p \to \overline{\mathbb{F}}_p$, $x \mapsto x^p$. Note that for $i \in \mathbb{Z}^+$ and $a \in \mathbb{Z}$, $\sigma^a|_{\mathbb{F}_{p^i}}$ depends only on the image of a in $\mathbb{Z}/i\mathbb{Z}$. For $i, j \in \mathbb{Z}^+$ with $i \mid j$, let

$$\begin{array}{rccc} \rho_{ij} : & \mathbb{Z}/j\mathbb{Z} & \longrightarrow & \mathbb{Z}/i\mathbb{Z} \\ & n + j\mathbb{Z} & \longmapsto & n + i\mathbb{Z}, \quad n \in \mathbb{Z}, \end{array}$$

be the natural homomorphism. Let

$$G = \Big\{ (a_i)_{i\in\mathbb{Z}^+} \in \prod_{i\in\mathbb{Z}^+} \mathbb{Z}/i\mathbb{Z} : \rho_{ij}(a_j) = a_i \text{ for all } i \mid j \Big\},$$

which is a subgroup of $\prod_{i\in\mathbb{Z}^+} \mathbb{Z}/i\mathbb{Z}$. Define

$$\begin{array}{rccc} \psi : & G & \longrightarrow & \mathrm{Aut}(\overline{\mathbb{F}}_p) \\ & (a_i)_{i\in\mathbb{Z}^+} & \longmapsto & \alpha, \end{array}$$

where

$$\begin{array}{rccc} \alpha : & \overline{\mathbb{F}}_p & \longrightarrow & \overline{\mathbb{F}}_p \\ & x & \longmapsto & \sigma^{a_i}(x) \quad \text{if } x \in \mathbb{F}_{p^i}. \end{array}$$

Prove that ψ is a group isomorphism. (Note: G is the limit of the inverse system $((\mathbb{Z}/i\mathbb{Z})_{i\in\mathbb{Z}^+}, (\rho_{ij})_{i|j})$ over the poset $(\mathbb{Z}^+, \mid\ )$; that is, $G = \varprojlim \mathbb{Z}/i\mathbb{Z}$.)

1.15. Let p be a prime and let F be a subfield of $\overline{\mathbb{F}}_p$ such that $[\overline{\mathbb{F}}_p : F] < \infty$. Prove that $F = \overline{\mathbb{F}}_p$. (This is a special case of a theorem of Artin and Schreier.)

1.16. Let ϕ be the Euler function and μ the Möbius function of $(\mathbb{Z}^+, \mid\ )$. Prove that

$$\phi(n) = \sum_{d|n} d\, \mu(d, n) \quad \text{for all } n \in \mathbb{Z}^+.$$

1.17. Let F be a field, and let $f = \sum_{i \geq 1} a_i \mathtt{X}^i$ and $g = \sum_{j \geq 1} b_j \mathtt{X}^j \in F[[\mathtt{X}]]$ be two formal power series with constant term 0. Define the *multiplicative convolution* of f and g as

$$f * g = \sum_{n \geq 1} \Big(\sum_{ij=n} a_i b_j \Big) \mathtt{X}^n.$$

Prove that

$$\Big(\sum_{i \geq 1} \mathtt{X}^i \Big) * \Big(\sum_{j \geq 1} \mu(1,j) \mathtt{X}^j \Big) = \mathtt{X}.$$

1.18. Let $\zeta(s) = \sum_{n \geq 1} 1/n^s$, $\operatorname{Re} s > 1$, be the Riemann zeta function. Prove that

$$\zeta(s)^{-1} = \sum_{n \geq 1} \frac{\mu(1,n)}{n^s}.$$

(Note: Traditionally, $\mu(1,n)$ is denoted by $\mu(n)$.)

Chapter 2

Polynomials over Finite Fields

2.1. Number of Irreducible Polynomials

Let $q > 1$ be a prime power and $n > 0$ an integer. Denote by $\mathcal{I}_q(n)$ the set of all monic irreducible polynomials of degree n in $\mathbb{F}_q[\mathtt{X}]$. We will derive an explicit formula for $|\mathcal{I}_q(n)|$.

Lemma 2.1. *We have*

$$\mathtt{X}^{q^n} - \mathtt{X} = \prod_{d \mid n} \prod_{f \in \mathcal{I}_q(d)} f. \tag{2.1}$$

Proof. Let $F = \mathtt{X}^{q^n} - \mathtt{X}$. Since $\gcd(F, F') = 1$, the factorization of F does not have repeated irreducible factors. Thus, to prove (2.1), it suffices to show that $\bigcup_{d \mid n} \mathcal{I}_q(d)$ is precisely the set of monic irreducible factors of F.

First, let $f \in \mathcal{I}_q(d)$ for some $d \mid n$. Let a be any root of f (in some extension of $\mathbb{F}_q$). Then $\mathbb{F}_q(a) = \mathbb{F}_{q^d}$, and f is the minimal polynomial of a over $\mathbb{F}_q$. Since $a \in \mathbb{F}_{q^d} \subset \mathbb{F}_{q^n}$, a is a root of F. Therefore $f \mid F$.

Now assume that $f \in \mathcal{I}_q(d)$ is a monic irreducible factor of F. Since $\mathbb{F}_{q^n}$ is the splitting field of F over $\mathbb{F}_q$, f splits in $\mathbb{F}_{q^n}$. Let $a \in \mathbb{F}_{q^n}$ be any root of f. Then we have $d = [\mathbb{F}_q(a) : \mathbb{F}_q] \mid [\mathbb{F}_{q^n} : \mathbb{F}_q] = n$. □

By comparing the degrees of both sides of (2.1), we have the following corollary.

Corollary 2.2. *We have*

$$q^n = \sum_{d|n} d\,|\mathcal{I}_q(d)|.$$

In Example 1.23, we determined the Möbius function μ of the poset $(\mathbb{Z}^+, |\,)$. By (1.22), for $x, y \in \mathbb{Z}^+$ with $x \mid y$, $\mu(x, y)$ depends only on y/x; we denote $\mu(x, y)$ by $\mu(y/x)$.

Theorem 2.3. *We have*

$$|\mathcal{I}_q(n)| = \frac{1}{n}\sum_{d|n} \mu(n/d) q^d.$$

Proof. For each $n \in \mathbb{Z}^+$, let

$$N(n) = n|\mathcal{I}_q(n)|$$

and

$$N_{\le}(n) = \sum_{d|n} N(d) = \sum_{d|n} d|\mathcal{I}_q(d)|.$$

By Corollary 2.2, $N_{\le}(n) = q^n$. Hence, by the Möbius inversion (Theorem 1.18, equation (1.18)),

$$n|\mathcal{I}_q(n)| = N(n) = \sum_{d|n} \mu(n/d) N_{\le}(d) = \sum_{d|n} \mu(n/d) q^d,$$

i.e.,

$$|\mathcal{I}_q(n)| = \frac{1}{n}\sum_{d|n} \mu(n/d) q^d.$$

□

In the next two propositions, we collect some useful facts about irreducible polynomials in $\mathbb{F}_q[\mathtt{X}]$.

Proposition 2.4.

(i) *Every irreducible polynomial $f \in \mathbb{F}_q[\mathtt{X}]$ is separable, i.e., f has no multiple roots (in its splitting field).*

(ii) *If $f \in \mathbb{F}_q[\mathtt{X}]$ is irreducible with* $\deg f = n$, *then f splits in $\mathbb{F}_{q^n}$.*

(iii) *For each $a \in \mathbb{F}_{q^n}$, let f_a be the minimal polynomial of a over $\mathbb{F}_q$. We have*

(2.2) $$f_a = \prod_{b \in [a]} (\mathtt{X} - b),$$

where $[a] = \{\gamma(a) : \gamma \in \mathrm{Aut}(\mathbb{F}_{q^n}/\mathbb{F}_q)\}$ *is the* $\mathrm{Aut}(\mathbb{F}_{q^n}/\mathbb{F}_q)$*-orbit of a. Equivalently,*

(2.3) $$f_a = (\mathtt{X} - a^{q^0})(\mathtt{X} - a^{q^1}) \cdots (\mathtt{X} - a^{q^{m-1}}),$$

where m is the smallest positive integer such that $a^{q^m} = a$.

Proof. (i) Since f has a root $\mathtt{X}+(f) \in \mathbb{F}_q[\mathtt{X}]/(f)$ and $\mathbb{F}_q[\mathtt{X}]/(f) \cong \mathbb{F}_{q^n}$, where $n = \deg f$, f has a root, say a, in $\mathbb{F}_{q^n}$. Thus, f is the minimal polynomial of a over $\mathbb{F}_q$. Since $\mathbb{F}_{q^n}/\mathbb{F}_q$ is Galois, f is separable.

(ii) By the proof of (i), f has a root in $\mathbb{F}_{q^n}$. Since $\mathbb{F}_{q^n}/\mathbb{F}_q$ is Galois and f is irreducible over $\mathbb{F}_q$, f splits in $\mathbb{F}_{q^n}$.

(iii) Since $\mathbb{F}_{q^n}/\mathbb{F}_q$ is Galois, f_a splits in $\mathbb{F}_{q^n}$ with no multiple roots, and $\mathrm{Aut}(\mathbb{F}_{q^n}/\ \mathbb{F}_q)$ acts transitively on the set of roots of f_a. Thus $[a]$ consists of all the roots of f_a, and hence (2.2) follows. For (2.3), note that $[a] = \{a^{q^0}, a^{q^1}, \dots, a^{q^{m-1}}\}$. □

Proposition 2.5. *Let $f \in \mathbb{F}_q[\mathtt{X}]$ be monic and irreducible with $\deg f = n$. Let $m > 0$ be an integer and $d = \gcd(m,n)$. In $\mathbb{F}_{q^m}[\mathtt{X}]$, f is a product of d distinct monic irreducibles of degree n/d. More precisely, let $a \in \mathbb{F}_{q^n}$ be a root of f, and let*

$$f_i = \prod_{j=0}^{\frac{n}{d}-1}\Big(\mathtt{X} - \tau^{i+jd}(a)\Big), \qquad 0 \le i \le d-1,$$

where $\langle\tau\rangle = \mathrm{Aut}(\mathbb{F}_{q^n}/\mathbb{F}_q)$. Then $f_0, \dots, f_{d-1} \in \mathbb{F}_{q^d}[\mathtt{X}]$ are distinct and irreducible over $\mathbb{F}_{q^m}$ and $f = f_0 \cdots f_{d-1}$.

Remark.

(i) Let F be a field. The action of $\mathrm{Aut}(F)$ on $F[\mathtt{X}]$ is defined as follows: For $f = \sum_i a_i\mathtt{X}^i \in F[\mathtt{X}]$ and $\sigma \in \mathrm{Aut}(F)$, $\sigma(f) = \sum_i \sigma(a_i)\mathtt{X}^i$.

(ii) In Proposition 2.5, $\{f_0, \dots, f_{d-1}\}$ is the $\langle\tau\rangle$-orbit of f_0.

Proof of Proposition 2.5. First, we have

$$f = \prod_{i=0}^{n-1}(\mathtt{X} - \tau^i(a)) = \prod_{i=0}^{d-1}\prod_{j=0}^{\frac{n}{d}-1}(\mathtt{X} - \tau^{i+jd}(a)) = f_0 \cdots f_{d-1}.$$

Since $\mathbb{F}_{q^n} = \mathbb{F}_q(a)$, $\tau^0(a), \dots, \tau^{n-1}(a)$ are all distinct, and hence $f_0, \dots, f_{d-1}$ are all distinct. For each $0 \le i \le d-1$, we have $\tau^d(f_i) = f_i$, where $\langle\tau^d\rangle = \mathrm{Aut}(\mathbb{F}_{q^n}/\mathbb{F}_{q^d})$. Hence $f_i \in \mathbb{F}_{q^d}[\mathtt{X}]$. To prove that $f_0, \dots, f_{d-1}$ are irreducible in $\mathbb{F}_{q^m}[\mathtt{X}]$, it suffices to show that f_0 is, since $f_0, \dots, f_{d-1}$ are $\langle\tau\rangle$-conjugates. Note that a is a root of $f_0 \in \mathbb{F}_{q^m}[\mathtt{X}]$ and that

$$[\mathbb{F}_{q^m}(a) : \mathbb{F}_{q^m}] = [\mathbb{F}_{q^{\mathrm{lcm}(m,n)}} : \mathbb{F}_{q^m}] = \frac{\mathrm{lcm}(m,n)}{m} = \frac{n}{d} = \deg f_0.$$

Therefore, f_0 is the minimal polynomial of a over $\mathbb{F}_{q^m}$ and hence is irreducible over $\mathbb{F}_{q^m}$. □

Definition 2.6. An irreducible polynomial $f \in \mathbb{F}_q[\mathtt{X}]$ of degree n is called a *primitive polynomial* over $\mathbb{F}_q$ if f is the minimal polynomial of a primitive element of $\mathbb{F}_{q^n}$.

The number of monic primitive polynomials of a given degree is easily counted.

Theorem 2.7. *The number of monic primitive polynomials of degree n over $\mathbb{F}_q$ is $\phi(q^n - 1)/n$, where ϕ is the Euler function.*

Proof. Let P be the set of all primitive elements of $\mathbb{F}_{q^n}$. Since $\mathbb{F}_{q^n}^*$ is cyclic of degree $q^n - 1$, $|P| = \phi(q^n - 1)$. The group $\mathrm{Aut}(\mathbb{F}_{q^n}/\mathbb{F}_q)$ acts on P. Since each $a \in P$ is of degree n over $\mathbb{F}_q$, by Proposition 2.4 (iii), the $\mathrm{Aut}(\mathbb{F}_{q^n}/\mathbb{F}_q)$-orbit $[a]$ of a has n elements. Therefore, P is partitioned into $\phi(q^n-1)/n$ orbits by the $\mathrm{Aut}(\mathbb{F}_{q^n}/\mathbb{F}_q)$ action. By Proposition 2.4 (iii) again, each $\mathrm{Aut}(\mathbb{F}_{q^n}/\mathbb{F}_q)$-orbit in P corresponds to a primitive polynomial of degree n over $\mathbb{F}_q$. Therefore, there are precisely $\phi(q^n - 1)/n$ primitive polynomials of degree n over $\mathbb{F}_q$. □

2.2. Berlekamp's Factorization Algorithm

Let $f \in \mathbb{F}_q[\mathtt{X}]$ be a polynomial with $\deg f > 0$. How can f be factored into irreducibles? In general, this is a difficult problem. In this section, we describe an algorithm by Berlekamp [**4**] for factoring polynomials in $\mathbb{F}_q[\mathtt{X}]$. The algorithm works efficiently when q is small.

Berlekamp's algorithm is an iterative method. Given $f \in \mathbb{F}_q[\mathtt{X}]$ with $\deg f > 0$, if f is not irreducible, we try to find a nontrivial factorization of f and repeat the process for the factors.

Lemma 2.8. *Let $f \in \mathbb{F}_q[\mathtt{X}]$ be monic and let $h \in \mathbb{F}_q[\mathtt{X}]$. Then $h^q \equiv h \pmod{f}$ if and only if*

$$f = \prod_{a \in \mathbb{F}_q} \gcd(f,\, h - a). \tag{2.4}$$

Proof. ($\Rightarrow$) Since $\mathbb{F}_q$ consists of all roots of $\mathtt{X}^q - \mathtt{X}$, we have

$$\mathtt{X}^q - \mathtt{X} = \prod_{a \in \mathbb{F}_q} (\mathtt{X} - a). \tag{2.5}$$

Substituting h for $\mathtt{X}$ in (2.5) gives

$$h^q - h = \prod_{a \in \mathbb{F}_q} (h - a).$$

Since $f \mid (h^q - h)$ and $h - a$, $a \in \mathbb{F}_q$, are pairwise coprime, we have

$$f = \gcd(f,\, h^q - h) = \prod_{a \in \mathbb{F}_q} \gcd(f,\, h - a).$$

($\Leftarrow$) We have

$$f = \prod_{a \in \mathbb{F}_q} \gcd(f,\, h - a) \,\Big|\, \prod_{a \in \mathbb{F}_q} (h - a) = h^q - h.$$

□

Remark. In Lemma 2.8, if $0 < \deg h < \deg f$, then $\deg \gcd(f, h-a) < \deg f$ for all $a \in \mathbb{F}_q$. Thus (2.4) is a nontrivial factorization of f.

Definition 2.9. Let $f \in \mathbb{F}_q[\mathtt{X}]$ be a polynomial with $\deg f > 0$. A polynomial $h \in \mathbb{F}_q[\mathtt{X}]$ is called an *f-reducing polynomial* if $0 < \deg h < \deg f$ and $h^q \equiv h \pmod{f}$.

Let $n = \deg f$ and let A be the $n \times n$ matrix over $\mathbb{F}_q$ defined by

$$\begin{bmatrix} \mathtt{X}^{0q} \\ \mathtt{X}^{1q} \\ \vdots \\ \mathtt{X}^{(n-1)q} \end{bmatrix} \equiv A \begin{bmatrix} \mathtt{X}^{0} \\ \mathtt{X}^{1} \\ \vdots \\ \mathtt{X}^{n-1} \end{bmatrix} \pmod{f}. \tag{2.6}$$

Then, for $h = a_0 + \cdots + a_{n-1}\mathtt{X}^{n-1} \in \mathbb{F}_q[\mathtt{X}]$,

$$\begin{aligned} h^q - h &= (a_0, \dots, a_{n-1}) \begin{bmatrix} \mathtt{X}^{0q} \\ \vdots \\ \mathtt{X}^{(n-1)q} \end{bmatrix} - (a_0, \dots, a_{n-1}) \begin{bmatrix} \mathtt{X}^{0} \\ \vdots \\ \mathtt{X}^{n-1} \end{bmatrix} \\ &\equiv (a_0, \dots, a_{n-1})(A - I) \begin{bmatrix} \mathtt{X}^{0} \\ \vdots \\ \mathtt{X}^{n-1} \end{bmatrix} \pmod{f}, \end{aligned}$$

where I is the identity matrix. Hence $h^q \equiv h \pmod{f}$ if and only if

$$(a_0, \dots, a_{n-1})(A - I) = 0. \tag{2.7}$$

By (2.6), the first row of A is $(1, 0, \dots, 0)$. Thus $(a_0, \dots, a_{n-1}) = (1, 0, \dots, 0)$ is always a solution of (2.7). The solutions $(a_0, \dots, a_{n-1})$ of (2.7) with $(a_1, \dots, a_{n-1}) \neq (0, \dots, 0)$ are precisely the coefficients of f-reducing polynomials. The existence of f-reducing polynomials, when f has at least two distinct irreducible factors, is guaranteed by the following theorem.

Theorem 2.10. *In the above notation,*

$$\mathrm{null}(A-I) = \textit{the number of distinct irreducible factors of } f,$$

where null( ) *denotes the nullity of a square matrix.*

Proof. Consider the $\mathbb{F}_q$-isomorphism

$$\begin{array}{rccc} \alpha: & \mathbb{F}_q^n & \longrightarrow & \mathbb{F}_q[\mathtt{X}]/(f) \\ & (a_0,\dots,a_{n-1}) & \longmapsto & a_0+\cdots+a_{n-1}\mathtt{X}^{n-1}+(f). \end{array}$$

From the above, we have

$$\begin{aligned} &\alpha\big(\{(a_0,\dots,a_{n-1})\in\mathbb{F}_q^n : (a_0,\dots,a_{n-1})(A-I)=0\}\big) \\ &= \{H\in\mathbb{F}_q[\mathtt{X}]/(f) : H^q=H\}. \end{aligned}$$

Therefore

$$\mathrm{null}(A-I) = \dim_{\mathbb{F}_q}\{H\in\mathbb{F}_q[\mathtt{X}]/(f) : H^q=H\}. \tag{2.8}$$

Let $f = f_1^{e_1}\cdots f_k^{e_k}$, where $f_1,\dots,f_k\in\mathbb{F}_q[\mathtt{X}]$ are distinct irreducibles and $e_i>0$, $1\le i\le k$. By the Chinese remainder theorem, there is an $\mathbb{F}_q$-algebra isomorphism

$$\mathbb{F}_q[\mathtt{X}]/(f)\cong\big(\mathbb{F}_q[\mathtt{X}]/(f_1^{e_1})\big)\times\cdots\times\big(\mathbb{F}_q[\mathtt{X}]/(f_k^{e_k})\big).$$

Hence

$$\text{the right side of (2.8)} = \sum_{i=1}^{k}\dim_{\mathbb{F}_q}\{H\in\mathbb{F}_q[\mathtt{X}]/(f_i^{e_i}) : H^q=H\}. \tag{2.9}$$

Let $H = h+(f_i^{e_i})\in\mathbb{F}_q[\mathtt{X}]/(f_i^{e_i})$, where $h\in\mathbb{F}_q[\mathtt{X}]$. By Lemma 2.8, $H^q=H$ if and only if

$$f_i^{e_i} = \prod_{a\in\mathbb{F}_q}\gcd(f_i^{e_i}, h-a). \tag{2.10}$$

Since $h-a$, $a\in\mathbb{F}_q$, are pairwise coprime, (2.10) holds if and only if $f_i^{e_i}\mid h-a_0$ for some $a_0\in\mathbb{F}_q$, which happens if and only if $H=a_0+(f_i^{e_i})$. Therefore,

$$\begin{aligned} &\dim_{\mathbb{F}_q}\{H\in\mathbb{F}_q[\mathtt{X}]/(f_i^{e_i}) : H^q=H\} \\ &= \dim_{\mathbb{F}_q}\{a+(f_i^{e_i})\in\mathbb{F}_q[\mathtt{X}]/(f_i^{e_i}) : a\in\mathbb{F}_q\} = 1. \end{aligned} \tag{2.11}$$

Combining (2.8), (2.9) and (2.11) gives $\mathrm{null}(A-I)=k$. □

Berlekamp's algorithm. Given any polynomial $f\in\mathbb{F}_q[\mathtt{X}]$ with $\deg f = n>0$, the following steps either produce a nontrivial factorization of f or confirm that f is irreducible.

STEP 1. Compute $d=\gcd(f,f')$. If $\deg d>0$, then $f=d\cdot(f/d)$ is a nontrivial factorization. If $\deg d=0$, go to step 2.

STEP 2. Find the matrix A defined by (2.6).

STEP 3. If $\operatorname{null}(A-I)=1$, f is irreducible. If $\operatorname{null}(A-I)>1$, find $(a_1,\dots,a_{n-1})\neq(0,\dots,0)$ such that $(0,a_1,\dots,a_{n-1})(A-I)=0$. Let $h=a_1\mathtt{X}+\cdots+a_{n-1}\mathtt{X}^{n-1}$.

STEP 4. Compute $\gcd(f,h-a)$ for all $a\in\mathbb{F}_q$. Then $f=\prod_{a\in\mathbb{F}_q}\gcd(f,h-a)$ is a nontrivial factorization of f.

Example 2.11. We factor

$$f=\mathtt{X}^{11}+2\mathtt{X}^7+2\mathtt{X}^5+2\mathtt{X}^4+\mathtt{X}+2\in\mathbb{F}_3[\mathtt{X}]$$

using Berlekamp's algorithm. We have the following congruences modulo f:

$$\begin{aligned}
\mathtt{X}^{0\cdot 3}&\equiv 1,\\
\mathtt{X}^{1\cdot 3}&\equiv \mathtt{X}^3,\\
\mathtt{X}^{2\cdot 3}&\equiv \mathtt{X}^6,\\
\mathtt{X}^{3\cdot 3}&\equiv \mathtt{X}^9,\\
\mathtt{X}^{4\cdot 3}&\equiv \mathtt{X}+2\mathtt{X}^2+\mathtt{X}^5+\mathtt{X}^6+\mathtt{X}^8,\\
\mathtt{X}^{5\cdot 3}&\equiv 1+2\mathtt{X}+2\mathtt{X}^4+\mathtt{X}^7+\mathtt{X}^8+\mathtt{X}^9,\\
\mathtt{X}^{6\cdot 3}&\equiv 1+2\mathtt{X}^2+\mathtt{X}^3+2\mathtt{X}^5+\mathtt{X}^6+\mathtt{X}^8+\mathtt{X}^{10},\\
\mathtt{X}^{7\cdot 3}&\equiv 1+2\mathtt{X}+\mathtt{X}^2+\mathtt{X}^4+2\mathtt{X}^6+2\mathtt{X}^7+2\mathtt{X}^8+2\mathtt{X}^9,\\
\mathtt{X}^{8\cdot 3}&\equiv 2+\mathtt{X}^2+\mathtt{X}^3+\mathtt{X}^4+2\mathtt{X}^5+2\mathtt{X}^6+2\mathtt{X}^8+2\mathtt{X}^9+2\mathtt{X}^{10},\\
\mathtt{X}^{9\cdot 3}&\equiv 2+2\mathtt{X}^4+2\mathtt{X}^5+2\mathtt{X}^6+2\mathtt{X}^7+\mathtt{X}^8+\mathtt{X}^9,\\
\mathtt{X}^{10\cdot 3}&\equiv 1+2\mathtt{X}^2+2\mathtt{X}^3+\mathtt{X}^4+2\mathtt{X}^5+\mathtt{X}^6+2\mathtt{X}^9+2\mathtt{X}^{10}.
\end{aligned}$$

Therefore,

$$A=\begin{bmatrix}
1&0&0&0&0&0&0&0&0&0&0\\
0&0&0&1&0&0&0&0&0&0&0\\
0&0&0&0&0&0&1&0&0&0&0\\
0&0&0&0&0&0&0&0&0&1&0\\
0&1&2&0&0&1&1&0&1&0&0\\
1&2&0&0&2&0&0&1&1&1&0\\
1&0&2&1&0&2&1&0&1&0&1\\
1&2&1&0&1&0&2&2&2&2&0\\
2&0&1&1&1&2&2&0&2&2&2\\
2&0&0&0&2&2&2&2&1&1&0\\
1&0&2&2&1&2&1&0&0&2&2
\end{bmatrix}.$$

It turns out that $\operatorname{rank}(A-I)=8$. Hence f has three distinct irreducible factors. The vector space $\{(a_1,\dots,a_{10})\in\mathbb{F}_3^{10}:(0,a_1,\dots,a_{10})(A-I)=0\}$

has a basis

$$(2,2,2,2,2,0,1,0,0,0), \quad (2,1,2,2,0,0,0,1,0,1),$$

whose corresponding polynomials are

$$h_1 = 2\mathtt{X} + 2\mathtt{X}^2 + 2\mathtt{X}^3 + 2\mathtt{X}^4 + 2\mathtt{X}^5 + \mathtt{X}^7,$$
$$h_2 = 2\mathtt{X} + \mathtt{X}^2 + 2\mathtt{X}^3 + 2\mathtt{X}^4 + \mathtt{X}^8 + \mathtt{X}^{10}.$$

Using the Euclidean algorithm, we find that

$$\begin{cases} \gcd(f,\, h_1) = 1, \\ \gcd(f,\, h_1 - 1) = h_1 - 1, \\ \gcd(f,\, h_1 - 2) = \mathtt{X}^4 + \mathtt{X}^2 + \mathtt{X} + 1, \end{cases}$$

$$\begin{cases} \gcd(f,\, h_2) = \mathtt{X}^3 + 2\mathtt{X} + 1, \\ \gcd(f,\, h_2 - 1) = \mathtt{X}^4 + \mathtt{X} + 2, \\ \gcd(f,\, h_2 - 2) = \mathtt{X}^4 + \mathtt{X}^2 + \mathtt{X} + 1. \end{cases}$$

Therefore, the complete factorization of f is

$$f = (\mathtt{X}^3 + 2\mathtt{X} + 1)(\mathtt{X}^4 + \mathtt{X} + 2)(\mathtt{X}^4 + \mathtt{X}^2 + \mathtt{X} + 1).$$

Theorem 2.10 can be generalized as follows.

Theorem 2.12. *Let $f = f_1^{e_1} \cdots f_k^{e_k}$, where $f_1, \dots, f_k$ are distinct irreducible polynomials in $\mathbb{F}_q[\mathtt{X}]$ with $\deg f_i = n_i$ and $e_i > 0$, $1 \le i \le k$. Let $n = \deg f = e_1 n_1 + \cdots + e_k n_k$ and let A be the $n \times n$ matrix defined by (2.6). Then for each integer $m > 0$,*

$$\text{(2.12)} \qquad \operatorname{null}(A^m - I) = \sum_{i=1}^{k} \gcd(m, n_i).$$

Proof. Using (2.6) repeatedly, we have

$$\begin{bmatrix} \mathtt{X}^{0\cdot q^m} \\ \mathtt{X}^{1\cdot q^m} \\ \vdots \\ \mathtt{X}^{(n-1)q^m} \end{bmatrix} \equiv A^m \begin{bmatrix} \mathtt{X}^0 \\ \mathtt{X}^1 \\ \vdots \\ \mathtt{X}^{n-1} \end{bmatrix} \pmod{f}.$$

Therefore, by Theorem 2.10,

$$\operatorname{null}(A^m - I) = \text{the number of distinct irreducible factors of } f \text{ in } \mathbb{F}_{q^m}[\mathtt{X}].$$

By Proposition 2.5, in $\mathbb{F}_{q^m}[\mathtt{X}]$, f_i splits into $\gcd(m, n_i)$ distinct irreducibles. Hence, the number of distinct irreducible factors of f in $\mathbb{F}_{q^m}[\mathtt{X}]$ is

$$\sum_{i=1}^{k} \gcd(m, n_i),$$

and (2.12) is proved. □

Using Theorem 2.12, we can derive a formula for the number of irreducible factors of a given degree of f in terms of the matrix A.

Lemma 2.13. *Let $m, n \in \mathbb{Z}^+$. We have*

$$\sum_{d|m} \mu\Big(\frac{m}{d}\Big)\gcd(d,n) = \begin{cases} \phi(m) & \text{if } m \mid n, \\ 0 & \text{if } m \nmid n, \end{cases}$$

where μ is the Möbius function of $(\mathbb{Z}^+, |)$ and ϕ is the Euler function.

Proof. If $m \mid n$, by Exercise 1.16, we have

$$\sum_{d|m} \mu\Big(\frac{m}{d}\Big)\gcd(d,n) = \sum_{d|m} \mu\Big(\frac{m}{d}\Big) d = \phi(m).$$

If $m \nmid n$, write $m = p_1^{a_1} \cdots p_t^{a_t}$ and $n = p_1^{b_1} \cdots p_t^{b_t}$, where $p_1, \dots, p_t$ are distinct primes and $a_1 > b_1$. Then

$$\begin{aligned} &\sum_{d|m} \mu\Big(\frac{m}{d}\Big)\gcd(d,n) \\ &= \Big(\sum_{d_1|p_1^{a_1}} \mu\Big(\frac{p_1^{a_1}}{d_1}\Big)\gcd(d_1, p_1^{b_1})\Big) \cdots \Big(\sum_{d_t|p_t^{a_t}} \mu\Big(\frac{p_t^{a_t}}{d_t}\Big)\gcd(d_t, p_t^{b_t})\Big), \end{aligned}$$

where

$$\sum_{d_1|p_1^{a_1}} \mu\Big(\frac{p_1^{a_1}}{d_1}\Big)\gcd(d_1, p_1^{b_1}) = \gcd(p_1^{a_1}, p_1^{b_1}) - \gcd(p_1^{a_1-1}, p_1^{b_1}) = 0.$$

□

Theorem 2.14. *Let $f \in \mathbb{F}_q[\mathtt{X}]$ be a polynomial with $\deg f = n > 0$ and let A be the matrix defined in (2.6). For each integer $m > 0$, the number of distinct irreducible factors of f of degree m is given by*

$$\sum_{\substack{s \le n \\ m|s}} \mu\Big(\frac{s}{m}\Big)\frac{1}{\phi(s)}\sum_{d|s} \mu\Big(\frac{s}{d}\Big)\,\mathrm{null}(A^d - I).$$

Proof. Let $n_1, \dots, n_k$ be the degrees of the distinct irreducible factors of f. For each $m \in \{1, \dots, n\}$, let

$$N(m) = |\{1 \le i \le k : n_i = m\}|$$

and

$$N_{\ge}(m) = \sum_{\substack{s \le n \\ m|s}} N(s) = |\{1 \le i \le k : m \mid n_i\}|.$$

By (2.12), we have

$$\begin{aligned}\sum_{d|m}\mu\Big(\frac{m}{d}\Big)\,\mathrm{null}(A^d-I)&=\sum_{i=1}^{k}\sum_{d|m}\mu\Big(\frac{m}{d}\Big)\gcd(d,n_i)\\&=\big|\{1\le i\le k: m\mid n_i\}\big|\,\phi(m)\qquad\text{(by Lemma 2.13)}\\&=N_{\ge}(m)\,\phi(m),\end{aligned}$$

that is,

$$N_{\ge}(m)=\frac{1}{\phi(m)}\sum_{d|m}\mu\Big(\frac{m}{d}\Big)\,\mathrm{null}(A^d-I).$$

Thus, by the Möbius inversion (1.17) for the sub-poset $(\{1,\dots,n\},\,|\,)$,

$$N(m)=\sum_{\substack{s\le n\\ m|s}}\mu\Big(\frac{s}{m}\Big)N_{\ge}(s)=\sum_{\substack{s\le n\\ m|s}}\mu\Big(\frac{s}{m}\Big)\frac{1}{\phi(s)}\sum_{d|s}\mu\Big(\frac{s}{d}\Big)\,\mathrm{null}(A^d-I).$$

□

2.3. Functions from $\mathbb{F}_q^n$ to $\mathbb{F}_q$

For any two sets X and Y, $\mathcal{F}(X,Y)$ denotes the set of all functions from X to Y.

In this section, we investigate the $\mathbb{F}_q$-algebra $\mathcal{F}(\mathbb{F}_q^n,\mathbb{F}_q)$ of all functions from $\mathbb{F}_q^n$ to $\mathbb{F}_q$, where $n>0$ is an integer. A property peculiar to finite fields is that every function in $\mathcal{F}(\mathbb{F}_q^n,\mathbb{F}_q)$ is a polynomial function.

Let $\mathbb{F}_q[\mathtt{X}_1,\dots,\mathtt{X}_n]$ be the polynomial ring in $\mathtt{X}_1,\dots,\mathtt{X}_n$ over $\mathbb{F}_q$. Each element $f(\mathtt{X}_1,\dots,\mathtt{X}_n)\in\mathbb{F}_q[\mathtt{X}_1,\dots,\mathtt{X}_n]$ induces a function

$$\begin{array}{cccc}\bar f: & \mathbb{F}_q^n & \longrightarrow & \mathbb{F}_q\\ & (a_1,\dots,a_n) & \longmapsto & f(a_1,\dots,a_n).\end{array}$$

Clearly, $\overline{(\ )}: f\mapsto\bar f$ is an $\mathbb{F}_q$-algebra homomorphism from $\mathbb{F}_q[\mathtt{X}_1,\dots,\mathtt{X}_n]$ to $\mathcal{F}(\mathbb{F}_q^n,\mathbb{F}_q)$. For each $(a_1,\dots,a_n)\in\mathbb{F}_q^n$, define

$$f_{(a_1,\dots,a_n)}=\prod_{i=1}^{n}\prod_{b\in\mathbb{F}_q\setminus\{a_i\}}\frac{\mathtt{X}_i-b}{a_i-b}\in\mathbb{F}_q[\mathtt{X}_1,\dots,\mathtt{X}_n].\tag{2.13}$$

Then for $(b_1,\dots,b_n)\in\mathbb{F}_q^n$,

$$\bar f_{(a_1,\dots,a_n)}(b_1,\dots,b_n)=\begin{cases}1 & \text{if } (b_1,\dots,b_n)=(a_1,\dots,a_n),\\ 0 & \text{if } (b_1,\dots,b_n)\ne(a_1,\dots,a_n).\end{cases}$$

Therefore, $\bar f_{(a_1,\dots,a_n)}$, $(a_1,\dots,a_n)\in\mathbb{F}_q^n$, form a basis of $\mathcal{F}(\mathbb{F}_q^n,\mathbb{F}_q)$. Consequently, $\overline{(\ )}:\mathbb{F}_q[\mathtt{X}_1,\dots,\mathtt{X}_n]\to\mathcal{F}(\mathbb{F}_q^n,\mathbb{F}_q)$ is onto.

Remark. The polynomials $f_{(a_1,\dots,a_n)}$ in (2.13) are the fundamental polynomials of the multivariate Lagrange interpolation of $\mathbb{F}_q^n$.

Theorem 2.15. *The homomorphism* $\overline{(\ )} : \mathbb{F}_q[\mathtt{X}_1,\dots,\mathtt{X}_n] \to \mathcal{F}(\mathbb{F}_q^n,\mathbb{F}_q)$ *induces an* $\mathbb{F}_q$*-algebra isomorphism*

$$\mathbb{F}_q[\mathtt{X}_1,\dots,\mathtt{X}_n]/(\mathtt{X}_1^q-\mathtt{X}_1,\dots,\mathtt{X}_n^q-\mathtt{X}_n) \cong \mathcal{F}(\mathbb{F}_q^n,\mathbb{F}_q), \tag{2.14}$$

where $(\mathtt{X}_1^q-\mathtt{X}_1,\dots,\mathtt{X}_n^q-\mathtt{X}_n)$ *is the ideal of* $\mathbb{F}_q[\mathtt{X}_1,\dots,\mathtt{X}_n]$ *generated by* $\mathtt{X}_1^q-\mathtt{X}_1$, $\dots$, $\mathtt{X}_n^q-\mathtt{X}_n$.

Proof. Since $a^q-a=0$ for all $a\in\mathbb{F}_q$, we have $(\mathtt{X}_1^q-\mathtt{X}_1,\dots,\mathtt{X}_n^q-\mathtt{X}_n)\subset\ker\overline{(\ )}$. Thus $\overline{(\ )}$ induces an onto homomorphism

$$\epsilon : \mathbb{F}_q[\mathtt{X}_1,\dots,\mathtt{X}_n]/(\mathtt{X}_1^q-\mathtt{X}_1,\dots,\mathtt{X}_n^q-\mathtt{X}_n) \longrightarrow \mathcal{F}(\mathbb{F}_q^n,\mathbb{F}_q).$$

However,

$$\dim_{\mathbb{F}_q}\mathbb{F}_q[\mathtt{X}_1,\dots,\mathtt{X}_n]/(\mathtt{X}_1^q-\mathtt{X}_1,\dots,\mathtt{X}_n^q-\mathtt{X}_n) \le q^n = \dim_{\mathbb{F}_q}\mathcal{F}(\mathbb{F}_q^n,\mathbb{F}_q).$$

(The inequality holds in the above since $\mathtt{X}_1^{e_1}\cdots\mathtt{X}_n^{e_n}$, $0\le e_i\le q-1$, $1\le i\le n$, span $\mathbb{F}_q[\mathtt{X}_1,\dots,\mathtt{X}_n]/(\mathtt{X}_1^q-\mathtt{X}_1,\dots,\mathtt{X}_n^q-\mathtt{X}_n)$.) Therefore ϵ is an isomorphism. □

The meaning of (2.14) is concrete: Every function from $\mathbb{F}_q^n$ to $\mathbb{F}_q$ is uniquely represented by a polynomial in $\mathbb{F}_q[\mathtt{X}_1,\dots,\mathtt{X}_n]$ in which the degree of each $\mathtt{X}_i$ is at most $q-1$. In particular, every function from $\mathbb{F}_q$ to $\mathbb{F}_q$ is uniquely represented by a polynomial of degree $\le q-1$ in $\mathbb{F}_q[\mathtt{X}]$.

Let

$$\mathcal{P}_{q,n} = \mathbb{F}_q[\mathtt{X}_1,\dots,\mathtt{X}_n]/(\mathtt{X}_1^q-\mathtt{X}_1,\dots,\mathtt{X}_n^q-\mathtt{X}_n).$$

We identify the two $\mathbb{F}_q$-algebras $\mathcal{P}_{q,n}$ and $\mathcal{F}(\mathbb{F}_q^n,\mathbb{F}_q)$ by the isomorphism in Theorem 2.15. When it is convenient and causes no confusion, elements in $\mathcal{P}_{q,n}$ and $\mathcal{F}(\mathbb{F}_q^n,\mathbb{F}_q)$ are simply denoted by polynomials in $\mathbb{F}_q[\mathtt{X}_1,\dots,\mathtt{X}_n]$. Every element f in $\mathcal{P}_{q,n}$ or $\mathcal{F}(\mathbb{F}_q^n,\mathbb{F}_q)$ is uniquely of the form

$$f = \sum_{(e_1,\dots,e_n)\in[0,\,q-1]^n} a_{e_1,\dots,e_n}\mathtt{X}_1^{e_1}\cdots\mathtt{X}_n^{e_n}, \tag{2.15}$$

where $[0,q-1]=\{0,1,\dots,q-1\}$ and $a_{e_1,\dots,e_n}\in\mathbb{F}_q$. We define $\deg f$ to be the total degree of the polynomial on the right side of (2.15), i.e.,

$$\deg f = \max\{e_1+\cdots+e_n : a_{e_1,\dots,e_n}\ne 0\}.$$

Unless specified otherwise, $\deg 0$ is defined to be $-\infty$.

For each $-1\le r\le n(q-1)$, let

$$R_q(r,n) = \{f\in\mathcal{P}_{q,n} : \deg f\le r\}.$$

Then $R_q(r,n)$ is an $\mathbb{F}_q$-subspace of $\mathcal{P}_{q,n}$ and is called the q-ary *Reed-Muller code* of order r and length q^n. For the coding-theoretic background of $R_q(r,n)$, see [**9**].

The quotient space $R_q(r,n)/R_q(r-1,n)$ is the space of homogeneous polynomial functions of degree r in $\mathcal{P}_{q,n}$. (Here we allow 0 to be treated as a homogeneous polynomial function of degree r.) Since $\mathtt{X}_1^{e_1}\cdots\mathtt{X}_n^{e_n}$, $0\le e_i\le q-1$ with $e_1+\cdots+e_n=r$, form a basis of $R_q(r,n)/R_q(r-1,n)$, we have

$$\begin{aligned}
&\dim_{\mathbb{F}_q} R_q(r,n)/R_q(r-1,n)\\
&=\big|\{(e_1,\dots,e_n)\in[0,q-1]^n : e_1+\cdots+e_n=r\}\big|\\
&=\text{the coefficient of } \mathtt{X}^r \text{ in } (1+\mathtt{X}+\cdots+\mathtt{X}^{q-1})^n\\
&=\text{the coefficient of } \mathtt{X}^r \text{ in } (1-\mathtt{X}^q)^n(1-\mathtt{X})^{-n}\\
&=\text{the coefficient of } \mathtt{X}^r \text{ in } \Big(\sum_{i=0}^{n}\binom{n}{i}(-1)^i\mathtt{X}^{qi}\Big)\Big(\sum_{j=0}^{\infty}\binom{n+j-1}{j}\mathtt{X}^j\Big)\\
&=\sum_{i\le\lfloor r/q\rfloor}(-1)^i\binom{n}{i}\binom{r-qi+n-1}{r-qi}.
\end{aligned}$$

Consequently,

$$\begin{aligned}
&\dim_{\mathbb{F}_q} R_q(r,n)\\
&=\sum_{s=0}^{r}\dim_{\mathbb{F}_q} R_q(s,n)/R_q(s-1,n)=\sum_{s=0}^{r}\sum_{i\le\lfloor s/q\rfloor}(-1)^i\binom{n}{i}\binom{s-qi+n-1}{s-qi}\\
&=\sum_{i\le\lfloor r/q\rfloor}(-1)^i\binom{n}{i}\sum_{j=0}^{r-qi}\binom{j+n-1}{n-1}=\sum_{i\le\lfloor r/q\rfloor}(-1)^i\binom{n}{i}\binom{r-qi+n}{n}.
\end{aligned}$$

When $q=2$, the above dimension formulas are much simpler. We have

$$\begin{aligned}
&\dim_{\mathbb{F}_2} R_2(r,n)/R_2(r-1,n)\\
&=\big|\{(e_1,\dots,e_n)\in[0,1]^n : e_1+\cdots+e_n=r\}\big|=\binom{n}{r}
\end{aligned}$$

and

$$\dim_{\mathbb{F}_2} R_2(r,n)=\sum_{s=0}^{r}\binom{n}{s}.$$

The method of representing functions in $\mathcal{F}(\mathbb{F}_q^n,\mathbb{F}_q)$ as polynomials in $\mathbb{F}_q[\mathtt{X}_1,\dots,\mathtt{X}_n]$ is referred to as the *multivariate* approach. We now describe another method, called the *univariate* approach, which identifies $\mathcal{F}(\mathbb{F}_q^n,\mathbb{F}_q)$ with $\mathcal{F}(\mathbb{F}_{q^n},\mathbb{F}_q)$ and represents functions in $\mathcal{F}(\mathbb{F}_{q^n},\mathbb{F}_q)$ as compositions of $\mathrm{Tr}_{q^n/q}$ and polynomials in $\mathbb{F}_{q^n}[\mathtt{X}]$.

Since $\mathrm{Tr}:\mathbb{F}_{q^n}\to\mathbb{F}_q$ is onto (Theorem 1.9 (i)), every function in $\mathcal{F}(\mathbb{F}_{q^n},\mathbb{F}_q)$ is a composition $\mathrm{Tr}_{q^n/q}\circ f$ for some function $f:\mathbb{F}_{q^n}\to\mathbb{F}_{q^n}$. We also treat $\mathrm{Tr}_{q^n/q}$ as the polynomial

$$\mathrm{Tr}_{q^n/q}=\mathtt{X}+\mathtt{X}^q+\cdots+\mathtt{X}^{q^{n-1}}\in\mathbb{F}_q[\mathtt{X}].$$

Thus we have an onto $\mathbb{F}_q$-map

$$(2.16)\qquad \begin{array}{rccc}\psi: & \mathbb{F}_{q^n}[\mathtt{X}] & \longrightarrow & \mathcal{F}(\mathbb{F}_{q^n},\mathbb{F}_q)\\ & f & \longmapsto & \mathrm{Tr}_{q^n/q}(f),\end{array}$$

where $\mathrm{Tr}_{q^n/q}(f)\in\mathbb{F}_q[\mathtt{X}]$ is considered as a function from $\mathbb{F}_{q^n}$ to $\mathbb{F}_q$. The $\mathbb{F}_q$-map ψ induces an $\mathbb{F}_q$-isomorphism

$$\bar{\psi}:\mathbb{F}_{q^n}[\mathtt{X}]/\ker\psi\xrightarrow{\cong}\mathcal{F}(\mathbb{F}_{q^n},\mathbb{F}_q).$$

We claim that

$$(2.17)\qquad \ker\psi=(\mathtt{X}^{q^n}-\mathtt{X})+K,$$

where $(\mathtt{X}^{q^n}-\mathtt{X})$ is the ideal of $\mathbb{F}_{q^n}[\mathtt{X}]$ generated by $\mathtt{X}^{q^n}-\mathtt{X}$ and K is the $\mathbb{F}_q$-span of

$$\Big\{a\mathtt{X}^{(e_0,\dots,e_{n-1})(q^0,\dots,q^{n-1})^T}-a^q\mathtt{X}^{(e_{n-1},e_0,\dots,e_{n-2})(q^0,\dots,q^{n-1})^T}:\\ a\in\mathbb{F}_{q^n},\ e_i\in[0,q-1]\Big\}.$$

For all $x\in\mathbb{F}_{q^n}$, we have $x^{q^n}-x=0$ and

$$\mathrm{Tr}_{q^n/q}\Big(a^q x^{(e_{n-1},e_0,\dots,e_{n-2})(q^0,\dots,q^{n-1})^T}\Big)=\mathrm{Tr}_{q^n/q}\Big(a x^{(e_0,\dots,e_{n-1})(q^0,\dots,q^{n-1})^T}\Big)\Big),$$

so $(\mathtt{X}^{q^n}-\mathtt{X})+K\subset\ker\psi$. On the other hand, if $f\in\ker\psi$, by Theorem 1.12, $f\equiv g-g^q \pmod{\mathtt{X}^{q^n}-\mathtt{X}}$ for some $g\in\mathbb{F}_{q^n}[\mathtt{X}]$ with $\deg g<q^n$. To prove that $g-g^q\in(\mathtt{X}^{q^n}-\mathtt{X})+K$, it suffices to consider $g=a\mathtt{X}^{(e_0,\dots,e_{n-1})(q^0,\dots,q^{n-1})^T}$, where $a\in\mathbb{F}_{q^n}$ and $0\le e_i\le q-1$, $0\le i\le n-1$, in which case we have

$$\begin{aligned}g-g^q&=a\mathtt{X}^{(e_0,\dots,e_{n-1})(q^0,\dots,q^{n-1})^T}-a^q\mathtt{X}^{(e_0,\dots,e_{n-1})(q^0,\dots,q^{n-1})^Tq}\\&\equiv a\mathtt{X}^{(e_0,\dots,e_{n-1})(q^0,\dots,q^{n-1})^T}-a^q\mathtt{X}^{(e_{n-1},e_0,\dots,e_{n-2})(q^0,\dots,q^{n-1})^T}\pmod{\mathtt{X}^{q^n}-\mathtt{X}}\\&\in K.\end{aligned}$$

Hence $\ker\psi\subset(\mathtt{X}^{q^n}-\mathtt{X})+K$, and (2.17) is proved.

Fix a basis $u_1,\dots,u_n$ of $\mathbb{F}_{q^n}$ over $\mathbb{F}_q$. We have an $\mathbb{F}_q$-module isomorphism

$$(2.18)\qquad \begin{array}{rccc}\alpha: & \mathbb{F}_q^n & \longrightarrow & \mathbb{F}_{q^n}\\ & (a_1,\dots,a_n) & \longmapsto & a_1u_1+\cdots+a_nu_n\end{array}$$

and an $\mathbb{F}_q$-algebra isomorphism

$$(2.19)\qquad \begin{array}{rccc} \alpha^*: & \mathcal{F}(\mathbb{F}_{q^n},\mathbb{F}_q) & \longrightarrow & \mathcal{F}(\mathbb{F}_q^n,\mathbb{F}_q) \\ & F & \longmapsto & F\circ\alpha. \end{array}$$

We identify $\mathcal{F}(\mathbb{F}_{q^n},\mathbb{F}_q)$ with $\mathcal{F}(\mathbb{F}_q^n,\mathbb{F}_q)$ through the isomorphism α^*. A natural question arises: how to choose polynomials $f\in\mathbb{F}_{q^n}[\mathtt{X}]$ so that $\mathrm{Tr}_{q^n/q}(f)$, treated as functions in $\mathcal{F}(\mathbb{F}_{q^n},\mathbb{F}_q)$, form an $\mathbb{F}_q$-basis of $R_q(r,n)/R_q(r-1,n)$? We answer this question in the rest of this section.

Lemma 2.16. *Let $a\in\mathbb{F}_{q^n}$ and $e_0,\dots,e_{n-1}\in[0,q-1]$. Then for all $x\in\mathbb{F}_{q^n}$ and $(x_1,\dots,x_n)=\alpha(x)\in\mathbb{F}_q^n$,*

$$(2.20)\qquad \mathrm{Tr}(ax^{e_0+e_1q+\cdots+e_{n-1}q^{n-1}})=\sum_{s_1+\cdots+s_n=r} c_{s_1,\dots,s_n}x_1^{s_1}\cdots x_n^{s_n},$$

where $r=e_0+\cdots+e_{n-1}$ and $c_{s_1,\dots,s_n}\in\mathbb{F}_q$ are constants.

Proof. Since $x=\sum_{i=1}^n x_iu_i$, we have

$$\begin{aligned} x^{e_0+e_1q+\cdots+e_{n-1}q^{n-1}} &= \Big(\sum_{i=1}^n x_iu_i\Big)^{e_0+e_1q+\cdots+e_{n-1}q^{n-1}} \\ &= \Big(\sum_{i=1}^n x_iu_i\Big)^{e_0}\Big(\sum_{i=1}^n x_iu_i^q\Big)^{e_1}\cdots\Big(\sum_{i=1}^n x_iu_i^{q^{n-1}}\Big)^{e_{n-1}}. \end{aligned}$$

The above is a homogeneous polynomial of degree $e_0+\cdots+e_{n-1}=r$ in $x_1,\dots,x_n$, so we can write

$$x^{e_0+e_1q+\cdots+e_{n-1}q^{n-1}}=\sum_{s_1+\cdots+s_n=r} b_{s_1,\dots,s_n}x_1^{s_1}\cdots x_n^{s_n},$$

where $b_{s_1,\dots,s_n}\in\mathbb{F}_{q^n}$ are constants. Therefore,

$$\mathrm{Tr}_{q^n/q}(ax^{e_0+e_1q+\cdots+e_{n-1}q^{n-1}})=\sum_{s_1+\cdots+s_n=r}\mathrm{Tr}_{q^n/q}(ab_{s_1,\dots,s_n})\,x_1^{s_1}\cdots x_n^{s_n}.$$

□

Remark. $\mathrm{Tr}_{q^n/q}(ax^{e_0+e_1q+\cdots+e_{n-1}q^{n-1}})$ is a homogeneous polynomial in x_1, $\dots,x_n$ of degree $e_0+\cdots+e_{n-1}$, which is the base-q weight of $e_0+e_1q+\cdots+e_{n-1}q^{n-1}$.

Let τ be the Frobenius map of $\mathbb{F}_{q^n}$ over $\mathbb{F}_q$ and let

$$E_r=\big\{(e_0,\dots,e_{n-1})\in[0,q-1]^n : e_0+\cdots+e_{n-1}=r\big\}.$$

Define a $\langle\tau\rangle$-action on E_r by

$$\tau^m(e_0,\dots,e_{n-1})=(e_{-m},e_{-m+1},\dots,e_{-m+n-1}),$$

where the subscript i of e_i is taken modulo n. The action of τ^m on $(e_0, \dots, e_{n-1})$ is the cyclic shift of the components m positions to the right. Therefore, two elements in E_r are in the same $\langle\tau\rangle$-orbit if and only if one can be obtained from the other through a cyclic shift. Observe that for all $x \in \mathbb{F}_{q^n}$ and $(e_0, \dots, e_{n-1}) \in [0, q-1]^n$,

$$\begin{aligned}(2.21)\qquad &\tau^m\Big(x^{(e_0,\dots,e_{n-1})(q^0,\dots,q^{n-1})^T}\Big)\\ &= \tau^m\Big(x^{e_0q^0+\cdots+e_{n-1}q^{n-1}}\Big) = x^{(e_0q^0+\cdots+e_{n-1}q^{n-1})q^m}\\ &= x^{e_{-m}q^0+\cdots+e_{-m+n-1}q^{n-1}} \qquad\qquad (\text{since } x^{q^n} = x)\\ &= x^{\tau^m(e_0,\dots,e_{n-1})(q^0,\dots,q^{n-1})^T}.\end{aligned}$$

Let $\epsilon_1, \dots, \epsilon_k \in E_r$ be a set of representatives of the $\langle\tau\rangle$-orbits in E_r and let $[\epsilon_i]$ be the $\langle\tau\rangle$-orbit of ϵ_i, $1 \le i \le k$. The stabilizer of ϵ_i in $\langle\tau\rangle$ is $\langle\tau^{|[\epsilon_i]|}\rangle = \mathrm{Aut}(\mathbb{F}_{q^n}/\mathbb{F}_{q^{|[\epsilon_i]|}})$.

Theorem 2.17. *In the above notation, a basis of $R_q(r,n)/R_q(r-1,n)$ is given by*

$$(2.22)\qquad \mathrm{Tr}_{q^n/q}\Big(a\mathtt{X}^{\epsilon_i(q^0,\dots,q^{n-1})^T}\Big),\quad 1 \le i \le k,\ a \in A_i,$$

where A_i is any subset of $\mathbb{F}_{q^n}$ such that $\mathrm{Tr}_{q^n/q^{|[\epsilon_i]|}}(a)$, $a \in A_i$, form a basis $\mathbb{F}_{q^{|[\epsilon_i]|}}$ over $\mathbb{F}_q$.

Proof. We use induction on r. The statement is obviously true for $r = 0$. Assume that $r > 0$. By Lemma 2.16, all functions in (2.22) belong to $R_q(r,n)$. The number of functions listed in (2.22) is

$$\sum_{i=1}^{k} |A_i| = \sum_{i=1}^{k} |[\epsilon_i]| = |E_r| = \dim_{\mathbb{F}_q} R_q(r,n)/R_q(r-1,n).$$

Therefore, it suffices to show that the functions in (2.22), modulo $R_q(r-1,n)$, are linearly independent over $\mathbb{F}_q$.

Assume that

$$(2.23)\qquad \sum_{i=1}^{k}\sum_{a\in A_i} b_{i,a}\mathrm{Tr}_{q^n/q}\Big(a\mathtt{X}^{\epsilon_i(q^0,\dots,q^{n-1})^T}\Big) = f$$

for some $f \in R_q(r-1,n)$, where $b_{i,a} \in \mathbb{F}_q$ for all $1 \le i \le k$ and $a \in A_i$. For all $x \in \mathbb{F}_{q^n}$, by (2.21),

$$\tau^{|[\epsilon_i]|}\Big(x^{\epsilon_i(q^0,\dots,q^{n-1})^T}\Big) = x^{\tau^{|[\epsilon_i]|}(\epsilon_i)(q^0,\dots,q^{n-1})^T} = x^{\epsilon_i(q^0,\dots,q^{n-1})^T},$$

so $x^{\epsilon_i(q^0,\dots,q^{n-1})^T} \in \mathbb{F}_{q^{|[\epsilon_i]|}}$. Therefore (2.23) gives

$$\begin{aligned}
(2.24)\qquad f(x) &= \sum_{i=1}^{k}\sum_{a\in A_i} b_{i,a}\mathrm{Tr}_{q^{|[\epsilon_i]|}/q}\Big(\mathrm{Tr}_{q^n/q^{|[\epsilon_i]|}}(a)\,x^{\epsilon_i(q^0,\dots,q^{n-1})^T}\Big)\\
&= \sum_{i=1}^{k}\mathrm{Tr}_{q^{|[\epsilon_i]|}/q}\Big[\Big(\sum_{a\in A_i} b_{i,a}\mathrm{Tr}_{q^n/q^{|[\epsilon_i]|}}(a)\Big)x^{\epsilon_i(q^0,\dots,q^{n-1})^T}\Big]\\
&= \sum_{i=1}^{k}\sum_{j=0}^{|[\epsilon_i]|-1}\tau^j\Big(\sum_{a\in A_i} b_{i,a}\mathrm{Tr}_{q^n/q^{|[\epsilon_i]|}}(a)\Big)x^{\tau^j(\epsilon_i)(q^0,\dots,q^{n-1})^T}
\end{aligned}$$

for all $x \in \mathbb{F}_{q^n}$. The right side of (2.24) is a polynomial in x of degree $\le q^n-1$, and the exponents of x all have base-q weight r and are all distinct. (In fact, $\tau^j(\epsilon_i)$, $1\le i\le k$, $0\le j\le |[\epsilon_i]|-1$, are the base-$q$ digit vectors of all integers in $[0, q^n-1]$ with base-q weight r.) By the induction hypothesis, the left side of (2.24) is a polynomial (over $\mathbb{F}_{q^n}$) in x of degree $\le q^n-1$, and the exponents of x all have base-q weight $\le r-1$. Therefore, for all $1\le i\le k$,

$$\sum_{a\in A_i} b_{i,a}\mathrm{Tr}_{q^n/q^{|[\epsilon_i]|}}(a) = 0.$$

From the definition of A_i, we must have $b_{i,a} = 0$ for all $a \in A_i$. Therefore, the proof of the theorem is complete. □

A basis of $R_q(r,n)$ can be obtained by taking the union of the bases of $R_q(r,n)/R_q(r-1,n)$, $R_q(r-1,n)/R_q(r-2,n),\dots,R_q(0,n)$ given by Theorem 2.17.

Example 2.18. We determine a basis of $R_q(2,n)/R_q(1,n)$ using Theorem 2.17. For $1\le i\le n-1$, let

$$\epsilon_i = (\underset{0}{1}, 0, \cdots, 0, \underset{i}{1}, 0, \cdots, 0) \in E_2.$$

Case 1. Assume that $q = 2$.

Case 1.1. Assume that n is odd. Then the $\langle\tau\rangle$-orbits of E_2 are $[\epsilon_i]$, $1\le i\le (n-1)/2$, where $|[\epsilon_i]| = n$ for all $1\le i\le (n-1)/2$. A basis of $R_2(2,n)/R_2(1,n)$ is given by

$$\mathrm{Tr}_{2^n/2}\Big(a\mathtt{X}^{\epsilon_i(2^0,\dots,2^{n-1})^T}\Big) = \mathrm{Tr}_{2^n/2}\Big(a\mathtt{X}^{1+2^i}\Big),\quad 1\le i\le \frac{n-1}{2},\ a\in A_i,$$

where A_i is any basis of $\mathbb{F}_{2^n}$ over $\mathbb{F}_2$.

Case 1.2. Assume that n is even. The $\langle\tau\rangle$-orbits of E_2 are $[\epsilon_i]$, $1 \le i \le n/2$, where

$$||[\epsilon_i]|| = \begin{cases} n & \text{if } 1 \le i \le \frac{n}{2} - 1, \\ \frac{n}{2} & \text{if } i = \frac{n}{2}. \end{cases}$$

A basis of $R_2(2,n)/R_2(1,n)$ is given by

$$\mathrm{Tr}_{2^n/2}\Big(a\mathtt{X}^{\epsilon_i(2^0,\dots,2^{n-1})^T}\Big) = \mathrm{Tr}_{2^n/2}\Big(a\mathtt{X}^{1+2^i}\Big), \quad 1 \le i \le \frac{n}{2}, \ a \in A_i,$$

where A_i, $1 \le i \le n/2 - 1$, is any basis of $\mathbb{F}_{2^n}$ over $\mathbb{F}_2$ and $A_{n/2}$ is any subset of $\mathbb{F}_{2^n}$ such that $\mathrm{Tr}_{2^n/2^{n/2}}(a)$, $a \in A_{n/2}$, form a basis of $\mathbb{F}_{2^{n/2}}$ over $\mathbb{F}_2$. Since $\ker(\mathrm{Tr}_{2^n/2^{n/2}}) = \mathbb{F}_{2^{n/2}}$, the condition on $A_{n/2}$ simply means that the image of $A_{n/2}$ in $\mathbb{F}_{2^n}/\mathbb{F}_{2^{n/2}}$ forms an $\mathbb{F}_2$-basis of $\mathbb{F}_{2^n}/\mathbb{F}_{2^{n/2}}$. (A general fact: If $g : V \to W$ is an onto homomorphism of vector spaces, then $g(v_i)$, $i \in I$, form a basis of W if and only if $v_i + \ker g$, $i \in I$, form a basis of $V/\ker g$.)

Case 2. Assume that $q > 2$. Let $\delta = (2, 0, \dots, 0) \in E_2$.

Case 2.1. Assume that n is odd. The $\langle\tau\rangle$-orbits of E_2 are $[\delta]$ and $[\epsilon_i]$, $1 \le i \le (n-1)/2$, where $||[\delta]|| = n$ and $||[\epsilon_i]|| = n$, $1 \le i \le (n-1)/2$. A basis of $R_q(2,n)/R_q(1,n)$ is given by

$$\mathrm{Tr}_{q^n/q}\Big(a\mathtt{X}^{\delta(q^0,\dots,q^{n-1})^T}\Big) = \mathrm{Tr}_{q^n/q}(a\mathtt{X}^2), \quad a \in A,$$

and

$$\mathrm{Tr}_{q^n/q}\Big(a\mathtt{X}^{\epsilon_i(q^0,\dots,q^{n-1})^T}\Big) = \mathrm{Tr}_{q^n/q}\Big(a\mathtt{X}^{1+q^i}\Big), \quad 1 \le i \le \frac{n-1}{2}, \ a \in A_i,$$

where each of $A, A_1, \dots, A_{(n-1)/2}$ is a basis of $\mathbb{F}_{q^n}$ over $\mathbb{F}_q$.

Case 2.2. Assume that n is even. The $\langle\tau\rangle$-orbits of E_2 are $[\delta]$ and $[\epsilon_i]$, $1 \le i \le n/2$, where $||[\delta]|| = n$ and

$$||[\epsilon_i]|| = \begin{cases} n & \text{if } 1 \le i \le \frac{n}{2} - 1, \\ \frac{n}{2} & \text{if } i = \frac{n}{2}. \end{cases}$$

A basis of $R_q(2,n)/R_q(1,n)$ is given by

$$\mathrm{Tr}_{q^n/q}\Big(a\mathtt{X}^{\delta(q^0,\dots,q^{n-1})^T}\Big) = \mathrm{Tr}_{q^n/q}(a\mathtt{X}^2), \quad a \in A,$$

and

$$\mathrm{Tr}_{q^n/q}\Big(a\mathtt{X}^{\epsilon_i(q^0,\dots,q^{n-1})^T}\Big) = \mathrm{Tr}_{q^n/q}\Big(a\mathtt{X}^{1+q^i}\Big), \quad 1 \le i \le \frac{n}{2}, \ a \in A_i,$$

where each of $A, A_1, \dots, A_{n/2-1}$ is a basis of $\mathbb{F}_{q^n}$ over $\mathbb{F}_q$ and $A_{n/2}$ is any subset of $\mathbb{F}_{q^n}$ such that $\mathrm{Tr}_{q^n/q^{n/2}}(a)$, $a \in A_{n/2}$, form a basis of $\mathbb{F}_{q^{n/2}}$ over $\mathbb{F}_q$. If q is even, the condition on $A_{n/2}$ means that the image of $A_{n/2}$ in

$\mathbb{F}_{q^n}/\mathbb{F}_{q^{n/2}}$ forms an $\mathbb{F}_q$-basis of $\mathbb{F}_{q^n}/\mathbb{F}_{q^{n/2}}$. If q is odd, one can choose for $A_{n/2}$ any basis of $\mathbb{F}_{q^{n/2}}$ over $\mathbb{F}_q$.

2.4. Permutation Polynomials

Definition 2.19. A polynomial $f \in \mathbb{F}_q[\mathtt{X}]$ is called a *permutation polynomial* of $\mathbb{F}_q$ if the function $a \mapsto f(a)$ is a permutation of $\mathbb{F}_q$.

By Theorem 2.15, every function from $\mathbb{F}_q$ to $\mathbb{F}_q$ is uniquely represented by a polynomial of degree $\leq q-1$ in $\mathbb{F}_q[\mathtt{X}]$. Therefore, the number of permutation polynomials of $\mathbb{F}_q$ of degree $\leq q-1$ is $q!$.

A main question concerning permutation polynomials of $\mathbb{F}_q$ is how to recognize them. The following is a useful criterion for this purpose.

Theorem 2.20 (Hermite's criterion). *A polynomial $f \in \mathbb{F}_q[\mathtt{X}]$ is a permutation polynomial of $\mathbb{F}_q$ if and only if the following two conditions are both satisfied.*

(i) *f has exactly one root in $\mathbb{F}_q$.*

(ii) *For each integer $1 \leq s \leq q-2$, $f^s \equiv f_s \pmod{\mathtt{X}^q - \mathtt{X}}$ for some $f_s \in \mathbb{F}_q[\mathtt{X}]$ with $\deg f_s \leq q-2$.*

Lemma 2.21. *Let $a_0, a_1, \ldots, a_{q-1} \in \mathbb{F}_q$. Then the following two conditions are equivalent.*

(i) *$a_0, a_1, \ldots, a_{q-1}$ are distinct, i.e., $\mathbb{F}_q = \{a_0, a_1, \ldots, a_{q-1}\}$.*

(ii)

$$\sum_{j=0}^{q-1} a_j^s = \begin{cases} 0 & \textit{if } 0 \leq s \leq q-2, \\ -1 & \textit{if } s = q-1. \end{cases}$$

Proof. For each $a \in \mathbb{F}_q$, let

$$h_a = 1 - (\mathtt{X} - a)^{q-1} \in \mathbb{F}_q[\mathtt{X}]. \tag{2.25}$$

Clearly, h_a maps a to 1 and all $x \in \mathbb{F}_q \setminus \{a\}$ to 0. Thus, $a_0, \ldots, a_{q-1} \in \mathbb{F}_q$ are distinct if and only if

$$f = \sum_{j=0}^{q-1} h_{a_j}$$

maps all $x \in \mathbb{F}_q$ to 1. The latter condition is equivalent to $f \equiv 1 \pmod{\mathtt{X}^q - \mathtt{X}}$, i.e., $f = 1$, since $\deg f < q$. It remains to show that condition (ii) is equivalent to the equation $f = 1$.

Since $(\mathtt{X}-a)^q = \mathtt{X}^q - a^q = (\mathtt{X}-a)\sum_{i=0}^{q-1} a^{q-1-i}\mathtt{X}^i$, we have $(\mathtt{X}-a)^{q-1} = \sum_{i=0}^{q-1} a^{q-1-i}\mathtt{X}^i$. Thus,

$$h_a = 1 - \sum_{i=0}^{q-1} a^{q-1-i}\mathtt{X}^i \tag{2.26}$$

and

$$f = -\sum_{j=0}^{q-1}\sum_{i=0}^{q-1} a_j^{q-1-i}\mathtt{X}^i = -\sum_{i=0}^{q-1}\Big(\sum_{j=0}^{q-1} a_j^{q-1-i}\Big)\mathtt{X}^i. \tag{2.27}$$

From (2.27), it is clear that $f = 1$ if and only if (ii) holds. □

Proof of Theorem 2.20. ($\Rightarrow$) (i) is obviously true. For $1 \le s \le q-2$, write

$$f^s \equiv \sum_{i=0}^{q-1} c_i \mathtt{X}^i \pmod{\mathtt{X}^q - \mathtt{X}},$$

where $c_i \in \mathbb{F}_q$. By Lemma 2.21,

$$0 = \sum_{a\in\mathbb{F}_q} f(a)^s = \sum_{i=0}^{q-1} c_i \sum_{a\in\mathbb{F}_q} a^i = -c_{q-1},$$

that is, $c_{q-1} = 0$. Hence (ii) holds.

($\Leftarrow$) For $1 \le s \le q-2$, by (ii) and Lemma 2.21,

$$\sum_{a\in\mathbb{F}_q} f(a)^s = \sum_{a\in\mathbb{F}_q} f_s(a) = 0.$$

For $s = q-1$, by (i),

$$\sum_{a\in\mathbb{F}_q} f(a)^{q-1} = -1.$$

It follows from Lemma 2.21 that $\{f(a) : a \in \mathbb{F}_q\} = \mathbb{F}_q$, i.e., f is a permutation polynomial of $\mathbb{F}_q$. □

In Theorem 2.20 (ii), for each $1 \le s \le q-2$, as seen in the above proof, the reduction of f^s modulo $\mathtt{X}^p - \mathtt{X}$ has degree $< q-1$ if and only if

$$\sum_{a\in\mathbb{F}_q} f(a)^s = 0. \tag{2.28}$$

Let $q = p^n$, where p is a prime, and let $t \in \{1, \dots, q-2\}$ be such that $t \equiv ps \pmod{q-1}$. Then (2.28) is equivalent to $\sum_{a\in\mathbb{F}_q} f(a)^t = 0$ since $\sum_{a\in\mathbb{F}_q} f(a)^t = (\sum_{a\in\mathbb{F}_q} f(a)^s)^p$. We say that two integers $s, t \in \{1, \dots, q-2\}$ are equivalent if $t \equiv p^i s \pmod{q-1}$ for some integer $i \ge 0$, that is, if the base-p digits of t are obtained from those of s through a cyclic shift. The equivalence classes so defined are called the *p-cyclotomic cosets* of

$\{1, \ldots, q-2\}$ modulo $q-1$. Condition (ii) of Theorem 2.20 is satisfied for all $1 \le s \le q-2$ if and only if it is satisfied for a representative from each p-cyclotomic coset of $\{1, \ldots, q-2\}$ modulo $q-1$. This fact may reduce the amount of computation to some extent when Hermite's criterion is used to determine if a given polynomial is a permutation polynomial.

Corollary 2.22. *Let $f \in \mathbb{F}_q[\mathtt{X}]$. For each $1 \le s \le q-1$, let $f^s \equiv f_s \pmod{\mathtt{X}^q - \mathtt{X}}$, where $f_s \in \mathbb{F}_q[\mathtt{X}]$ with $\deg f_s \le q-1$. Then f is a permutation polynomial of $\mathbb{F}_q$ if and only if*

$$\deg f_s \begin{cases} \le q-2 & \text{if } 1 \le s \le q-2, \\ = q-1 & \text{if } s = q-1. \end{cases} \tag{2.29}$$

Proof. ($\Rightarrow$) By Theorem 2.20, $\deg f_s \le q-2$ for $1 \le s \le q-2$. Write $f_{q-1} = a_0 + \cdots + a_{q-1}\mathtt{X}^{q-1}$, $a_i \in \mathbb{F}_q$. Then

$$-1 = \sum_{a \in \mathbb{F}_q} f(a)^{q-1} = \sum_{a \in \mathbb{F}_q} f_{q-1}(a) = -a_{q-1}.$$

Thus $a_{q-1} = 1$ and hence $\deg f_{q-1} = q-1$.

($\Leftarrow$) Clearly, (2.29) is equivalent to

$$\sum_{a \in \mathbb{F}_q} f(a)^s = \begin{cases} 0 & \text{if } 1 \le s \le q-2, \\ c & \text{if } s = q-1 \end{cases} \tag{2.30}$$

for some $0 \ne c \in \mathbb{F}_q$. Let

$$F = \sum_{a \in \mathbb{F}_q} h_{f(a)},$$

where $h_{f(a)}$ is defined in (2.25). By (2.27) and (2.30),

$$F = -\sum_{i=0}^{q-1} \Big(\sum_{a \in \mathbb{F}_q} f(a)^{q-1-i} \Big) \mathtt{X}^i = -c.$$

Assume to the contrary that f is not a permutation polynomial of $\mathbb{F}_q$. Then there exists $z \in \mathbb{F}_q \setminus f(\mathbb{F}_q)$. It follows that

$$0 = \sum_{a \in \mathbb{F}_q} h_{f(a)}(z) = F(z) = -c,$$

which is a contradiction. □

Corollary 2.23. *If $f \in \mathbb{F}_q[\mathtt{X}]$ is such that $\deg f > 1$ and $\deg f \mid q-1$, then f is not a permutation polynomial of $\mathbb{F}_q$.*

Proof. Let $s = (q-1)/\deg f$. Then $1 \le s < q-1$ and $\deg f^s = q-1$. By Theorem 2.20 (ii), f is not a permutation polynomial of $\mathbb{F}_q$. □

There are many known families of permutation polynomials, the simplest one being $\mathtt{X}^k \in \mathbb{F}_q[\mathtt{X}]$ where $(k, q-1) = 1$. The linearized polynomials to be discussed in the next section are another source of permutation polynomials. In the rest of this section, we introduce a remarkable family of permutation polynomials known as the Dickson polynomials.

For $n \in \mathbb{N}$, $\mathtt{X}^n + \mathtt{Y}^n \in \mathbb{Z}[\mathtt{X}, \mathtt{Y}]$ is a symmetric polynomial in $\mathtt{X}$ and $\mathtt{Y}$, so it can be expressed as a polynomial with coefficients in $\mathbb{Z}$ in the elementary symmetric polynomials in $\mathtt{X}$ and $\mathtt{Y}$; that is, there exists $D_n(\mathtt{X}, \mathtt{Y}) \in \mathbb{Z}[\mathtt{X}, \mathtt{Y}]$ such that

$$\mathtt{X}^n + \mathtt{Y}^n = D_n(\mathtt{X}+\mathtt{Y}, \mathtt{XY}). \tag{2.31}$$

For $n \geq 2$, we have

$$\mathtt{X}^n + \mathtt{Y}^n = (\mathtt{X}+\mathtt{Y})(\mathtt{X}^{n-1} + \mathtt{Y}^{n-1}) - \mathtt{XY}(\mathtt{X}^{n-2} + \mathtt{Y}^{n-2}),$$

i.e.,

$$D_n(\mathtt{X}+\mathtt{Y}, \mathtt{XY}) = (\mathtt{X}+\mathtt{Y})D_{n-1}(\mathtt{X}+\mathtt{Y}, \mathtt{XY}) - \mathtt{XY}D_{n-2}(\mathtt{X}+\mathtt{Y}, \mathtt{XY}).$$

Hence

$$D_n(\mathtt{X}, \mathtt{Y}) = \mathtt{X}D_{n-1}(\mathtt{X}, \mathtt{Y}) - \mathtt{Y}D_{n-2}(\mathtt{X}, \mathtt{Y}).$$

Obviously, $D_0(\mathtt{X}, \mathtt{Y}) = 2$ and $D_1(\mathtt{X}, \mathtt{Y}) = \mathtt{X}$. Therefore $D_n(\mathtt{X}, \mathtt{Y})$ is given by the recursive relation

$$\begin{cases} D_0(\mathtt{X}, \mathtt{Y}) = 2, \\ D_1(\mathtt{X}, \mathtt{Y}) = \mathtt{X}, \\ D_n(\mathtt{X}, \mathtt{Y}) = \mathtt{X}D_{n-1}(\mathtt{X}, \mathtt{Y}) - \mathtt{Y}D_{n-2}(\mathtt{X}, \mathtt{Y}), \qquad n \geq 2. \end{cases} \tag{2.32}$$

One can also easily derive (2.31) from (2.32) by induction.

From (2.32) we can determine the generating function $\sum_{n\geq 0} D_n(\mathtt{X}, \mathtt{Y})\mathtt{T}^n$. We have

$$\begin{aligned}
&(1 - \mathtt{XT} + \mathtt{YT}^2)\sum_{n\geq 0} D_n(\mathtt{X}, \mathtt{Y})\mathtt{T}^n \\
&= \sum_{n\geq 0} D_n(\mathtt{X}, \mathtt{Y})\mathtt{T}^n - \sum_{n\geq 0} \mathtt{X}D_n(\mathtt{X}, \mathtt{Y})\mathtt{T}^{n+1} + \sum_{n\geq 0} \mathtt{Y}D_n(\mathtt{X}, \mathtt{Y})\mathtt{T}^{n+2} \\
&= \sum_{n\geq 2} D_n(\mathtt{X}, \mathtt{Y})\mathtt{T}^n - \sum_{n\geq 2} \mathtt{X}D_{n-1}(\mathtt{X}, \mathtt{Y})\mathtt{T}^n + \sum_{n\geq 2} \mathtt{Y}D_{n-2}(\mathtt{X}, \mathtt{Y})\mathtt{T}^n + 2 + \mathtt{XT} - \mathtt{X}2\mathtt{T} \\
&= \sum_{n\geq 2} \big[D_n(\mathtt{X}, \mathtt{Y}) - \mathtt{X}D_{n-1}(\mathtt{X}, \mathtt{Y}) + \mathtt{Y}D_{n-2}(\mathtt{X}, \mathtt{Y})\big]\mathtt{T}^n + 2 - \mathtt{XT} \\
&= 2 - \mathtt{XT}.
\end{aligned}$$

Thus

$$\sum_{n\geq 0} D_n(\mathtt{X}, \mathtt{Y})\mathtt{T}^n = \frac{2 - \mathtt{XT}}{1 - \mathtt{XT} + \mathtt{YT}^2}. \tag{2.33}$$

An explicit formula for $D_n(\mathtt{X},\mathtt{Y})$ can be derived from (2.33) as follows. We have

$$\begin{aligned}
\frac{2-\mathtt{XT}}{1-\mathtt{XT}+\mathtt{YT}^2} &= (2-\mathtt{XT})\sum_{k\geq 0}(\mathtt{XT}-\mathtt{YT}^2)^k\\
&= (2-\mathtt{XT})\sum_{k\geq 0}\sum_{i}\binom{k}{i}(\mathtt{XT})^{k-i}(-\mathtt{YT}^2)^i\\
&= (2-\mathtt{XT})\sum_{k\geq 0}\sum_{i}\binom{k}{i}(-1)^i\mathtt{X}^{k-i}\mathtt{Y}^i\mathtt{T}^{k+i}\\
&= (2-\mathtt{XT})\sum_{n\geq 0}\Big(\sum_{i=0}^{\lfloor n/2\rfloor}\binom{n-i}{i}(-1)^i\mathtt{X}^{n-2i}\mathtt{Y}^i\Big)\mathtt{T}^n\\
&= \sum_{n\geq 0}\Big(\sum_{i=0}^{\lfloor n/2\rfloor}2\binom{n-i}{i}(-1)^i\mathtt{X}^{n-2i}\mathtt{Y}^i\Big)\mathtt{T}^n\\
&\quad -\sum_{n\geq 0}\Big(\sum_{i=0}^{\lfloor n/2\rfloor}\binom{n-i}{i}(-1)^i\mathtt{X}^{n+1-2i}\mathtt{Y}^i\Big)\mathtt{T}^{n+1}\\
&= \sum_{n\geq 0}\Big(\sum_{i=0}^{\lfloor n/2\rfloor}2\binom{n-i}{i}(-1)^i\mathtt{X}^{n-2i}\mathtt{Y}^i\Big)\mathtt{T}^n\\
&\quad -\sum_{n\geq 1}\Big(\sum_{i=0}^{\lfloor n/2\rfloor}\binom{n-1-i}{i}(-1)^i\mathtt{X}^{n-2i}\mathtt{Y}^i\Big)\mathtt{T}^n\\
&= 2+\sum_{n\geq 1}\Big(\sum_{i=0}^{\lfloor n/2\rfloor}\Big[2\binom{n-i}{i}-\binom{n-1-i}{i}\Big](-1)^i\mathtt{X}^{n-2i}\mathtt{Y}^i\Big)\mathtt{T}^n\\
&= 2+\sum_{n\geq 1}\Big(\sum_{i=0}^{\lfloor n/2\rfloor}\frac{n}{n-i}\binom{n-i}{i}(-1)^i\mathtt{X}^{n-2i}\mathtt{Y}^i\Big)\mathtt{T}^n.
\end{aligned}$$

Therefore, for $n\geq 1$,

$$D_n(\mathtt{X},\mathtt{Y}) = \sum_{i=0}^{\lfloor n/2\rfloor}\frac{n}{n-i}\binom{n-i}{i}(-1)^i\mathtt{X}^{n-2i}\mathtt{Y}^i. \tag{2.34}$$

Definition 2.24. Let $a\in\mathbb{F}_q$ and $n\in\mathbb{N}$. The polynomial $D_n(\mathtt{X},a)\in\mathbb{F}_q[\mathtt{X}]$ is called the nth *Dickson polynomial* with parameter a.

By (2.32), $D_n(\mathtt{X}, a)$ is monic of degree n for $n \geq 1$. For the first few n, we have

$$\begin{aligned} D_0(\mathtt{X}, a) &= 2, \\ D_1(\mathtt{X}, a) &= \mathtt{X}, \\ D_2(\mathtt{X}, a) &= \mathtt{X}^2 - 2a, \\ D_3(\mathtt{X}, a) &= \mathtt{X}^3 - 3a\mathtt{X}, \\ D_4(\mathtt{X}, a) &= \mathtt{X}^4 - 4a\mathtt{X}^2 + 2a^2. \end{aligned}$$

If $a = 0$, $D_n(\mathtt{X}, 0) = \mathtt{X}^n$, which is a permutation polynomial of $\mathbb{F}_q$ if and only if $\gcd(n, q-1) = 1$. For $a \in \mathbb{F}_q^*$, we have the following result.

Theorem 2.25. *Let $a \in \mathbb{F}_q^*$ and $n \in \mathbb{Z}^+$. Then $D_n(\mathtt{X}, a) \in \mathbb{F}_q[\mathtt{X}]$ is a permutation polynomial of $\mathbb{F}_q$ if and only if* $\gcd(n, q^2 - 1) = 1$.

Before proving Theorem 2.25, we observe some simple facts. Let F be any field and let $y_1, y_2, a \in F^*$. Then

$$y_1 + \frac{a}{y_1} = y_2 + \frac{a}{y_2} \quad \text{if and only if} \quad y_1 = y_2 \text{ or } \frac{a}{y_2}. \tag{2.35}$$

In fact, $y_1 + a/y_1 - (y_2 + a/y_2) = (y_1 - y_2)(1 - a/(y_1 y_2))$.

Lemma 2.26. *Let $a \in \mathbb{F}_q^*$ and $Y = \{y \in \mathbb{F}_{q^2}^* : y^{q-1} = 1 \text{ or } y^{q+1} = a\}$. Then*

$$\mathbb{F}_q = \Big\{y + \frac{a}{y} : y \in Y\Big\}.$$

Proof. For each $y \in Y$, we have $y^q = y$ or $y^q = a/y$. Thus, $(y + a/y)^q = y^q + a/y^q = y + a/y$, so $y + a/y \in \mathbb{F}_q$. On the other hand, for each $b \in \mathbb{F}_q$, the equation $y + a/y = b$, i.e., $y^2 - by + a = 0$, has a solution $y \in \mathbb{F}_{q^2}^*$. Since $y^q + a/y^q = b^q = b = y + a/y$, by (2.35), $y^q = y$ or a/y, i.e., $y \in Y$. □

Proof of Theorem 2.25. ($\Leftarrow$) Assume that $D_n(x_1, a) = D_n(x_2, a)$, where $x_1, x_2 \in \mathbb{F}_q$. We show that $x_1 = x_2$. From the above, we can write $x_i = y_i + a/y_i$ for some $y_i \in \mathbb{F}_{q^2}^*$, $i = 1, 2$. By (2.31),

$$y_1^n + \frac{a^n}{y_1^n} = D_n(x_1, a) = D_n(x_2, a) = y_2^n + \frac{a^n}{y_2^n}.$$

Thus $y_1^n = y_2^n$ or a^n/y_2^n. Since $\gcd(n, q^2 - 1) = 1$, we have $y_1 = y_2$ or a/y_2, implying that $x_1 = x_2$.

($\Rightarrow$) Assume to the contrary that $\gcd(n, q^2 - 1) > 1$. We show that $D_n(\mathtt{X}, a)$ is not a one-to-one function on $\mathbb{F}_q$.

If $\gcd(n, q^2 - 1)$ is even, then n is even and q is odd. By (2.34), $D_n(-1, a) = D_n(1, a)$, where $-1 \neq 1$ in $\mathbb{F}_q$.

If $\gcd(n, q^2 - 1)$ is odd, it has a prime factor $r \geq 3$. It suffices to show that there exist $y_1, y_2 \in Y$, where Y is defined in Lemma 2.26, such that $y_1 \notin \{y_2, a/y_2\}$ but $y_1^n = y_2^n$. In fact, from this we have

$$D_n\Big(y_1 + \frac{a}{y_1}, a\Big) = y_1^n + \frac{a^n}{y_1^n} = y_2^n + \frac{a^n}{y_2^n} = D_n\Big(y_2 + \frac{a}{y_2}, a\Big),$$

where $y_i + a/y_i \in \mathbb{F}_q$, $i = 1, 2$, are distinct.

Case 1. Assume that $r \mid q - 1$. Let $y_2 = 1 \in Y$. Since the equation $y^r = 1$ has $r \geq 3$ solutions in $\mathbb{F}_q^*$, there exists $y_1 \in \mathbb{F}_q^* \subset Y$ such that $y_1^r = 1$ and $y_1 \notin \{y_2, a/y_2\}$. Clearly, $y_1^n = 1 = y_2^n$.

Case 2. Assume that $r \mid q + 1$. Write $a = y_2^{q+1} = \mathrm{N}_{q^2/q}(y_2)$ for some $y_2 \in \mathbb{F}_{q^2}^*$. (Recall that $\mathrm{N}_{q^2/q} : \mathbb{F}_{q^2}^* \to \mathbb{F}_q^*$ is onto.) Let $\epsilon \in \mathbb{F}_{q^2}^*$ be a root of the equation $\epsilon^r = 1$ such that $\epsilon \notin \{1, a/y_2^2\}$, and set $y_1 = \epsilon y_2$. Then $y_1^{q+1} = y_2^{q+1} = a$ and hence $y_1, y_2 \in Y$. Clearly, $y_1^n = y_2^n$ and $y_1 \notin \{y_2, a/y_2\}$. □

2.5. Linearized Polynomials

Definition 2.27. Let $q > 1$ be a prime power and $n \in \mathbb{Z}^+$. An $\mathbb{F}_q$-*linearized polynomial* (or simply a *q-polynomial*) over $\mathbb{F}_{q^n}$ is a polynomial of the form

$$f = \sum_{i=0}^{k} a_i \mathtt{X}^{q^i} \in \mathbb{F}_{q^n}[\mathtt{X}]. \tag{2.36}$$

The polynomial function $\bar{f} : \mathbb{F}_{q^n} \to \mathbb{F}_{q^n}$ defined by the polynomial f in (2.36) is an $\mathbb{F}_q$-map. In fact, $\bar{f} = \sum_{i=0}^{k} a_i \tau^i \in \mathrm{Hom}_{\mathbb{F}_q}(\mathbb{F}_{q^n}, \mathbb{F}_{q^n})$, where τ is the Frobenius map of $\mathbb{F}_{q^n}/\mathbb{F}_q$. Denote by $\mathcal{L}(q, n)$ the set of all q-polynomials in $\mathbb{F}_{q^n}[\mathtt{X}]$ and by $\mathcal{L}_k(q, n)$ the set of all q-polynomials of degree $\leq q^k$ in $\mathbb{F}_{q^n}[\mathtt{X}]$. The following proposition shows that every element in $\mathrm{Hom}_{\mathbb{F}_q}(\mathbb{F}_{q^n}, \mathbb{F}_{q^n})$ is uniquely represented by a q-polynomial in $\mathcal{L}_{n-1}(q, n)$.

Proposition 2.28. *The mapping* $\overline{(\ )} : \mathcal{L}_{n-1}(q, n) \to \mathrm{Hom}_{\mathbb{F}_q}(\mathbb{F}_{q^n}, \mathbb{F}_{q^n})$ *is an* $\mathbb{F}_q$*-isomorphism.*

Proof. If $f \in \mathcal{L}_{n-1}(q, n)$ is such that $\bar{f} = 0$, then $f(a) = 0$ for all $a \in \mathbb{F}_q$. Since $\deg f \leq q^{n-1}$, we have $f = 0$. Therefore $\overline{(\ )} : \mathcal{L}_{n-1}(q, n) \to \mathrm{Hom}_{\mathbb{F}_q}(\mathbb{F}_{q^n}, \mathbb{F}_{q^n})$ is one-to-one. On the other hand, since $\mathtt{X}^{q^0}, \dots, \mathtt{X}^{q^{n-1}}$ is an $\mathbb{F}_{q^n}$-basis of $\mathcal{L}_{n-1}(q, n)$, we have

$$\dim_{\mathbb{F}_q} \mathcal{L}_{n-1}(q, n) = n \dim_{\mathbb{F}_{q^n}} \mathcal{L}_{n-1}(q, n) = n^2 = \dim_{\mathbb{F}_q} \mathrm{Hom}_{\mathbb{F}_q}(\mathbb{F}_{q^n}, \mathbb{F}_{q^n}).$$

Therefore $\overline{(\ )} : \mathcal{L}_{n-1}(q, n) \to \mathrm{Hom}_{\mathbb{F}_q}(\mathbb{F}_{q^n}, \mathbb{F}_{q^n})$ is an isomorphism. □

We now determine the dimension of the range of a q-polynomial. For

$$f = a_0\mathtt{X} + a_1\mathtt{X}^q + \cdots + a_{n-1}\mathtt{X}^{q^{n-1}} \in \mathcal{L}_{n-1}(q,n),$$

define

$$A(f) = \begin{bmatrix} a_0 & a_1 & \cdots & a_{n-1} \\ a_{n-1}^q & a_0^q & \cdots & a_{n-2}^q \\ \vdots & \vdots & \ddots & \vdots \\ a_1^{q^{n-1}} & a_2^{q^{n-1}} & \cdots & a_0^{q^{n-1}} \end{bmatrix}.$$

Theorem 2.29. *In the above notation, we have*

$$\operatorname{rank} A(f) = \dim_{\mathbb{F}_q} f(\mathbb{F}_{q^n}).$$

Lemma 2.30. *k elements $b_1, \dots, b_k \in \mathbb{F}_{q^n}$ are linearly independent over $\mathbb{F}_q$ if and only if the columns of*

$$B = \begin{bmatrix} b_1 & \cdots & b_k \\ b_1^q & \cdots & b_k^q \\ \vdots & & \vdots \\ b_1^{q^{n-1}} & \cdots & b_k^{q^{n-1}} \end{bmatrix}$$

are linearly independent over $\mathbb{F}_{q^n}$.

Proof. ($\Leftarrow$) Assume to the contrary that $b_1, \dots, b_k$ are linearly dependent over $\mathbb{F}_q$, that is, $\sum_{i=1}^k \beta_i b_i = 0$ for some $0 \neq (\beta_1, \dots, \beta_k) \in \mathbb{F}_q^k$. Then $B(\beta_1, \dots, \beta_k)^T = 0$, which is a contradiction.

($\Rightarrow$) 1° We first assume that $k = n$. The (i,j)-entry of $B^T B$ is

$$\sum_{l=0}^{n-1} (b_i b_j)^{q^l} = \operatorname{Tr}_{q^n/q}(b_i b_j).$$

Since $(x, y) \mapsto \operatorname{Tr}_{q^n/q}(xy)$ is a *nondegenerate* $\mathbb{F}_q$-bilinear form on $\mathbb{F}_{q^n}$ (Exercise 1.4), $\det(B^T B) \neq 0$. Hence B is nonsingular.

2° Now assume $1 \leq k \leq n$. Extend $b_1, \dots, b_k$ to an $\mathbb{F}_q$-basis $b_1, \dots, b_n$ of $\mathbb{F}_{q^n}$. By 1°, the columns of

$$\begin{bmatrix} b_1 & \cdots & b_n \\ b_1^q & \cdots & b_n^q \\ \vdots & & \vdots \\ b_1^{q^{n-1}} & \cdots & b_n^{q^{n-1}} \end{bmatrix}$$

are linearly independent over $\mathbb{F}_{q^n}$. $\square$

Lemma 2.30 can be stated as follows: For $b_1, \dots, b_k \in \mathbb{F}_{q^n}$,

$$\text{rank}\begin{bmatrix} b_1 & \cdots & b_k \\ b_1^q & \cdots & b_k^q \\ \vdots & & \vdots \\ b_1^{q^{n-1}} & \cdots & b_k^{q^{n-1}} \end{bmatrix} = \dim_{\mathbb{F}_q} \sum_{i=1}^{k} \mathbb{F}_q b_i. \tag{2.37}$$

Proof of Theorem 2.29. In the following, the subscript i of a_i is taken modulo n. For each $z \in \mathbb{F}_{q^n}$, we have

(2.38)

$$A(f)\begin{bmatrix} z \\ z^q \\ \vdots \\ z^{q^{n-1}} \end{bmatrix} = \begin{bmatrix} \sum_{i=0}^{n-1} a_i z^{q^i} \\ \sum_{i=0}^{n-1} a_{i-1}^q z^{q^i} \\ \vdots \\ \sum_{i=0}^{n-1} a_{i-(n-1)}^{q^{n-1}} z^{q^i} \end{bmatrix} = \begin{bmatrix} \sum_{i=0}^{n-1} a_i z^{q^i} \\ \sum_{i=0}^{n-1} a_i^q z^{q^{i+1}} \\ \vdots \\ \sum_{i=0}^{n-1} a_i^{q^{n-1}} z^{q^{i+n-1}} \end{bmatrix} = \begin{bmatrix} f(z) \\ f(z)^q \\ \vdots \\ f(z)^{q^{n-1}} \end{bmatrix}.$$

Let $b_1, \dots, b_n$ be an $\mathbb{F}_q$-basis of $\mathbb{F}_{q^n}$. By (2.38),

$$A(f)\begin{bmatrix} b_1 & \cdots & b_n \\ b_1^q & \cdots & b_n^q \\ \vdots & & \vdots \\ b_1^{q^{n-1}} & \cdots & b_n^{q^{n-1}} \end{bmatrix} = \begin{bmatrix} f(b_1) & \cdots & f(b_n) \\ f(b_1)^q & \cdots & f(b_n)^q \\ \vdots & & \vdots \\ f(b_1)^{q^{n-1}} & \cdots & f(b_n)^{q^{n-1}} \end{bmatrix}. \tag{2.39}$$

Therefore, by (2.39) and (2.37),

$$\text{rank}\, A(f) = \dim_{\mathbb{F}_q} \sum_{i=1}^{n} \mathbb{F}_q f(b_i) = \dim_{\mathbb{F}_q} f(\mathbb{F}_{q^n}).$$

□

Corollary 2.31. *A q-polynomial*

$$f = a_0\mathtt{X} + a_1\mathtt{X}^q + \cdots + a_{n-1}\mathtt{X}^{q^{n-1}} \in \mathcal{L}_{n-1}(q, n)$$

is a permutation polynomial of $\mathbb{F}_{q^n}$ if and only if $A(f)$ is nonsingular.

Proof. Immediate from Theorem 2.29. □

The set $\mathcal{L}(q, n)$ of all q-polynomials in $\mathbb{F}_{q^n}[\mathtt{X}]$ is closed under addition and composition. We now show that $(\mathcal{L}(q, n), +, \circ)$ is a ring which is isomorphic to a skew polynomial ring over $\mathbb{F}_{q^n}$.

Definition 2.32. Let F be a field and $\sigma \in \text{Aut}(F)$. The *skew polynomial ring* $F[\mathtt{X}; \sigma]$ is a ring such that

(i) $(F[\mathtt{X}; \sigma], +) = (F[\mathtt{X}], +)$;

(ii) the multiplication in $F[\mathtt{X};\sigma]$ is defined by distributivity, associativity and the rule that

$$\mathtt{X}a = \sigma(a)\mathtt{X} \quad \text{for all } a \in F.$$

Clearly, $F[\mathtt{X};\sigma]$ is commutative if and only if $\sigma = \mathrm{id}_F$ (the identity map of F). When $\sigma = \mathrm{id}_F$, $F[\mathtt{X};\sigma] = F[\mathtt{X}]$.

Example 2.33. Let σ be the Frobenius map of $\mathbb{F}_{q^n}/\mathbb{F}_q$, and let $f = \sum_{i=0}^{s} a_i\mathtt{X}^i$ and $g = \sum_{j=0}^{t} b_j\mathtt{X}^j \in \mathbb{F}_{q^n}[\mathtt{X};\sigma]$. Then

$$fg = \sum_{i,j} a_i\mathtt{X}^i b_j\mathtt{X}^j = \sum_{i,j} a_i b_j^{q^i}\mathtt{X}^{i+j} = \sum_{k=0}^{s+t}\Big(\sum_{i+j=k} a_i b_j^{q^i}\Big)\mathtt{X}^k.$$

Theorem 2.34. *Let σ be the Frobenius map of $\mathbb{F}_{q^n}/\mathbb{F}_q$. The map*

$$\begin{array}{rccc} \phi: & \mathbb{F}_{q^n}[\mathtt{X};\sigma] & \longrightarrow & \mathcal{L}(q,n) \\ & \displaystyle\sum_{i=0}^{s} a_i\mathtt{X}^i & \longmapsto & \displaystyle\sum_{i=0}^{s} a_i\mathtt{X}^{q^i} \end{array}$$

is a ring isomorphism from $\mathbb{F}_{q^n}[\mathtt{X};\sigma]$ to $(\mathcal{L}(q,n),+,\circ)$.

Proof. We only have to prove that $\phi(fg) = \phi(f)\circ\phi(g)$ for all $f,g \in \mathbb{F}_{q^n}[\mathtt{X};\sigma]$. Let $f = \sum_{i=0}^{s} a_i\mathtt{X}^i$ and $g = \sum_{j=0}^{t} b_j\mathtt{X}^j \in \mathbb{F}_{q^n}[\mathtt{X};\sigma]$. By Example 2.33, we have

$$\begin{aligned} \phi(fg) &= \phi\Big(\sum_{k=0}^{s+t}\Big(\sum_{i+j=k} a_i b_j^{q^i}\Big)\mathtt{X}^k\Big) = \sum_{k=0}^{s+t}\Big(\sum_{i+j=k} a_i b_j^{q^i}\Big)\mathtt{X}^{q^k} \\ &= \sum_{i,j} a_i b_j^{q^i}\mathtt{X}^{q^{i+j}} = \sum_{i=0}^{s} a_i\Big(\sum_{j=0}^{t} b_j\mathtt{X}^{q^j}\Big)^{q^i} = \phi(f)\circ\phi(g). \end{aligned}$$

□

Corollary 2.35. *The map*

$$\begin{array}{rccc} \phi: & \mathbb{F}_q[\mathtt{X}] & \longrightarrow & \mathcal{L}(q,1) \\ & \displaystyle\sum_{i=0}^{s} a_i\mathtt{X}^i & \longmapsto & \displaystyle\sum_{i=0}^{s} a_i\mathtt{X}^{q^i} \end{array}$$

is a ring isomorphism.

Proof. Let $n = 1$ in Theorem 2.34. The Frobenius map of $\mathbb{F}_q/\mathbb{F}_q$ is $\mathrm{id}_{\mathbb{F}_q}$ and $\mathbb{F}_q[\mathtt{X};\mathrm{id}_{\mathbb{F}_q}] = \mathbb{F}_q[\mathtt{X}]$. □

2.6. Payne's Theorem

In this section, we use Corollary 2.31 to prove (a generalization of) a theorem by Payne [**29**]. Let $f = \sum_{i=0}^{n-1} a_i\mathtt{X}^{q^i} \in \mathbb{F}_{q^n}[\mathtt{X}]$ be a q-polynomial which permutes $\mathbb{F}_{q^n}$. Consider

$$\frac{f(\mathtt{X})}{\mathtt{X}} = \sum_{i=0}^{n-1} a_i\mathtt{X}^{q^i-1}.$$

For each $x \in \mathbb{F}_{q^n}^*$, let $\bar{x}$ denote the image of x in $\mathbb{F}_{q^n}^*/\mathbb{F}_q^*$. Since $f : \mathbb{F}_{q^n} \to \mathbb{F}_{q^n}$ is $\mathbb{F}_q$-linear, the map

$$\begin{array}{ccc} \mathbb{F}_{q^n}^*/\mathbb{F}_q^* & \longrightarrow & \mathbb{F}_{q^n}^* \\ \bar{x} & \longmapsto & f(x)/x, \quad x \in \mathbb{F}_{q^n}^*, \end{array}$$

is well-defined. Note that $\mathbb{F}_{q^n}^*/\mathbb{F}_q^*$ is the $(n-1)$-dimensional projective space $\mathbb{P}^{n-1}(\mathbb{F}_q)$ over $\mathbb{F}_q$. If the map

$$\begin{array}{ccc} \mathbb{F}_{q^n}^*/\mathbb{F}_q^* & \longrightarrow & \mathbb{F}_{q^n}^*/\mathbb{F}_q^* \\ \bar{x} & \longmapsto & \overline{f(x)/x}, \quad x \in \mathbb{F}_{q^n}^*, \end{array}$$

is a permutation of $\mathbb{F}_{q^n}^*/\mathbb{F}_q^*$, we say that the polynomial $f(\mathtt{X})/\mathtt{X}$ is a permutation of $\mathbb{F}_{q^n}^*/\mathbb{F}_q^*$. Thus, $f(\mathtt{X})/\mathtt{X}$ is a permutation of $\mathbb{F}_{q^n}^*/\mathbb{F}_q^*$ if and only if for each $b \in \mathbb{F}_{q^n}^*$, there exist $z \in \mathbb{F}_{q^n}^*$ and $\epsilon \in \mathbb{F}_q^*$ such that $f(z)/z = \epsilon b$.

Theorem 2.36 (Generalized Payne's theorem [**15**]). *Let $f = \sum_{i=0}^{n-1} a_i\mathtt{X}^{q^i} \in \mathbb{F}_{q^n}[\mathtt{X}]$ be a q-polynomial. Then $f(\mathtt{X})/\mathtt{X}$ is a permutation of $\mathbb{F}_{q^n}^*/\mathbb{F}_q^*$ if and only if $\gcd(n, q-1) = 1$ and $f = a\mathtt{X}^{q^k}$, where $a \in \mathbb{F}_{q^n}^*$ and k is an integer such that $0 \le k \le n-1$ and $\gcd(k, n) = 1$.*

Lemma 2.37. *Let $f = \sum_{i=0}^{n-1} a_i\mathtt{X}^{q^i} \in \mathbb{F}_{q^n}[\mathtt{X}]$ be a q-polynomial such that $f(\mathtt{X})/\mathtt{X}$ is a permutation of $\mathbb{F}_{q^n}^*/\mathbb{F}_q^*$. Then the determinants of the principal submatrices of*

$$A(f) = \begin{bmatrix} a_0 & a_1 & \cdots & a_{n-1} \\ a_{n-1}^q & a_0^q & \cdots & a_{n-2}^q \\ \vdots & \vdots & \ddots & \vdots \\ a_1^{q^{n-1}} & a_2^{q^{n-1}} & \cdots & a_0^{q^{n-1}} \end{bmatrix}.$$

of size $m \times m$ $(1 \le m \le n-1)$ are all 0.

Proof. Let

$$D(\mathtt{X}) = \begin{vmatrix} a_0 + \mathtt{X} & a_1 & \cdots & a_{n-1} \\ a_{n-1}^q & (a_0 + \mathtt{X})^q & \cdots & a_{n-2}^q \\ \vdots & \vdots & \ddots & \vdots \\ a_1^{q^{n-1}} & a_2^{q^{n-1}} & \cdots & (a_0 + \mathtt{X})^{q^{n-1}} \end{vmatrix} \in \mathbb{F}_{q^n}[\mathtt{X}].$$

For each $b \in \mathbb{F}_{q^n}^*$, since $f(\mathtt{X})/\mathtt{X}$ is a permutation of $\mathbb{F}_{q^n}^*/\mathbb{F}_q^*$, there exist $z \in \mathbb{F}_{q^n}^*$ and $\epsilon \in \mathbb{F}_q^*$ such that $f(z)/z = -\epsilon b$. Thus z is a root of

$$(a_0 + \epsilon b)\mathtt{X} + a_1\mathtt{X}^q + \cdots + a_{n-1}\mathtt{X}^{q^{n-1}}. \tag{2.40}$$

Since the polynomial in (2.40) has at least two roots 0 and z, it is not a permutation polynomial of $\mathbb{F}_{q^n}$. It follows from Corollary 2.31 that $D(\epsilon b) = 0$. Therefore, for every $b \in \mathbb{F}_{q^n}^*$, $\prod_{\epsilon\in\mathbb{F}_q^*} D(\epsilon b) = 0$. Since $\deg \prod_{\epsilon\in\mathbb{F}_q^*} D(\epsilon\mathtt{X}) = q^n - 1$, we must have

$$\prod_{\epsilon\in\mathbb{F}_q^*} D(\epsilon\mathtt{X}) = \delta(\mathtt{X}^{q^n-1} - 1) \tag{2.41}$$

for some $\delta \in \mathbb{F}_{q^n}^*$. (In fact, $\delta = -1$, although this fact is not needed in the proof. Because $D(0)$ is invariant under the Frobenius map of $\mathbb{F}_{q^n}/\mathbb{F}_q$, (2.41) gives $-\delta = (D(0))^{q-1} = 1$.)

Let $0 \le i_1 < i_2 < \cdots < i_m \le n-1$, where $1 \le m \le n-1$. Write $\{0,\dots,n-1\}\setminus\{i_1,\dots,i_m\} = \{j_1,\dots,j_s\}$ with $0 \le j_1 < \cdots < j_s \le n-1$. We can determine the coefficient of $\mathtt{X}^{(q-1)q^{j_1}+\cdots+(q-1)q^{j_s}}$ in $\prod_{\epsilon\in\mathbb{F}_q^*} D(\epsilon\mathtt{X})$ through the expansion of $D(\epsilon\mathtt{X})$. We have

$$\begin{aligned} D(\epsilon\mathtt{X}) &= \begin{vmatrix} a_0 + \epsilon\mathtt{X} & a_1 & \cdots & a_{n-1} \\ a_{n-1}^q & a_0^q + \epsilon\mathtt{X}^q & \cdots & a_{n-2}^q \\ \vdots & \vdots & \ddots & \vdots \\ a_1^{q^{n-1}} & a_2^{q^{n-1}} & \cdots & a_0^{q^{n-1}} + \epsilon\mathtt{X}^{q^{n-1}} \end{vmatrix} \\ &= \sum_{c_0,\dots,c_{n-1}\in\{0,1\}} a_{c_0q^0+\cdots+c_{n-1}q^{n-1}}(\epsilon)\, \mathtt{X}^{c_0q^0+\cdots+c_{n-1}q^{n-1}}, \end{aligned} \tag{2.42}$$

where $a_{c_0q^0+\cdots+c_{n-1}q^{n-1}}(\epsilon) \in \mathbb{F}_{q^n}$. The coefficient of $\mathtt{X}^{q^{j_1}+\cdots+q^{j_s}}$ in (2.42) is

$$\epsilon^s \det\big(A(f)(i_1,\dots,i_m)\big),$$

where $A(f)(i_1,\dots,i_m)$ is the principal submatrix of $A(f)$ with row and column indices $i_1,\dots,i_m$, i.e., the submatrix of $A(f)$ obtained by deleting the rows and columns with indices $j_1,\dots,j_s$. Thus, $a_{q^{j_1}+\cdots+q^{j_s}}(\epsilon) = \epsilon^s \det\big(A(f)(i_1,\dots,i_m)\big)$. By the uniqueness of the base-q expansion of the integer $(q-1)q^{j_1} + \cdots + (q-1)q^{j_s}$, we see that the coefficient of $\mathtt{X}^{(q-1)q^{j_1}+\cdots+(q-1)q^{j_s}}$ in

$$\prod_{\epsilon\in\mathbb{F}_q^*} D(\epsilon\mathtt{X}) = \prod_{\epsilon\in\mathbb{F}_q^*}\Big(\sum_{c_0,\dots,c_{n-1}\in\{0,1\}} a_{c_0q^0+\cdots+c_{n-1}q^{n-1}}(\epsilon)\, \mathtt{X}^{c_0q^0+\cdots+c_{n-1}q^{n-1}}\Big)$$

equals

$$\prod_{\epsilon\in\mathbb{F}_q^*} a_{q^{j_1}+\cdots+q^{j_s}}(\epsilon) = \big[\det\big(A(f)(i_1,\dots,i_m)\big)\big]^{q-1}\prod_{\epsilon\in\mathbb{F}_q^*}\epsilon^s$$
$$= \big[\det\big(A(f)(i_1,\dots,i_m)\big)\big]^{q-1}(-1)^s.$$

Comparing the coefficients of $\mathtt{X}^{(q-1)q^{j_1}+\cdots+(q-1)q^{j_s}}$ on the two sides of (2.41), we have $\det\big(A(f)(i_1,\dots,i_m)\big)=0$. □

Proof of Theorem 2.36. ($\Leftarrow$) Since $\mathbb{F}_{q^n}^*/\mathbb{F}_q^*$ is a cyclic group of order $(q^n-1)/(q-1)$ and since

(2.43)
$$\begin{aligned}\gcd\Big(q^k-1,\,\frac{q^n-1}{q-1}\Big) &= \gcd\Big(\gcd(q^k-1,\,q^n-1),\,\frac{q^n-1}{q-1}\Big)\\ &= \gcd\Big(q-1,\,\frac{q^n-1}{q-1}\Big) \qquad (\text{since } \gcd(k,n)=1)\\ &= \gcd(q-1,\,1+q+\cdots+q^{n-1})\\ &= \gcd(q-1,\,n)=1,\end{aligned}$$

the map $(\ )\mapsto(\ )^{q^k-1}$ is a permutation of $\mathbb{F}_{q^n}^*/\mathbb{F}_q^*$. Thus, $f(\mathtt{X})/\mathtt{X}=a\mathtt{X}^{q^k-1}$ is a permutation of $\mathbb{F}_{q^n}^*/\mathbb{F}_q^*$.

($\Rightarrow$) 1° We first show that f has exactly one nonzero coefficient. Assume that $n>1$. By Lemma 2.37, the determinants of the principal submatrices of

$$A(f)=\begin{bmatrix} a_0 & a_1 & \cdots & a_{n-1}\\ a_{n-1}^q & a_0^q & \cdots & a_{n-2}^q\\ \vdots & \vdots & \ddots & \vdots\\ a_1^{q^{n-1}} & a_2^{q^{n-1}} & \cdots & a_0^{q^{n-1}}\end{bmatrix}$$

of sizes 1×1, 2×2, …, $(n-1)\times(n-1)$ are all 0. Write

$$A(f)=[b_{ij}]_{0\le i,j\le n-1},$$

and note that

(2.44) $$b_{ij}=0 \text{ if and only if } a_{j-i}=0,$$

where the subscript $j-i$ of a_{j-i} is taken modulo n.

We claim that if $i_1,\dots,i_m\in\{0,\dots,n-1\}$, where $1\le m\le n-1$, are such that $i_1+\cdots+i_m\equiv 0 \pmod n$, then

(2.45) $$a_{i_1}\cdots a_{i_m}=0.$$

To prove (2.45), we use induction on m. The case $m=1$ is obvious. Assume that $m>1$ and suppose to the contrary that $i_1+\cdots+i_m\equiv 0 \pmod n$

but $a_{i_1} \cdots a_{i_m} \neq 0$. We may assume that $0, i_1, i_1 + i_2, \ldots, i_1 + \cdots + i_{m-1}$ are all distinct modulo n. (Otherwise, $i_s + \cdots + i_t \equiv 0 \pmod{n}$ for some $1 \leq s < t \leq m-1$. By the induction hypothesis, $a_{i_s} \cdots a_{i_t} = 0$, which is a contradiction.) Consider the principal submatrix of $A(f)$ with row and column indices $j_0 = 0, j_1 = i_1, j_2 = i_1 + i_2, \ldots, j_{m-1} = i_1 + \cdots + i_{m-1}$:

$$B = \begin{bmatrix} 0 & b_{0j_1} & b_{0j_2} & \cdots & b_{0j_{m-1}} \\ b_{j_10} & 0 & b_{j_1j_2} & \cdots & b_{j_1j_{m-1}} \\ b_{j_20} & b_{j_2j_1} & 0 & \cdots & b_{j_2j_{m-1}} \\ \vdots & \vdots & \vdots & \ddots & \vdots \\ b_{j_{m-1}0} & b_{j_{m-1}j_1} & b_{j_{m-1}j_2} & \cdots & 0 \end{bmatrix}.$$

Since $a_{i_1}, \ldots, a_{i_m}$ are all nonzero, by (2.44), $b_{0j_1}, b_{j_1j_2}, \ldots, b_{j_{m-2}j_{m-1}}, b_{j_{m-1}0}$ are all nonzero. Since all 2×2 principal submatrices of B have determinant 0, $b_{j_10} = b_{j_2j_1} = \cdots = b_{j_{m-1}j_{m-2}} = 0$. Since all 3×3 principal submatrices of B have determinant 0, it follows that $b_{j_20} = b_{j_3j_1} = \cdots = b_{j_{m-1}j_{m-3}} = 0$. For example,

$$0 = \begin{vmatrix} 0 & b_{0j_1} & b_{0j_2} \\ 0 & 0 & b_{j_1j_2} \\ b_{j_20} & 0 & 0 \end{vmatrix} = b_{0j_1} b_{j_1j_2} b_{j_20}$$

implies that $b_{j_20} = 0$. In the same way, by considering the principal submatrices of B up to size $(m-1) \times (m-1)$, we conclude that

$$B = \begin{bmatrix} 0 & b_{0j_1} & * & \cdots & * & * \\ 0 & 0 & b_{j_1j_2} & \cdots & * & * \\ 0 & 0 & 0 & \cdots & * & * \\ \vdots & \vdots & \vdots & \ddots & \vdots & \vdots \\ 0 & 0 & 0 & \cdots & 0 & b_{j_{m-2}j_{m-1}} \\ b_{j_{m-1}0} & 0 & 0 & \cdots & 0 & 0 \end{bmatrix}.$$

It follows that $b_{0j_1} b_{j_1j_2} \cdots b_{j_{m-2}j_{m-1}} b_{j_{m-1}0} = \det B = 0$, which is a contradiction. Thus (2.45) is proved.

Assume that $a_k \neq 0$ for some $0 \leq k \leq n-1$. We claim that $\gcd(k, n) = 1$. Otherwise, there is an integer $1 \leq l \leq n-1$ such that $lk \equiv 0 \pmod{n}$. By (2.45), we have

$$\underbrace{a_k \cdots a_k}_{l} = 0,$$

which is a contradiction. For each $0 \le i \le n-1$ with $i \ne k$, we can write $i \equiv -jk \pmod{n}$ with $0 \le j \le n-2$. By (2.45) again, we have

$$a_i \underbrace{a_k \cdots a_k}_{j} = 0,$$

which implies that $a_i = 0$. Thus a_k is the only nonzero coefficient of f.

2° By 1°, $f = a\mathtt{X}^{q^k}$ for some $0 \le k \le n-1$ and $a \in \mathbb{F}_{q^n}^*$. Since $f(\mathtt{X})/\mathtt{X} = a\mathtt{X}^{q^k-1}$ is a permutation of $\mathbb{F}_{q^n}^*/\mathbb{F}_q^*$, we have $\gcd(q^k-1, (q^n-1)/(q-1)) = 1$. It follows that $\gcd(n, q-1) = 1$ and $\gcd(k, n) = 1$; see (2.43). □

Corollary 2.38 (Payne's theorem). *Let $f = \sum_{i=0}^{n-1} a_i\mathtt{X}^{2^i} \in \mathbb{F}_{2^n}[\mathtt{X}]$ be a 2-polynomial. Then $f(\mathtt{X})/\mathtt{X} = \sum_{i=0}^{n-1} a_i\mathtt{X}^{2^i-1}$ is a permutation of $\mathbb{F}_{2^n}^*$ if and only if $f = a\mathtt{X}^{2^k}$, where $a \in \mathbb{F}_{2^n}^*$ and k is an integer such that $0 \le k \le n-1$ and* $\gcd(k, n) = 1$.

Proof. Since $\mathbb{F}_{2^n}^* = \mathbb{F}_{2^n}^*/\mathbb{F}_2^*$, the conclusion follows immediately from Theorem 2.36. □

Exercises

2.1. Let $f = \mathtt{X}^2 + b\mathtt{X} + c \in \mathbb{F}_q[\mathtt{X}]$.
   (i) If q is odd, prove that f is irreducible over $\mathbb{F}_q$ if and only if $b^2 - 4c$ is not a square in $\mathbb{F}_q$.
   (ii) If q is even, prove that f is irreducible over $\mathbb{F}_q$ if and only if $b \ne 0$ and $\mathrm{Tr}_{q/2}(b^{-2}c) = 1$.

2.2. Prove that

$$\lim_{n\to\infty} nq^{-n}|\mathcal{I}_q(n)| = 1,$$

that is, $|\mathcal{I}_q(n)| \sim q^n/n$ as $n \to \infty$.

2.3. (i) Let $f \in \mathbb{F}_q[\mathtt{X}]$ be a monic irreducible polynomial of degree n and $f \ne \mathtt{X}$. Prove that f is primitive if and only if $\mathtt{X}^d \not\equiv 1 \pmod{f}$ for all $d \mid q^n - 1$ with $d < q^n - 1$.
   (ii) Determine all monic irreducible cubic polynomials in $\mathbb{F}_3[\mathtt{X}]$. Among these irreducible cubics, identify the ones that are primitive.

2.4. Use Berlekamp's algorithm to factor $f = \mathtt{X}^7 + \mathtt{X}^6 + \mathtt{X}^5 + \mathtt{X}^4 + \mathtt{X}^2 + 1$ over $\mathbb{F}_2$.

2.5. Let m and n be positive integers with $\gcd(m, n) = d$ and $\mathrm{lcm}(m, n) = l$. Prove that

$$\mathbb{F}_{q^m} \otimes_{\mathbb{F}_q} \mathbb{F}_{q^n} \cong \underbrace{\mathbb{F}_{q^l} \times \cdots \times \mathbb{F}_{q^l}}_{d}$$

as $\mathbb{F}_q$-algebras.

2.6. (Dillon) Let $f = \sum_{1\le i<j\le n} \mathtt{X}_i\mathtt{X}_j \in \mathbb{F}_2[\mathtt{X}_1, \dots, \mathtt{X}_n]$. For $x = (x_1, \dots, x_n) \in \mathbb{F}_2^n$, define $|x| = |\{1 \le i \le n : x_i \ne 0\}|$. Prove that

$$f(x) = \begin{cases} 0 & \text{if } |x| \equiv 0, 1 \pmod 4, \\ 1 & \text{if } |x| \equiv 2, 3 \pmod 4. \end{cases}$$

2.7. Let $D_n(\mathtt{X}, \mathtt{Y})$ be the polynomial defined in (2.31). Prove that for integers $m, n \ge 0$,

$$D_m\big(D_n(\mathtt{X}, \mathtt{Y}), \mathtt{Y}^n\big) = D_{mn}(\mathtt{X}, \mathtt{Y}).$$

2.8. For any $f \in \mathbb{F}_q[\mathtt{X}]$,

$$f \equiv -\sum_{i=1}^{q-1}\Big(\sum_{a\in\mathbb{F}_q} a^{q-1-i} f(a)\Big)\mathtt{X}^i + f(0) \pmod{\mathtt{X}^q - \mathtt{X}}.$$

2.9. Let $q = p^n$, where p is a prime. Let $a_0, a_1, \dots, a_{q-1} \in \mathbb{F}_q$ be such that

$$\sum_{j=0}^{q-1} a_j^i = 0 \quad \text{for all } 1 \le i \le q-2.$$

(i) If $\sum_{j=0}^{q-1} a_j^{q-1} \ne 0$, prove that $a_0, \dots, a_{q-1}$ are all distinct.
(ii) If $\sum_{j=0}^{q-1} a_j^{q-1} = 0$, prove that each element of $\{a_0, \dots, a_{q-1}\}$ appears a multiple of p times in the sequence $a_0, \dots, a_{q-1}$.

2.10. Let F be any field and let A be a finite subgroup of F^* of order l. Let $a_1, \dots, a_l \in A$. Prove that $a_1, \dots, a_l$ are distinct if and only if

$$\sum_{j=1}^{l} a_j^s = 0 \quad \text{for all } 1 \le s \le l-1.$$

2.11. Let q be odd and $f = \mathtt{X}^{(q+1)/2} + a\mathtt{X} \in \mathbb{F}_q[\mathtt{X}]$. Prove that f is a permutation polynomial of $\mathbb{F}_q$ if and only if $a^2 - 1$ is a square in $\mathbb{F}_q^*$. (Hint: When x is a square in $\mathbb{F}_q^*$, $f(x) = (1+a)x$; when x is a nonsquare in $\mathbb{F}_q^*$, $f(x) = (-1+a)x$; when $x = 0$, $f(x) = 0$.)

2.12. Let $q = p^e$, where p is an odd prime. A function $f : \mathbb{F}_q \to \mathbb{F}_q$ is said to be *planar* if for every $a \in \mathbb{F}_q^*$, the function $f(x+a) - f(x)$, $x \in \mathbb{F}_q$, is a permutation of $\mathbb{F}_q$.
(i) Let n be a positive integer. Prove that $\mathtt{X}^n$ is a planar function on $\mathbb{F}_q$ if and only if $(\mathtt{X}+1)^n - \mathtt{X}^n$ is a permutation polynomial of $\mathbb{F}_q$.

(ii) (Dembowski-Ostrom) Let $i \geq 0$. Prove that $\mathtt{X}^{1+p^i}$ is a planar function on $\mathbb{F}_q$ if and only if $\nu_2(i) \geq \nu_2(e)$, where $\nu_2(\ )$ is the 2-adic order.

2.13. Let $n \geq 3$. Prove that $f = \mathtt{X}^{2^n} + \mathtt{X} + 1$ is reducible over $\mathbb{F}_2$. (Hint: First show that $f \mid \mathtt{X}^{2^{2n}} - \mathtt{X}$.)

2.14. Let q_1 and q_2 be powers of a prime p. Prove that if $f \in \mathbb{F}_{q_2}[\mathtt{X}]$ is a q_1-polynomial and $g \in \mathbb{F}_{q_1}[\mathtt{X}]$ is a q_2-polynomial, then $f \circ g = g \circ f$.

2.15. (The *Moore determinant*) Let q be a prime power and define

$$\Delta(\mathtt{X}_1, \dots, \mathtt{X}_n) = \begin{vmatrix} \mathtt{X}_1 & \cdots & \mathtt{X}_n \\ \mathtt{X}_1^q & \cdots & \mathtt{X}_n^q \\ \vdots & & \vdots \\ \mathtt{X}_1^{q^{n-1}} & \cdots & \mathtt{X}_n^{q^{n-1}} \end{vmatrix} \in \mathbb{F}_q[\mathtt{X}_1, \dots, \mathtt{X}_n].$$

(i) Prove that

$$\Delta(\mathtt{X}_1, \dots, \mathtt{X}_n) = \prod_{i=1}^{n} \prod_{a_1, \dots, a_{i-1} \in \mathbb{F}_q} \Big(\mathtt{X}_i - \sum_{j=1}^{i-1} a_j \mathtt{X}_j\Big).$$

(When $i = 1$, the inner product is understood to be $\mathtt{X}_1$.)

(ii) Let $\mathtt{X}_1, \dots, \mathtt{X}_n, \mathtt{Y}$ be algebraically independent indeterminates over $\mathbb{F}_q$. Prove that

$$\prod_{a_1, \dots, a_n \in \mathbb{F}_q} (\mathtt{Y} + a_1\mathtt{X}_1 + \cdots + a_n\mathtt{X}_n) = (-1)^n \frac{\Delta(\mathtt{Y}, \mathtt{X}_1, \dots, \mathtt{X}_n)}{\Delta(\mathtt{X}_1, \dots, \mathtt{X}_n)}.$$

2.16. (i) Let $f = \sum_{i=0}^{k} a_i \mathtt{X}^{q^i} \in \overline{\mathbb{F}}_q[\mathtt{X}]$. Prove that f is separable (i.e., without multiple roots) if and only if $a_0 \neq 0$.

(ii) Let $f \in \overline{\mathbb{F}}_q[\mathtt{X}]$. Prove that f is a monic separable q-polynomial if and only if $f = \prod_{a \in A} (\mathtt{X} - a)$ for some finite-dimensional $\mathbb{F}_q$-vector space $A \subset \overline{\mathbb{F}}_q$. (Hint: Use Exercise 2.15.)

2.17. In the notation of Corollary 2.35, prove that for $f, g \in \mathbb{F}_q[\mathtt{X}]$ not both 0, $\gcd(\phi(f), \phi(g)) = \phi(\gcd(f, g))$.

2.18. Let $\sigma \in \mathrm{Aut}(\overline{\mathbb{F}}_q/\mathbb{F}_q)$ be defined by $\sigma(\ ) = (\ )^q$. Prove that f is a monic separable q-polynomial over $\mathbb{F}_q$ if and only if $f = \prod_{a \in A} (\mathtt{X} - a)$, where A is a finite $\mathbb{F}_q[\langle\sigma\rangle]$-submodule of $\overline{\mathbb{F}}_q$. Find the number of $\mathbb{F}_q[\langle\sigma\rangle]$-submodules of $\overline{\mathbb{F}}_q$ of dimension n over $\mathbb{F}_q$.

Chapter 3

Gauss Sums

3.1. Characters of Finite Abelian Groups

Characters. Let A be a finite abelian group and let $\mathbb{C}^*$ be the multiplicative group of $\mathbb{C}$. A *character* of A is a group homomorphism $\chi : A \to \mathbb{C}^*$, i.e., an element of $\mathrm{Hom}_{\mathbb{Z}}(A, \mathbb{C}^*)$. The abelian group $\mathrm{Hom}_{\mathbb{Z}}(A, \mathbb{C}^*)$ is called the *character group* of A and is denoted by $\widehat{A}$. The identity of $\widehat{A}$ is called the *principal* character of A and is denoted by 1^A.

Proposition 3.1. *For a finite abelian group A, $\widehat{A} \cong A$.*

Proof. Write $A = (\mathbb{Z}/n_1\mathbb{Z}) \times \cdots \times (\mathbb{Z}/n_k\mathbb{Z})$, where $n_1, \dots, n_k$ are positive integers, and let $\zeta_n = e^{2\pi i/n}$ for $n \in \mathbb{Z}^+$. Define

$$\begin{array}{rccc} \phi: & (\mathbb{Z}/n_1\mathbb{Z}) \times \cdots \times (\mathbb{Z}/n_k\mathbb{Z}) & \longrightarrow & \widehat{A} \\ & (a_1, \dots, a_k) & \longmapsto & \chi_{a_1,\dots,a_k}, \end{array}$$

where

$$\begin{array}{rccc} \chi_{a_1,\dots,a_k}: & (\mathbb{Z}/n_1\mathbb{Z}) \times \cdots \times (\mathbb{Z}/n_k\mathbb{Z}) & \longrightarrow & \mathbb{C}^* \\ & (x_1, \dots, x_k) & \longmapsto & \zeta_{n_1}^{a_1x_1} \cdots \zeta_{n_k}^{a_kx_k}. \end{array} \tag{3.1}$$

Clearly, ϕ is a group homomorphism. If $\chi_{a_1,\dots,a_k} = 1^A$, we have

$$1 = \chi_{a_1,\dots,a_k}(0, \dots, 0, \underset{i}{1}, 0, \dots, 0) = \zeta_{n_i}^{a_i} \quad \text{for all } 1 \le i \le k,$$

that is, $(a_1, \dots, a_k) = (0, \dots, 0)$. Hence ϕ is one-to-one. For each $\chi \in \widehat{A}$, let $\chi(0, \dots, 0, \underset{i}{1}, 0, \dots, 0) = \zeta_{n_i}^{a_i}$, where $a_i \in \mathbb{Z}/n_i\mathbb{Z}$. Then

$$\chi(x_1, \dots, x_k) = \zeta_{n_1}^{a_1x_1} \cdots \zeta_{n_k}^{a_kx_k} \text{ for all } (x_1, \dots, x_k) \in (\mathbb{Z}/n_1\mathbb{Z}) \times \cdots \times (\mathbb{Z}/n_k\mathbb{Z}),$$

that is, $\chi = \chi_{a_1,\dots,a_k}$. Hence ϕ is also onto. □

The pairing between A and $\widehat{A}$. For a finite abelian group A, define a pairing map $\langle\cdot,\cdot\rangle : \widehat{A}\times A\to\mathbb{C}^*$ by $\langle\chi,a\rangle=\chi(a)$, where $\chi\in\widehat{A}$ and $a\in A$. Clearly, $\langle\cdot,\cdot\rangle$ is $\mathbb{Z}$-bilinear. The $\mathbb{Z}$-bilinear map $\langle\cdot,\cdot\rangle$ is nondegenerate, i.e., for each $\chi\in\widehat{A}\setminus\{1^A\}$ there exists $a\in A$ such that $\langle\chi,a\rangle\neq 1$, and for each $a\in A\setminus\{0\}$ there exists $\chi\in\widehat{A}$ such that $\langle\chi,a\rangle\neq 1$. The first claim is obviously true; the second claim follows from (3.1). Consequently, the maps $\widehat{A}\to\widehat{A}:\chi\mapsto\langle\chi,\cdot\rangle$ and $A\to\widehat{\widehat{A}}:a\mapsto\langle\cdot,a\rangle$ are both group monomorphisms. The first one is the identity map on $\widehat{A}$; the second map is an isomorphism from A to $\widehat{\widehat{A}}$ (since $|A|=|\widehat{\widehat{A}}|$ by Proposition 3.1), which, unlike the isomorphism between A and $\widehat{A}$ in Proposition 3.1, does not depend on the decomposition of A as a direct product of cyclic groups.

For each $B\subset A$ and $H\subset\widehat{A}$, define

$$B^\perp=\{\chi\in\widehat{A}:\langle\chi,a\rangle=0\text{ for all }a\in B\},$$
$$H^\perp=\{a\in A:\langle\chi,a\rangle=0\text{ for all }\chi\in H\}.$$

Note that $B^\perp<\widehat{A}$ and $H^\perp<A$. Obviously, $(\ )^\perp$ is an inclusion-reversing operator, i.e., $B_1\subset B_2\subset A$ implies $B_1^\perp\supset B_2^\perp$, and $H_1\subset H_2\subset\widehat{A}$ implies $H_1^\perp\supset H_2^\perp$.

Theorem 3.2. *Let A be a finite abelian group. Let $\mathcal{S}(A)$ and $\mathcal{S}(\widehat{A})$ be the sets of all subgroups of A and $\widehat{A}$, respectively.*

(i) *The maps $(\ )^\perp:\mathcal{S}(A)\to\mathcal{S}(\widehat{A})$ and $(\ )^\perp:\mathcal{S}(\widehat{A})\to\mathcal{S}(A)$ are both bijections and are inverses of each other. Moreover, for all $B<A$ and $H<\widehat{A}$,*

$$|B|\,|B^\perp|=|A|=|H||H^\perp|.$$

(ii) *For $B<C<A$, there is an isomorphism*

$$\begin{array}{rccc}\phi: & B^\perp/C^\perp & \longrightarrow & \widehat{C/B} \\ & \chi+C^\perp & \longmapsto & \langle\chi,\cdot\rangle, \qquad \chi\in B^\perp,\end{array}\tag{3.2}$$

where

$$\begin{array}{rccc}\langle\chi,\cdot\rangle: & C/B & \longrightarrow & \mathbb{C}^* \\ & a+B & \longmapsto & \langle\chi,a\rangle, \qquad a\in C.\end{array}\tag{3.3}$$

(iii) *For $H<K<\widehat{A}$, there is an isomorphism*

$$\begin{array}{rccc}\psi: & H^\perp/K^\perp & \longrightarrow & \widehat{K/H} \\ & a+K^\perp & \longmapsto & \langle\cdot,a\rangle, \qquad a\in H^\perp,\end{array}\tag{3.4}$$

where

$$\begin{aligned} \langle\,\cdot\,,a\rangle: \quad K/H &\longrightarrow \mathbb{C}^* \\ \chi+H &\longmapsto \langle\chi,a\rangle, \qquad \chi\in K. \end{aligned}$$

Proof. (i) 1° We first show that for each $B\in\mathcal{S}(A)$, $|B||B^\perp|=|A|$.

Let $\pi: A\to A/B$ be the canonical homomorphism. An element of $B^\perp$ is a homomorphism $\chi: A\to\mathbb{C}^*$ with $\ker\chi\supset B$. Such a homomorphism χ factors through π uniquely: $\chi=\tilde\chi\circ\pi$, where $\tilde\chi\in\widehat{A/B}$ is uniquely determined by χ. The one-to-one correspondence $\chi\leftrightarrow\tilde\chi$ shows that $|B^\perp|=|\widehat{A/B}|$. On the other hand, by Proposition 3.1, $|\widehat{A/B}|=|A/B|=|A|/|B|$. Hence $|B^\perp||B|=|A|$.

2° From the definition of $(\ )^\perp$, it is clear that

$$B\subset B^{\perp\perp} \quad \text{for all } B\in\mathcal{S}(A), \tag{3.5}$$

$$H\subset H^{\perp\perp} \quad \text{for all } B\in\mathcal{S}(\widehat{A}). \tag{3.6}$$

For $B\in\mathcal{S}(A)$, by (3.5), $B^\perp\supset B^{\perp\perp\perp}$. On the other hand, by (3.6) with $H=B^\perp$, $B^\perp\subset B^{\perp\perp\perp}$. Thus $B^\perp=B^{\perp\perp\perp}$. Therefore, by using 1° twice,

$$|B^{\perp\perp}|=\frac{|A|}{|B^{\perp\perp\perp}|}=\frac{|A|}{|B^\perp|}=|B|.$$

Consequently, $B^{\perp\perp}=B$. Now the composition

$$\mathcal{S}(A)\xrightarrow{(\ )^\perp}\mathcal{S}(\widehat{A})\xrightarrow{(\ )^\perp}\mathcal{S}(A)$$

is the identity map. Note that $|\mathcal{S}(A)|=|\mathcal{S}(\widehat{A})|<\infty$ since $A\cong\widehat{A}$. Thus, both $(\ )^\perp$ are bijections and they are inverses of each other.

3° For $H\in\mathcal{S}(\widehat{A})$, write $H=B^\perp$ for some $B\in\mathcal{S}(A)$. Then $|H||H^\perp|=|B^\perp||B|=|A|$ by 1°.

(ii) It is clear that the maps in (3.3) and (3.2) are well-defined homomorphisms. Assume that $\chi+C^\perp\in\ker\phi$, where $\chi\in B^\perp$. Then $\langle\chi,a\rangle=0$ for all $a\in C$, that is, $\chi\in C^\perp$. Thus $\ker\phi=\{0+C^\perp\}$ and hence ϕ is one-to-one. Since

$$|B^\perp/C^\perp|=|C|/|B|=|\widehat{C/B}|,$$

ϕ is an isomorphism.

(iii) Same as the proof of (ii). □

Theorem 3.3. *Let A be a finite abelian group written additively.*

(i) $\chi(-a)=\overline{\chi(a)}$ *for all $a\in A$ and $\chi\in\widehat{A}$.*

(ii) *Let $B < A$, $b \in A$ and $H < \widehat{A}$, $\lambda \in \widehat{A}$. Then*

(3.7)
$$\sum_{a \in B} \lambda(a) = \begin{cases} |B| & \text{if } \lambda \in B^{\perp}, \\ 0 & \text{if } \lambda \notin B^{\perp}, \end{cases}$$

(3.8)
$$\sum_{\chi \in H} \chi(b) = \begin{cases} |H| & \text{if } b \in H^{\perp}, \\ 0 & \text{if } b \notin H^{\perp}. \end{cases}$$

In particular, with $B = A$ and $H = \widehat{A}$,

(3.9)
$$\sum_{a \in A} \lambda(a) = \begin{cases} |A| & \text{if } \lambda = 1^A, \\ 0 & \text{if } \lambda \neq 1^A, \end{cases}$$

(3.10)
$$\sum_{\chi \in \widehat{A}} \chi(b) = \begin{cases} |A| & \text{if } b = 0, \\ 0 & \text{if } b \neq 0. \end{cases}$$

(iii) *(Orthogonal relations) Let $a, b \in A$ and $\alpha, \beta \in \widehat{A}$. Then*

(3.11)
$$\sum_{c \in A} \alpha(c)\overline{\beta(c)} = \begin{cases} |A| & \text{if } \alpha = \beta, \\ 0 & \text{if } \alpha \neq \beta, \end{cases}$$

(3.12)
$$\sum_{\chi \in \widehat{A}} \chi(a)\overline{\chi(b)} = \begin{cases} |A| & \text{if } a = b, \\ 0 & \text{if } a \neq b. \end{cases}$$

Proof. (i) Since $\chi(a) \in \mathbb{C}^*$ is a root of unity, $\chi(-a) = \chi(a)^{-1} = \overline{\chi(a)}$.

(ii) If $\lambda \in B^{\perp}$, then $\sum_{a \in B} \lambda(a) = \sum_{a \in B} 1 = |B|$. If $\lambda \notin B^{\perp}$, choose $c \in B$ such that $\lambda(c) \neq 1$. Then

$$\sum_{a \in B} \lambda(a) = \sum_{a \in B} \lambda(c + a) = \sum_{a \in B} \lambda(c)\lambda(a) = \lambda(c) \sum_{a \in B} \lambda(a),$$

which forces $\sum_{a \in B} \lambda(a) = 0$. Thus (3.7) is proved. The proof of (3.8) is identical.

(iii) By (3.9), we have

$$\sum_{c \in A} \alpha(c)\overline{\beta(c)} = \sum_{c \in A} (\alpha\bar{\beta})(c) = \begin{cases} |A| & \text{if } \alpha\bar{\beta} = 1^A, \text{ i.e., } \alpha = \beta, \\ 0 & \text{if } \alpha\bar{\beta} \neq 1^A, \text{ i.e., } \alpha \neq \beta. \end{cases}$$

The proof of (3.12) is the same. □

Fourier transform. Let A be a finite abelian group and let $f : A \to \mathbb{C}$ be a function. The *Fourier transform* of f is a function $\widetilde{f} : \widehat{A} \to \mathbb{C}$ defined by

$$\widetilde{f}(\chi) = \sum_{a \in A} f(a)\bar{\chi}(a), \qquad \chi \in \widehat{A}. \tag{3.13}$$

If f and g are two functions from A to $\mathbb{C}$, the *convolution* of f and g is a function $f * g : A \to \mathbb{C}$ defined by

$$(f * g)(a) = \sum_{\substack{b,c \in A \\ b+c=a}} f(b)g(c), \qquad a \in A.$$

The next proposition is a collection of some basic identities about the Fourier transform.

Proposition 3.4. *Let A be a finite abelian group and $B < A$, and let $f, g : A \to \mathbb{C}$ be two functions. Then we have the following identities.*

$$f(a) = \frac{1}{|A|} \sum_{\chi \in \widehat{A}} \widetilde{f}(\chi)\chi(a) \quad \text{for all } a \in A \qquad \text{(the inversion formula)}. \tag{3.14}$$

$$\widetilde{\widetilde{f}}(\langle \cdot, a\rangle) = |A| f(-a) \quad \text{for all } a \in A. \tag{3.15}$$

(Recall that $\langle \cdot, a\rangle \in \widehat{\widehat{A}}$ is defined by $\chi \mapsto \langle \chi, a\rangle$.)

$$\widetilde{f * g} = \widetilde{f}\,\widetilde{g}. \tag{3.16}$$

$$\widetilde{fg} = \frac{1}{|A|} \widetilde{f} * \widetilde{g}. \tag{3.17}$$

$$\sum_{\chi \in B^{\perp}} \widetilde{f}(\chi) = \frac{|A|}{|B|} \sum_{a \in B} f(a) \qquad \text{(Poisson summation formula)}. \tag{3.18}$$

$$\sum_{\chi \in \widehat{A}} |\widetilde{f}(\chi)|^2 = |A| \sum_{a \in A} |f(a)|^2 \qquad \text{(Parseval's identity)}. \tag{3.19}$$

Proof. *Proof of* (3.14): We have

$$\begin{aligned}\frac{1}{|A|}\sum_{\chi\in\widehat{A}}\widetilde{f}(\chi)\chi(a) &= \frac{1}{|A|}\sum_{\chi\in\widehat{A}}\sum_{b\in A} f(b)\overline{\chi(b)}\chi(a) = \frac{1}{|A|}\sum_{b\in A} f(b)\sum_{\chi\in\widehat{A}}\chi(a)\overline{\chi(b)}\\ &= \frac{1}{|A|}\sum_{b=a} f(b)|A| \qquad\qquad \text{(by (3.12))}\\ &= f(a).\end{aligned}$$

Proof of (3.15): We have

$$\begin{aligned}\widetilde{\widetilde{f}}(\langle\,\cdot\,,a\rangle) &= \sum_{\chi\in\widehat{A}}\widetilde{f}(\chi)\overline{\langle\,\cdot\,,a\rangle}(\chi) = \sum_{\chi\in\widehat{A}}\sum_{b\in A} f(b)\bar{\chi}(b)\bar{\chi}(a)\\ &= \sum_{b\in A} f(b)\sum_{\chi\in\widehat{A}}\bar{\chi}(b+a) = \sum_{b=-a} f(b)|A| = |A|f(-a).\end{aligned}$$

Proof of (3.16): For each $\chi\in\widehat{A}$,

$$\begin{aligned}\widetilde{(f*g)}(\chi) &= \sum_{a\in A}(f*g)(a)\,\bar{\chi}(a) = \sum_{a\in A}\sum_{b\in A} f(b)g(a-b)\bar{\chi}(a)\\ &= \sum_{a\in A}\sum_{b\in A} f(b)g(a)\bar{\chi}(a+b) \qquad\qquad (a\mapsto a+b)\\ &= \Big(\sum_{b\in A} f(b)\bar{\chi}(b)\Big)\Big(\sum_{a\in A} g(a)\bar{\chi}(a)\Big) = \widetilde{f}(\chi)\,\widetilde{g}(\chi).\end{aligned}$$

Proof of (3.17): For each $a\in A$, by (3.15),

$$\widetilde{\widetilde{(fg)}}(\langle\,\cdot\,,a\rangle) = |A|(fg)(-a).$$

On the other hand, by (3.16) with $\widetilde{f}$ and $\widetilde{g}$ in place of f and g, respectively, and by (3.15),

$$\Big(\frac{1}{|A|}\widetilde{f}*\widetilde{g}\Big)^{\sim}(\langle\,\cdot\,,a\rangle) = \frac{1}{|A|}(\widetilde{\widetilde{f}}\,\widetilde{\widetilde{g}})(\langle\,\cdot\,,a\rangle) = |A|(fg)(-a).$$

Thus $\widetilde{\widetilde{fg}} = \big((\widetilde{f}*\widetilde{g})/|A|\big)^{\sim}$, and hence $\widetilde{fg} = (\widetilde{f}*\widetilde{g})/|A|$.

Proof of (3.18): We have

$$\begin{aligned}\sum_{\chi\in B^{\perp}}\widetilde{f}(\chi) &= \sum_{\chi\in B^{\perp}}\sum_{a\in A} f(a)\bar{\chi}(a) = \sum_{a\in A} f(a)\sum_{\chi\in B^{\perp}}\chi(-a)\\ &= \sum_{a\in B} f(a)|B^{\perp}| \qquad\qquad \text{(by (3.8))}\\ &= \frac{|A|}{|B|}\sum_{a\in B} f(a).\end{aligned}$$

Proof of (3.19): Let $g : A \to \mathbb{C}$ be defined by $g(a) = \overline{f(-a)}$, $a \in A$. Then for $\chi \in \widehat{A}$,

$$\widetilde{g}(\chi) = \sum_{a\in A} g(a)\bar{\chi}(a) = \overline{\Big(\sum_{a\in A} f(-a)\chi(a)\Big)} = \overline{\Big(\sum_{a\in A} f(a)\bar{\chi}(a)\Big)} = \overline{\widetilde{f}}(\chi).$$

Thus by (3.16) and by (3.18) with $B = \{0\}$,

$$\begin{aligned}\sum_{\chi\in\widehat{A}} |\widetilde{f}(\chi)|^2 &= \sum_{\chi\in\widehat{A}} (\widetilde{f}\,\widetilde{g})(\chi) = \sum_{\chi\in\widehat{A}} (\widetilde{f * g})(\chi) = |A|(f * g)(0) \\ &= |A|\sum_{a\in A} f(a)g(-a) = |A|\sum_{a\in A} |f(a)|^2.\end{aligned}$$

□

For a finite abelian group A, let $\mathbb{C}^A$ denote the set of all functions from A to $\mathbb{C}$. We have the following $\mathbb{C}$-maps of vector spaces over $\mathbb{C}$:

$$\mathbb{C}^A \xrightarrow{\widetilde{(\ )}} \mathbb{C}^{\widehat{A}} \xrightarrow{\widetilde{(\ )}} \mathbb{C}^{\widehat{\widehat{A}}} \xrightarrow{\phi^*} \mathbb{C}^A, \tag{3.20}$$

where

$$\begin{array}{rccc} \phi: & A & \xrightarrow{\cong} & \widehat{\widehat{A}} \\ & a & \longmapsto & \langle\,\cdot\,, a\rangle \end{array}$$

and

$$\begin{array}{rccc} \phi^* & \mathbb{C}^{\widehat{\widehat{A}}} & \longrightarrow & \mathbb{C}^A \\ & f & \longmapsto & f\circ\phi. \end{array}$$

By (3.15), the composition of the $\mathbb{C}$-maps in (3.20) is the map $\mathbb{C}^A \to \mathbb{C}^A$: $f(\ ) \mapsto |A|f(-(\ ))$, which is an isomorphism. Since $\dim_{\mathbb{C}} \mathbb{C}^A = \dim_{\mathbb{C}} \mathbb{C}^{\widehat{A}} = \dim_{\mathbb{C}} \mathbb{C}^{\widehat{\widehat{A}}} = |A| < \infty$, all the $\mathbb{C}$-maps in (3.20) are isomorphisms. Moreover, by (3.16), $\widetilde{(\ )} : (\mathbb{C}^A, +, *) \to (\mathbb{C}^{\widehat{A}}, +, \cdot)$ is a $\mathbb{C}$-algebra isomorphism, and by (3.17), $\widetilde{(\ )}/|A| : (\mathbb{C}^A, +, \cdot) \to (\mathbb{C}^{\widehat{A}}, +, *)$ is also a $\mathbb{C}$-algebra isomorphism.

Characters of $\mathbb{F}_q$ and $\mathbb{F}_q^*$. Characters of $(\mathbb{F}_q, +)$ and $(\mathbb{F}_q^*, \cdot)$ are called additive and multiplicative characters of $\mathbb{F}_q$, respectively. We write $\widehat{(\mathbb{F}_q, +)} = \widehat{\mathbb{F}_q}$ and $\widehat{(\mathbb{F}_q^*, \cdot)} = \widehat{\mathbb{F}_q^*}$.

For each $a \in \mathbb{F}_q$, define

$$\begin{array}{rccc} e_a: & \mathbb{F}_q & \longrightarrow & \mathbb{C}^* \\ & x & \longmapsto & \zeta_p^{\operatorname{Tr}_{q/p}(ax)}, \end{array}$$

where $p = \operatorname{char} \mathbb{F}_q$. Clearly, $e_a \in \widehat{\mathbb{F}_q}$. Moreover, $a \mapsto e_a$ is a one-to-one homomorphism, and hence an isomorphism, from $\mathbb{F}_q$ to $\widehat{\mathbb{F}_q}$. Therefore, $\widehat{\mathbb{F}_q} = \{e_a : a \in \mathbb{F}_q\}$.

Fix a generator α of $\mathbb{F}_q^*$. Then

$$\begin{aligned} \psi: \mathbb{F}_q^* &\longrightarrow \mathbb{C}^* \\ \alpha^i &\longmapsto \zeta_{q-1}^i \end{aligned}$$

is a character of $(\mathbb{F}_q^*, \cdot)$ of order $q-1$, and hence $\widehat{\mathbb{F}_q^*} = \langle \psi \rangle = \{\psi^i : 0 \le i \le q-2\}$. For $\chi \in \widehat{\mathbb{F}_q^*}$, we define

$$\chi(0) = \begin{cases} 1 & \text{if } \chi = 1^{\mathbb{F}_q^*}, \\ 0 & \text{if } \chi \neq 1^{\mathbb{F}_q^*}. \end{cases}$$

Thus

$$\sum_{x \in \mathbb{F}_q} \chi(x) = \begin{cases} q & \text{if } \chi = 1^{\mathbb{F}_q^*}, \\ 0 & \text{if } \chi \neq 1^{\mathbb{F}_q^*}. \end{cases}$$

3.2. Gauss Sums

For $a \in \mathbb{F}_q$ and $\chi \in \widehat{\mathbb{F}_q^*}$, the *Gauss sum* $g_a(\chi)$ is defined to be

$$g_a(\chi) = \sum_{x \in \mathbb{F}_q} \chi(x) e_a(x) = \sum_{x \in \mathbb{F}_q} \chi(x) \zeta_p^{\mathrm{Tr}_{q/p}(ax)}.$$

Let $\mathbb{C}^{\mathbb{F}_q}$ be the unitary space of all functions from $\mathbb{F}_q$ to $\mathbb{C}$ equipped with the inner product

$$\langle f, g \rangle = \frac{1}{q} \sum_{x \in \mathbb{F}_q} f(x)\overline{g(x)}.$$

Then $\{e_a : a \in \mathbb{F}_q\}$ is an orthonormal basis of $\mathbb{C}^{\mathbb{F}_q}$. Hence, for $\chi \in \widehat{\mathbb{F}_q^*}$, we have

$$\chi = \sum_{a \in \mathbb{F}_q} \langle \chi, e_a \rangle e_a = \frac{1}{q} \sum_{a \in \mathbb{F}_q} g_{-a}(\chi) e_a,$$

which gives an interpretation of the Gauss sum.

Fact 3.5. *Let $\chi \in \widehat{\mathbb{F}_q^*}$.*

(i)

$$g_0(\chi) = \begin{cases} q & \textit{if } \chi = 1^{\mathbb{F}_q^*}, \\ 0 & \textit{if } \chi \neq 1^{\mathbb{F}_q^*}. \end{cases}$$

(ii) *If $a \in \mathbb{F}_q^*$, then $g_a(\chi) = \chi(a^{-1}) g_1(\chi)$.*

Proof. (i) is obvious. For (ii), we have

$$\begin{aligned} g_a(\chi) &= \sum_{x\in\mathbb{F}_q} \chi(x)e_1(ax) = \sum_{x\in\mathbb{F}_q} \chi(a^{-1}x)e_1(x) \\ &= \chi(a^{-1}) \sum_{x\in\mathbb{F}_q} \chi(x)e_1(x) = \chi(a^{-1})g_1(\chi). \end{aligned}$$

□

Because of Fact 3.5, it suffices to consider the Gauss sum $g_1(\chi)$. We write $g(\chi) = g_1(\chi)$.

Fact 3.6. *Let* $\chi \in \widehat{\mathbb{F}_q^*}$.

(i) *If* $\chi = 1^{\mathbb{F}_q^*}$, $g(\chi) = 0$.
(ii) *If* $\chi \neq 1^{\mathbb{F}_q^*}$, $|g(\chi)| = q^{1/2}$.

Proof. (i) is obvious. For (ii), we have

$$\begin{aligned} |g(\chi)|^2 &= \sum_{x,y\in\mathbb{F}_q^*} \chi(x)e_1(x)\overline{\chi(y)e_1(y)} = \sum_{x,y\in\mathbb{F}_q^*} \chi(xy^{-1})e_1(x-y) \\ &= \sum_{x,y\in\mathbb{F}_q^*} \chi(x)e_1(xy-y) \qquad\qquad (x \mapsto xy) \\ &= \sum_{x\in\mathbb{F}_q^*} \chi(x) \sum_{y\in\mathbb{F}_q^*} e_{x-1}(y) = \sum_{x\in\mathbb{F}_q^*} \chi(x) \sum_{y\in\mathbb{F}_q} e_{x-1}(y) = \sum_{x=1} \chi(x)q = q. \end{aligned}$$

□

Jacobi sum. For $\chi_1, \ldots, \chi_k \in \widehat{\mathbb{F}_q^*}$, the *Jacobi sum* of $\chi_1, \ldots, \chi_k$ is defined as

$$J(\chi_1, \ldots, \chi_k) = \sum_{\substack{x_1,\ldots,x_k\in\mathbb{F}_q \\ x_1+\cdots+x_k=1}} \chi_1(x_1)\cdots\chi_k(x_k).$$

Proposition 3.7. *Assume that* $\chi_1, \ldots, \chi_k \in \widehat{\mathbb{F}_q^*}$.

(i)

$$J(\chi_1, \ldots, \chi_k) = \begin{cases} q^{k-1} & \text{if } \chi_1, \ldots, \chi_k \text{ are all principal,} \\ 0 & \text{if some, but not all, of } \chi_1, \ldots, \chi_k \text{ are principal.} \end{cases}$$

(ii) *If* $\chi_1, \ldots, \chi_k$ *are all nonprincipal, then*

$$J(\chi_1, \ldots, \chi_k) = \begin{cases} \dfrac{g(\chi_1)\cdots g(\chi_k)}{g(\chi_1\cdots\chi_k)} & \text{if } \chi_1\cdots\chi_k \neq 1^{\mathbb{F}_q^*}, \\ -\dfrac{1}{q}\, g(\chi_1)\cdots g(\chi_k) & \text{if } \chi_1\cdots\chi_k = 1^{\mathbb{F}_q^*}. \end{cases}$$

Proof. (i) If $\chi_1 = \cdots = \chi_k = 1^{\mathbb{F}_q^*}$,

$$J(\chi_1, \ldots, \chi_k) = \sum_{\substack{x_1,\ldots,x_k \in \mathbb{F}_q \\ x_1+\cdots+x_k=1}} 1 = q^{k-1}.$$

If $\chi_1 = 1^{\mathbb{F}_q^*}$ and $\chi_2 \neq 1^{\mathbb{F}_q^*}$,

$$\begin{aligned} J(\chi_1, \ldots, \chi_k) &= \sum_{\substack{x_1,\ldots,x_k \in \mathbb{F}_q \\ x_1+\cdots+x_k=1}} \chi_2(x_2)\cdots\chi_k(x_k) = \sum_{x_2,\ldots,x_k \in \mathbb{F}_q} \chi_2(x_2)\cdots\chi_k(x_k) \\ &= \prod_{i=2}^{k}\Big(\sum_{x_i \in \mathbb{F}_q} \chi_i(x_i)\Big) = 0. \end{aligned}$$

(ii) 1° Assume that $\chi_1 \cdots \chi_k \neq 1^{\mathbb{F}_q^*}$. Then

$$\begin{aligned} &g(\chi_1)\cdots g(\chi_k) \\ &= \sum_{x_1,\ldots,x_k \in \mathbb{F}_q} \chi_1(x_1)\cdots\chi_k(x_k)e_1(x_1+\cdots+x_k) \\ &= \sum_{a \in \mathbb{F}_q} e_1(a) \sum_{x_1+\cdots+x_k=a} \chi_1(x_1)\cdots\chi_k(x_k) \\ &= \sum_{a \in \mathbb{F}_q^*} e_1(a) \sum_{x_1+\cdots+x_k=a} \chi_1(x_1)\cdots\chi_k(x_k) + \sum_{x_1+\cdots+x_k=0} \chi_1(x_1)\cdots\chi_k(x_k) \\ &= \sum_{a \in \mathbb{F}_q^*} e_1(a)(\chi_1\cdots\chi_k)(a) \sum_{\frac{x_1}{a}+\cdots+\frac{x_k}{a}=1} \chi_1\Big(\frac{x_1}{a}\Big)\cdots\chi_k\Big(\frac{x_k}{a}\Big) \\ &\quad + \sum_{x_1+\cdots+x_k=0} \chi_1(x_1)\cdots\chi_k(x_k) \\ &= g(\chi_1\cdots\chi_k)J(\chi_1,\ldots,\chi_k) + \sum_{x_1+\cdots+x_k=0} \chi_1(x_1)\cdots\chi_k(x_k). \end{aligned}$$

In the above,

$$\begin{aligned} &\sum_{x_1+\cdots+x_k=0} \chi_1(x_1)\cdots\chi_k(x_k) \\ &= \sum_{x_1 \in \mathbb{F}_q^*} \sum_{\frac{x_2}{x_1}+\cdots+\frac{x_k}{x_1}=-1} (\chi_1\cdots\chi_k)(x_1)\chi_2\Big(\frac{x_2}{x_1}\Big)\cdots\chi_k\Big(\frac{x_k}{x_1}\Big) \\ &= \sum_{x_1 \in \mathbb{F}_q^*} (\chi_1\cdots\chi_k)(x_1) \sum_{x_2+\cdots+x_k=-1} \chi_2(x_2)\cdots\chi_k(x_k) = 0. \end{aligned} \tag{3.21}$$

Hence

$$J(\chi_1, \ldots, \chi_k) = \frac{g(\chi_1)\cdots g(\chi_k)}{g(\chi_1\cdots\chi_k)}.$$

2° Assume that $\chi_1 \cdots \chi_k = 1^{\mathbb{F}_q^*}$. We have

$$J(\chi_1, \dots, \chi_k) = \sum_{a \in \mathbb{F}_q^*} \chi_1(1-a) \sum_{x_2+\cdots+x_k=a} \chi_2(x_2)\cdots\chi_k(x_k) + \sum_{x_2+\cdots+x_k=0} \chi_2(x_2)\cdots\chi_k(x_k).$$

In the above, the second sum on the right side is 0 by (3.21) and the fact that $\chi_2 \cdots \chi_k \neq 1^{\mathbb{F}_q^*}$. Therefore,

$$\begin{aligned} J(\chi_1, \dots, \chi_k) &= \sum_{a \in \mathbb{F}_q^*} \chi_1(1-a)(\chi_2\cdots\chi_k)(a) \sum_{\frac{x_2}{a}+\cdots+\frac{x_k}{a}=1} \chi_2\Big(\frac{x_2}{a}\Big)\cdots\chi_k\Big(\frac{x_k}{a}\Big) \\ &= J(\chi_2, \dots, \chi_k) \sum_{a \in \mathbb{F}_q^*} \chi_1\Big(\frac{1-a}{a}\Big) \\ &= -J(\chi_2, \dots, \chi_k)\chi_1(-1) \qquad \Big(\text{since } \sum_{x \in \mathbb{F}_q} \chi_1(x) = 0\Big) \\ &= -\chi_1(-1)\frac{g(\chi_2)\cdots g(\chi_k)}{g(\chi_2\cdots\chi_k)} \qquad (\text{by } 1^\circ) \\ &= -\frac{g(\chi_1)\cdots g(\chi_k)}{\chi_1(-1)g(\chi_1^{-1})g(\chi_1)} \qquad (\text{since } \chi_2\cdots\chi_k = \chi_1^{-1}). \end{aligned}$$

In the above,

$$\begin{aligned} \chi_1(-1)g(\chi_1^{-1}) &= \sum_{x \in \mathbb{F}_q^*} \chi_1(-x^{-1})e_1(x) = \sum_{x \in \mathbb{F}_q^*} \chi_1(x^{-1})e_1(-x) \\ &= \sum_{x \in \mathbb{F}_q^*} \overline{\chi_1(x)e_1(x)} = \overline{g(\chi_1)}. \end{aligned} \tag{3.22}$$

Thus

$$J(\chi_1, \dots, \chi_k) = -\frac{g(\chi_1)\cdots g(\chi_k)}{\overline{g(\chi_1)}g(\chi_1)} = -\frac{1}{q}\, g(\chi_1)\cdots g(\chi_k).$$

□

3.3. The Davenport-Hasse Theorem

Let n be a positive integer. For each $\chi \in \widehat{\mathbb{F}_q^*}$, define $\chi^{(n)} = \chi \circ \mathrm{N}_{q^n/q} \in \widehat{\mathbb{F}_{q^n}^*}$, which is called the lift of χ from $\mathbb{F}_q$ to $\mathbb{F}_{q^n}$. According to our convention, the Gauss sums of χ and $\chi^{(n)}$ are defined as $g(\chi) = \sum_{x \in \mathbb{F}_q^*} \chi(x)\zeta_p^{\mathrm{Tr}_{q/p}(x)}$ and $g(\chi^{(n)}) = \sum_{x \in \mathbb{F}_{q^n}^*} \chi^{(n)}(x)\zeta_p^{\mathrm{Tr}_{q^n/p}(x)}$, respectively. It is natural to ask how $g(\chi)$ and $g(\chi^{(n)})$ are related. The answer is given by the Davenport-Hasse theorem.

Theorem 3.8 (The Davenport-Hasse theorem). *We have*

$$-g(\chi^{(n)}) = \big[-g(\chi)\big]^n.$$

Proof. Let $e(\ ) = \zeta_p^{\mathrm{Tr}_{q/p}(\ )}$ and $e^{(n)}(\ ) = \zeta_p^{\mathrm{Tr}_{q^n/p}(\ )}$.

1° Let $\mathcal{M}$ be the set of all monic polynomials in $\mathbb{F}_q[\mathtt{X}]$ and let $\mathcal{I} = \{f \in \mathcal{M} : f \text{ is irreducible}\}$. For each $f \in \mathcal{M}$, write $f = (\mathtt{X} - a_1)\cdots(\mathtt{X} - a_k)$, $a_1, \dots, a_k \in \overline{\mathbb{F}}_q$, and define

$$\rho(f) = \chi(a_1 \cdots a_k)e(a_1 + \cdots + a_k).$$

If $g = (\mathtt{X} - b_1)\cdots(\mathtt{X} - b_l) \in \mathcal{M}$, where $b_1, \dots, b_l \in \overline{\mathbb{F}}_q$, we have

$$\begin{aligned}\rho(fg) &= \chi(a_1 \cdots a_k b_1 \cdots b_l)e(a_1 + \cdots + a_k + b_1 + \cdots + b_l)\\ &= \chi(a_1 \cdots a_k)e(a_1 + \cdots + a_k)\chi(b_1 \cdots b_l)e(b_1 + \cdots + b_l)\\ &= \rho(f)\rho(g).\end{aligned}$$

2° Let $\alpha \in \mathbb{F}_{q^n}$ with minimal polynomial f over $\mathbb{F}_q$. Then

$$\rho(f) = \chi\big(\mathrm{N}_{\mathbb{F}_q(\alpha)/\mathbb{F}_q}(\alpha)\big)e\big(\mathrm{Tr}_{\mathbb{F}_q(\alpha)/\mathbb{F}_q}(\alpha)\big).$$

Since $[\mathbb{F}_{q^n} : \mathbb{F}_q(\alpha)] = n/\deg f$, we have

$$\begin{aligned}\rho(f)^{n/\deg f} &= \chi\big(\mathrm{N}_{\mathbb{F}_q(\alpha)/\mathbb{F}_q}(\alpha)^{n/\deg f}\big)e\Big(\frac{n}{\deg f}\mathrm{Tr}_{\mathbb{F}_q(\alpha)/\mathbb{F}_q}(\alpha)\Big)\\ &= \chi\big(\mathrm{N}_{\mathbb{F}_{q^n}/\mathbb{F}_q}(\alpha)\big)e\big(\mathrm{Tr}_{\mathbb{F}_{q^n}/\mathbb{F}_q}(\alpha)\big) = \chi^{(n)}(\alpha)e^{(n)}(\alpha).\end{aligned}$$

Thus

$$\begin{aligned}g(\chi^{(n)}) &= \sum_{\alpha \in \mathbb{F}_{q^n}} \chi^{(n)}(\alpha)e^{(n)}(\alpha) = \sum_{\substack{f \in \mathcal{I}\\ \deg f \mid n}} \sum_{f(\alpha)=0} \chi^{(n)}(\alpha)e^{(n)}(\alpha)\\ &= \sum_{\substack{f \in \mathcal{I}\\ \deg f \mid n}} (\deg f)\rho(f)^{n/\deg f}.\end{aligned} \tag{3.23}$$

3° Let $\mathtt{t}$ be an indeterminate. We have

$$\begin{aligned}\prod_{f \in \mathcal{I}} \frac{1}{1 - \rho(f)\mathtt{t}^{\deg f}} &= \prod_{f \in \mathcal{I}} \Big(\sum_{j(f)=0}^{\infty} \rho(f^{j(f)})\mathtt{t}^{j(f)\deg f}\Big)\\ &= \sum_{i=0}^{\infty}\Big(\sum_{\substack{j:\mathcal{I}\to\mathbb{N}\\ \sum_{f\in\mathcal{I}} j(f)\deg f = i}} \rho\Big(\prod_{f\in\mathcal{I}} f^{j(f)}\Big)\Big)\mathtt{t}^i\\ &= \sum_{i=0}^{\infty}\Big(\sum_{\substack{f \in \mathcal{M}\\ \deg f = i}} \rho(f)\Big)\mathtt{t}^i.\end{aligned}$$

Note that

$$\sum_{\substack{f\in\mathcal{M}\\ \deg f=0}} \rho(f)=1,\qquad \sum_{\substack{f\in\mathcal{M}\\ \deg f=1}} \rho(f)=\sum_{a\in\mathbb{F}_q}\rho(\mathtt{X}-a)=\sum_{a\in\mathbb{F}_q}\chi(a)e(a)=g(\chi),$$

and for $i>1$,

$$\begin{aligned}\sum_{\substack{f\in\mathcal{M}\\ \deg f=i}} \rho(f) &= \sum_{a_0,\dots,a_{i-1}\in\mathbb{F}_q} \rho(\mathtt{X}^i-a_{i-1}\mathtt{X}^{i-1}+\cdots+(-1)^i a_0)\\ &= q^{i-2}\Big(\sum_{a_0\in\mathbb{F}_q}\chi(a_0)\Big)\Big(\sum_{a_{i-1}\in\mathbb{F}_q} e(a_{i-1})\Big)=0.\end{aligned}$$

Thus

$$\prod_{f\in\mathcal{I}}\frac{1}{1-\rho(f)\mathtt{t}^{\deg f}}=1+g(\chi)\mathtt{t}. \tag{3.24}$$

From (3.24) we have

$$-\sum_{f\in\mathcal{I}}\log\big(1-\rho(f)\mathtt{t}^{\deg f}\big)=\log\big(1+g(\chi)\mathtt{t}\big).$$

Differentiating the above equation with respect to $\mathtt{t}$ gives

$$\sum_{f\in\mathcal{I}}\frac{(\deg f)\rho(f)\mathtt{t}^{\deg f-1}}{1-\rho(f)\mathtt{t}^{\deg f}}=\frac{g(\chi)}{1+g(\chi)\mathtt{t}}.$$

In the above,

$$\begin{aligned}\text{the left side} &= \mathtt{t}^{-1}\sum_{f\in\mathcal{I}}\deg f\sum_{j=1}^{\infty}\rho(f)^j\mathtt{t}^{j\deg f}\\ &= \mathtt{t}^{-1}\sum_{n=1}^{\infty}\Big(\sum_{\substack{f\in\mathcal{I}\\ \deg f\mid n}}(\deg f)\rho(f)^{n/\deg f}\Big)\mathtt{t}^n\\ &= \sum_{n=1}^{\infty} g(\chi^{(n)})\mathtt{t}^{n-1} \qquad\qquad \text{(by (3.23))},\end{aligned}$$

and

$$\text{the right side}=\sum_{n=1}^{\infty}(-1)^{n-1}g(\chi)^n\mathtt{t}^{n-1}.$$

Comparing the coefficients of $\mathtt{t}^{n-1}$ gives

$$g(\chi^{(n)})=(-1)^{n-1}g(\chi)^n.$$

□

3.4. The Gauss Quadratic Sum

The reader is asked to prove the following easy fact (Exercise 3.4): Let $k \mid q-1$ and let A_k be the subgroup of $\widehat{\mathbb{F}_q^*}$ of order k. Then

$$\sum_{x\in\mathbb{F}_q} e_1(x^k) = \sum_{\chi\in A_k} g(\chi) = \sum_{\chi\in A_k\setminus\{1^{\mathbb{F}_q^*}\}} g(\chi). \tag{3.25}$$

Let q be odd and let $\eta \in \widehat{\mathbb{F}_q^*}$ be such that $o(\eta) = 2$; η is called the *quadratic character* of $\mathbb{F}_q^*$ and is defined by

$$\eta(x) = \begin{cases} 1 & \text{if } x \text{ is a square in } \mathbb{F}_q^*, \\ -1 & \text{if } x \text{ is a nonsquare in } \mathbb{F}_q^*. \end{cases} \tag{3.26}$$

Equation (3.25) with $k = 2$ gives

$$g(\eta) = \sum_{x\in\mathbb{F}_q} e_1(x^2), \tag{3.27}$$

which is called the *Gauss quadratic sum* over $\mathbb{F}_q$. By (3.22), $\overline{g(\eta)} = \eta(-1)g(\eta)$, where

$$\eta(-1) = \begin{cases} 1 & \text{if } q \equiv 1 \pmod 4, \\ -1 & \text{if } q \equiv -1 \pmod 4. \end{cases}$$

Thus, $g(\eta)$ is real if $q \equiv 1 \pmod 4$ and is purely imaginary if $q \equiv -1 \pmod 4$. Since $|g(\eta)| = q^{1/2}$, we must have

$$g(\eta) = \begin{cases} \pm q^{1/2} & \text{if } q \equiv 1 \pmod 4, \\ \pm i q^{1/2} & \text{if } q \equiv -1 \pmod 4. \end{cases} \tag{3.28}$$

The determination of the signs in (3.28) requires some additional effort. We first consider the case $q = p$. In fact, the determination of the signs in (3.28) with $q = p$ was a major problem considered by Gauss, and it took four years for him to find a solution [**11**]. For a survey of various proofs of Gauss's result, see [**5**]. The proof adopted here is due to Siegel. It appears to be the simplest analytic proof in the sense that it does not use Fourier analysis and the theta function employed in the classical proofs by Dirichlet and Cauchy, and as Siegel had noted, it generalizes to Gauss sums and quadratic reciprocity in number fields.

Analytic evaluation of the Gauss quadratic sum over $\mathbb{F}_p$.

Theorem 3.9 (Siegel [**33**]). *Let $a, b \in \mathbb{Z}^+$ and $t \in \mathbb{Q}$ be such that $ab + 2bt$ is an even integer, and let $r = b/a$. Then*

$$\sum_{k=0}^{a-1} e^{\pi i r(k+t)^2} = r^{-1/2} e^{\pi i/4} \sum_{l=0}^{b-1} e^{\pi i(-r^{-1}l^2 + 2tl)}. \tag{3.29}$$

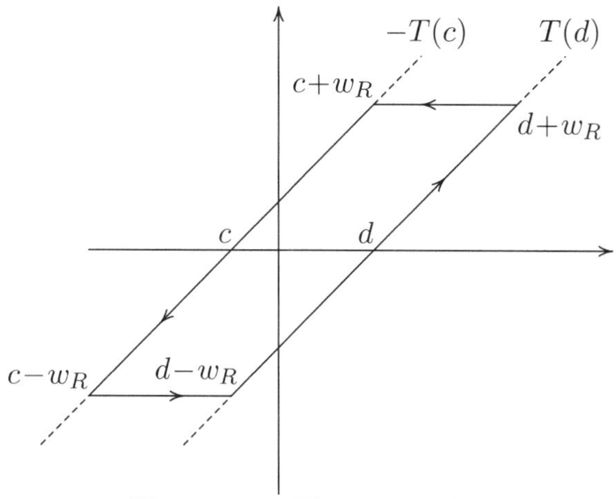

Figure 3.1. The contour C_R

Proof. 1° Let

$$f(z)=\frac{e^{\pi ir(z+t)^2}}{e^{2\pi iz}-1}.$$

This is a meromorphic function on $\mathbb{C}$ with poles at the integers and

$$\operatorname{Res}(f;k)=\frac{e^{\pi ir(k+t)^2}}{2\pi i},\qquad k\in\mathbb{Z}.$$

Let $w_R=Re^{\pi i/4}$, where $R>0$. For $c,d\in\mathbb{R}\setminus\mathbb{Z}$ with $c<d$, the integration of f on the perimeter C_R (oriented counterclockwise) of the parallelogram with vertices $c-w_R$, $d-w_R$, $d+w_R$, and $c+w_R$ gives

$$\int_{C_R} f(z)dz=2\pi i\sum_{c<k<d}\operatorname{Res}(f;k)=\sum_{c<k<d}e^{\pi ir(k+t)^2}. \tag{3.30}$$

For $z=x+iy$, $x,y\in\mathbb{R}$, $y\neq 0$, we have

$$|f(z)|=\frac{e^{-2\pi ry(x+t)}}{|e^{2\pi(-y+ix)}-1|}\leq\frac{e^{-2\pi ry(x+t)}}{|e^{-2\pi y}-1|}. \tag{3.31}$$

Letting $R\to\infty$ in (3.30) and using (3.31), we have

$$\int_{T(d)} f(z)dz-\int_{T(c)} f(z)dz=\sum_{c<k<d}e^{\pi ir(k+t)^2}, \tag{3.32}$$

where $T(d)$ is the line (oriented upward) which intersects the real axis at d with an angle of $\pi/4$. In the same way, we also have

$$\int_{T(d)} e^{\pi irz^2}dz-\int_{T(c)} e^{\pi irz^2}dz=0\quad\text{for all } c,d\in\mathbb{R}. \tag{3.33}$$

2° Denote the sums on the left and right sides of (3.29) by A and B, respectively. For $z \in \mathbb{C} \setminus \mathbb{Z}$,

(3.34)

$$\begin{aligned} f(z+a) - f(z) &= e^{\pi i r (z+t)^2} \cdot \frac{e^{\pi i r[a^2 + 2a(z+t)]} - 1}{e^{2\pi i z} - 1} \\ &= e^{\pi i r (z+t)^2} \cdot \frac{e^{2\pi i r z b} - 1}{e^{2\pi i z} - 1} \qquad \text{(since } ab + 2bt \in 2\mathbb{Z}) \\ &= e^{\pi i r (z+t)^2} \sum_{l=0}^{b-1} e^{2\pi i z l} = \sum_{l=0}^{b-1} e^{\pi i(-r^{-1}l^2 - 2tl)} e^{\pi i r (z+t+r^{-1}l)^2}. \end{aligned}$$

Therefore,

$$\begin{aligned} A &= \int_{T(-\frac{1}{2}+a)} f(z)dz - \int_{T(-\frac{1}{2})} f(z)dz \qquad \text{(by (3.32))} \\ &= \int_{T(-\frac{1}{2})} \big[f(z+a) - f(z)\big]dz \\ &= \sum_{l=0}^{b-1} e^{\pi i(-r^{-1}l^2 - 2tl)} \int_{T(-\frac{1}{2}+t+r^{-1}l)} e^{\pi i r z^2} dz \qquad \text{(by (3.34))} \\ &= B r^{-1/2} \int_{T(0)} e^{\pi i z^2} dz \qquad \text{(by (3.33))}. \end{aligned}$$

(For the last step, note that $e^{\pi i(-r^{-1}l^2 - 2tl)}$ depends only on l modulo b. Hence a substitution $l \mapsto -l$ gives $\sum_{l=0}^{b-1} e^{\pi i(-r^{-1}l^2 - 2tl)} = B$.) When $a = b = 1$ and $t = 1/2$, $A = e^{\pi i/4}$ and $B = 1$, which gives $\int_{T(0)} e^{\pi i z^2} dz = e^{\pi i/4}$. □

Theorem 3.10. *Let p be an odd prime and η the quadratic character of $\mathbb{F}_p^*$. Then*

$$g(\eta) = \sum_{j=0}^{p-1} e^{2\pi i j^2/p} = \begin{cases} p^{1/2} & \text{if } p \equiv 1 \pmod 4, \\ i p^{1/2} & \text{if } p \equiv -1 \pmod 4. \end{cases} \tag{3.35}$$

Proof. The first equality follows from (3.27); the second one follows from (3.29) with $a = p$, $b = 2$ and $t = 0$. □

Combining Theorems 3.8 and 3.10 gives an explicit formula for the Gauss quadratic sum over $\mathbb{F}_{p^n}$.

Theorem 3.11. *Let p be an odd prime and $q = p^n$, $n \in \mathbb{Z}^+$. Then $\eta^{(n)} = \eta \circ \mathrm{N}_{q/p}$ is the quadratic character of $\mathbb{F}_q^*$, and*

$$g(\eta^{(n)}) = \begin{cases} (-1)^{n-1} q^{1/2} & \text{if } p \equiv 1 \pmod 4, \\ (-1)^{n-1} i^n q^{1/2} & \text{if } p \equiv -1 \pmod 4. \end{cases}$$

Exercises

3.1. Define

$$\begin{array}{rccc} \alpha: & \mathrm{Hom}_{\mathbb{Z}}(\mathbb{Z}, \mathbb{C}^*) & \longrightarrow & \mathbb{C}^* \\ & f & \longmapsto & f(1). \end{array}$$

Prove that α is an isomorphism of abelian groups. (Note: Since $\mathbb{C}^* \not\cong \mathbb{Z}$ ($|\mathbb{C}^*| \neq |\mathbb{Z}|$), this exercise shows that $\mathrm{Hom}_{\mathbb{Z}}(\mathbb{Z}, \mathbb{C}^*) \not\cong \mathbb{Z}$.)

3.2. Let R be a finite ring. Since R is a $(\mathbb{Z}, R)$-bimodule, $\widehat{(R,+)} = \mathrm{Hom}_{\mathbb{Z}}(R, \mathbb{C}^*)$ is a left R-module. The left scalar multiplication of $\mathrm{Hom}_{\mathbb{Z}}(R, \mathbb{C}^*)$ by R is defined as follows: For $a \in R$ and $\chi \in \mathrm{Hom}_{\mathbb{Z}}(R, \mathbb{C}^*)$, $a\chi : R \to \mathbb{C}^*$ is defined by $(a\chi)(\cdot) = \chi(\cdot\, a)$. An element $\chi \in \widehat{(R,+)}$ is called a *generating character* of R if $\widehat{(R,+)} = R\chi$.

(i) Let $\chi \in \widehat{(R,+)}$. Prove that χ is a generating character of R if and only if $\ker \chi$ does not contain any nonzero left ideal of R.

(ii) If R has a unique minimal left ideal, prove that R has a generating character.

3.3. Let A be a finite abelian group and $B < A$, and let $f, g : A \to \mathbb{C}$ be two functions. Prove the following identities.

$$\sum_{\chi \in B^\perp} \widetilde{f}(\chi)\,\widetilde{g}(\chi) = \frac{|A|}{|B|} \sum_{a \in B} (f * g)(a).$$

$$\sum_{a \in B} f(a) g(a) = \frac{|B|}{|A|^2} \sum_{\chi \in B^\perp} (\widetilde{f} * \widetilde{g})(\chi).$$

$$\sum_{\chi \in \widehat{A}} |\widetilde{f}(\chi)|^4 = |A| \sum_{a \in A} |(f * f)(a)|^2.$$

3.4. Let $k \mid q-1$ and let A_k be the subgroup of $\widehat{\mathbb{F}_q^*}$ of order k. Prove that

$$\sum_{x \in \mathbb{F}_q} e_1(x^k) = \sum_{\chi \in A_k} g(\chi).$$

3.5. In the notation of Exercise 3.4, prove that for $a \in \mathbb{F}_q^*$,

$$\sum_{x \in \mathbb{F}_q} e_a(x^k) = \sum_{\chi \in A_k} \overline{\chi(a)} g(\chi).$$

3.6. Let $\chi \in \widehat{\mathbb{F}_q^*} \setminus \{1^{\mathbb{F}_q^*}\}$. Prove that $g(\chi) g(\chi^{-1}) = \chi(-1) q$.

3.7. Find the exact value of the Gauss quadratic sum over $\mathbb{F}_q$ with $q = 23, 25, 27, 29$.

3.8. Let η be the quadratic character of $\mathbb{F}_q^*$ defined in (3.26), where $q = p^n$, with p an odd prime and $n \in \mathbb{Z}^+$. Prove that
$$g(\eta) = (-1)^{(q+1)/2} i^{n(p^2+2p+5)/4} q^{1/2}.$$

3.9. For $\chi_1, \dots, \chi_k \in \widehat{\mathbb{F}_q^*}$, define
$$J_0(\chi_1, \dots, \chi_k) = \sum_{\substack{x_1,\dots,x_k \in \mathbb{F}_q \\ x_1+\cdots+x_k=0}} \chi_1(x_1) \cdots \chi_k(x_k).$$
Prove that
$$J_0(\chi_1, \dots, \chi_k) = \begin{cases} q^{k-1} & \text{if } \chi_1 = \cdots = \chi_k = 1^{\mathbb{F}_q^*}, \\ \dfrac{q-1}{q} g(\chi_1) \cdots g(\chi_k) & \text{if } \chi_i \neq 1^{\mathbb{F}_q^*},\ 1 \le i \le k, \text{ and } \chi_1 \cdots \chi_k = 1^{\mathbb{F}_q^*}, \\ 0 & \text{otherwise.} \end{cases}$$

3.10. Let $\chi \in \widehat{\mathbb{F}_q}$ and $a, b \in \mathbb{F}_q$. The *Kloosterman sum* $K(\chi; a, b)$ is defined as
$$K(\chi; a, b) = \sum_{x \in \mathbb{F}_q^*} \chi(ax + bx^{-1}).$$
If q is odd and $\chi \neq 1^{\mathbb{F}_q}$, and a and b are not both 0, prove that
$$K(\chi; a, b) = \sum_{x \in \mathbb{F}_q} \chi(x)\eta(x^2 - 4ab),$$
where η is the quadratic character of $\mathbb{F}_q^*$.

3.11. Prove that $f \in \mathbb{F}_q[\mathtt{X}]$ is a permutation polynomial of $\mathbb{F}_q$ if and only if
$$\sum_{x \in \mathbb{F}_q} e_a(f(x)) = 0 \quad \text{for all } a \in \mathbb{F}_q^*.$$

3.12. Let $f : X \to A$, where X is a finite set and A is a finite abelian group. Prove that f takes each value in A an equal number of times if and only if
$$\sum_{x \in X} \chi(f(x)) = 0 \quad \text{for all } \chi \in \widehat{A} \setminus \{1^A\}.$$

3.13. Let A and B be finite abelian groups. A function $f : A \to B$ is said to be *bent* if for all $\alpha \in \widehat{A}$ and $\beta \in \widehat{B} \setminus \{1^B\}$,
$$\Big|\sum_{x \in A} \alpha(x)\beta(f(x))\Big| = |A|^{1/2}.$$

Prove that $f : A \to B$ is bent if and only if for each $x \in A \setminus \{0\}$, the function $f_x : A \to B$, $f_x(y) = f(y + x) - f(y)$, takes each value of B an equal number of times.

3.14. Let p be a prime. Prove that $f : \mathbb{F}_p^n \to \mathbb{F}_p$ is bent if and only if

$$\left| \sum_{x \in \mathbb{F}_p^n} \zeta_p^{f(x)+xy^T} \right| = p^{n/2} \quad \text{for all } y \in \mathbb{F}_p^n.$$

Chapter 4

Algebraic Number Theory

This chapter is a brief but self-contained introduction to algebraic number theory. After exploring the basic facts about number fields, we focus on cyclotomic fields and prove the Stickelberger congruence for Gauss sums. For further reading in algebraic number theory, the reader is referred to [**20, 23, 31, 40**].

All rings in this chapter are commutative.

4.1. Number Fields

Integral elements. Let $R \subset S$ be rings. An element $\alpha \in S$ is said to be *integral* over R if there exists a monic polynomial $f \in R[\mathtt{X}]$ such that $f(\alpha) = 0$. If all elements of S are integral over S, S is said to be integral over R.

Proposition 4.1. *Let $R \subset S$ be rings and $\alpha \in S$. The following statements are equivalent.*

(i) *α is integral over R.*

(ii) *$R[\alpha]$ is a finitely generated R-module.*

(iii) *α is contained in a subring $T \subset S$ which is a finitely generated R-module.*

Proof. (i) $\Rightarrow$ (ii). There exists $f = \mathtt{X}^n + a_{n-1}\mathtt{X}^{n-1} + \cdots + a_0 \in R[\mathtt{X}]$ such that $f(\alpha) = 0$. Since $\alpha^n = -a_0 - \cdots - a_{n-1}\alpha^{n-1}$, $R[\alpha] = \langle 1, \alpha, \ldots, \alpha^{n-1}\rangle$ as an R-module.

(ii) ⇒ (iii). Let $T = R[\alpha]$.

(iii) ⇒ (i). Let $T = \langle \epsilon_1, \dots, \epsilon_n \rangle$ as an R-module. We may assume that $\epsilon_1 = 1$. Then

$$\alpha \begin{bmatrix} \epsilon_1 \\ \vdots \\ \epsilon_n \end{bmatrix} = A \begin{bmatrix} \epsilon_1 \\ \vdots \\ \epsilon_n \end{bmatrix},$$

where $A \in \mathrm{M}_{n\times n}(R)$, that is,

$$(\alpha I_n - A) \begin{bmatrix} \epsilon_1 \\ \vdots \\ \epsilon_n \end{bmatrix} = \begin{bmatrix} 0 \\ \vdots \\ 0 \end{bmatrix}.$$

The first column of $\alpha I_n - A$ is a linear combination of the other columns. Thus $\det(\alpha I_n - A) = 0$. Since $\det(\mathtt{X} I_n - A) \in R[\mathtt{X}]$ is monic, α is integral over R. □

Proposition 4.2. *Let $R \subset S$ be rings.*

(i) *Let $A \subset S$ be such that all elements of A are integral over R. Then $R[A]$ (the subring of S generated by R and A) is integral over R.*

(ii) *Let $\overline{R} = \{\alpha \in S : \alpha$ is integral over $R\}$. Then $\overline{R}$ is a subring of S. ($\overline{R}$ is called the* integral closure *of R in S.)*

Proof. (i) Let $\alpha \in R[A]$. There exist $\alpha_1, \dots, \alpha_n \in A$ such that $\alpha \in R[\alpha_1, \dots, \alpha_n]$. For each $1 \le i \le n$, α_i is integral over R, so $R[\alpha_1, \dots, \alpha_{i-1}, \alpha_i]$ is a finitely generated $R[\alpha_1, \dots, \alpha_{i-1}]$-module (Proposition 4.1). From the tower

$$R \subset R[\alpha_1] \subset R[\alpha_1, \alpha_2] \subset \cdots \subset R[\alpha_1, \dots, \alpha_n],$$

it follows that $R[\alpha_1, \dots, \alpha_n]$ is a finitely generated R-module. By Proposition 4.1, α is integral over R.

(ii) For any $\alpha, \beta \in \overline{R}$, $\alpha - \beta, \alpha\beta \in R[\alpha, \beta] \subset \overline{R}$ by (i). Hence $\overline{R}$ is a subring of S. □

Proposition 4.3. *Let $R \subset S \subset T$ be rings such that S is integral over R and T is integral over S. Then T is integral over R.*

Proof. Let $\alpha \in T$. There exists $f = \mathtt{X}^n + a_{n-1}\mathtt{X}^{n-1} + \cdots + a_0 \in S[\mathtt{X}]$ such that $f(\alpha) = 0$. Then α is integral over $R[a_0, \dots, a_{n-1}]$, and hence $R[a_0, \dots, a_{n-1}, \alpha]$ is a finitely generated $R[a_0, \dots, a_{n-1}]$-module. Since a_0, $\dots, a_{n-1}$ are integral over R, by the proof of Proposition 4.2 (i), $R[a_0, \dots, a_{n-1}]$ is a finitely generated R-module. Thus $R[a_0, \dots, a_{n-1}, \alpha]$ is a finitely generated R-module. Hence α is integral over R. □

Number fields. A subfield $F \subset \mathbb{C}$ with $[F : \mathbb{Q}] < \infty$ is called a *number field*. The integral closure of $\mathbb{Z}$ in F, denoted by $\mathfrak{o}_F$, is called the *ring of integers* of F.

Definition 4.4. An integral domain $\mathfrak{o}$ is called a *Dedekind domain* if every nonzero ideal of $\mathfrak{o}$ is a product of finitely many prime ideals of $\mathfrak{o}$.

Theorem 4.5. *Let F be a number field. Then $\mathfrak{o}_F$ is a Dedekind domain.*

The proof of Theorem 4.5 requires some preparation and will be given later in this section.

Trace and norm. Let $F \subset K \subset \overline{F}$ be fields, where $\overline{F}$ is the algebraic closure of F and $[K : F] < \infty$. Let $[K : F]_s$ and $[K : F]_i$ denote the separable degree and the inseparable degree of K/F, respectively. Then $r = [K : F]_s$ is the number of distinct F-embeddings (embeddings that are identity on F) of K into $\overline{F}$ ([**18**, Proposition V.6.12]). Let $\sigma_1, \dots, \sigma_r$ be such embeddings. For each $u \in K$, its *trace* and *norm* from K to F are defined, respectively, by

$$\begin{aligned}\mathrm{Tr}_{K/F}(u) &= [K : F]_i\big(\sigma_1(u) + \cdots + \sigma_r(u)\big),\\ \mathrm{N}_{K/F}(u) &= \big(\sigma_1(u) \cdots \sigma_r(u)\big)^{[K:F]_i}.\end{aligned}$$

Note that if K/F is not separable, $\mathrm{char}\, F \mid [K : F]_i$, and so $\mathrm{Tr}_{K/F}(u) = 0$ for all $u \in K$. If K/F is separable, since $\sigma_1, \dots, \sigma_r$ are linearly independent over $\overline{F}$ ([**24**, Theorem VI.4.1]), there exists $u \in K$ such that $\mathrm{Tr}_{K/F}(u) \neq 0$.

Proposition 4.6. *Let $F \subset K$ be fields with $[K : F] < \infty$ and $u \in K$. Let $f = \mathtt{X}^n + a_{n-1}\mathtt{X}^{n-1} + \cdots + a_0 \in F[\mathtt{X}]$ be the minimal polynomial of u over F.*

(i) $\mathrm{Tr}_{K/F}(u) = -[K : F(u)]a_{n-1}$ *and* $\mathrm{N}_{K/F}(u) = [(-1)^n a_0]^{[K:F(u)]}$. *(Note that* $\mathrm{Tr}_{K/F}(u), \mathrm{N}_{K/F}(u) \in F$.*)*

(ii) *Let $u_1, \dots, u_m$ be a basis of K/F, and assume that*

$$u \begin{bmatrix} u_1 \\ \vdots \\ u_m \end{bmatrix} = A \begin{bmatrix} u_1 \\ \vdots \\ u_m \end{bmatrix}, \qquad A \in \mathrm{M}_{m\times m}(F). \tag{4.1}$$

Then $\mathrm{Tr}_{K/F}(u) = \mathrm{Tr}(A)$ *and* $\mathrm{N}_{K/F}(u) = \det A$.

Proof. (i) Let $r = [F(u) : F]_s$ and $t = [K : F(u)]_s$. Let $\sigma_1, \dots, \sigma_r$ be the distinct F-embeddings of $F(u)$ in $\overline{F}$ and $\tau_1, \dots, \tau_t$ the distinct $F(u)$-embeddings of K in $\overline{F}$. We have ([**18**, Corollary V.6.14])

$$f = \big[(\mathtt{X} - \sigma_1(u)) \cdots (\mathtt{X} - \sigma_r(u))\big]^{[F(u):F]_i},$$

and hence

$$a_{n-1} = -[F(u):F]_i \sum_{j=1}^{r} \sigma_j(u), \qquad a_0 = \left[(-1)^r \prod_{j=1}^{r} \sigma_j(u)\right]^{[F(u):F]_i}.$$

Let $\bar{\sigma}_j \in \mathrm{Aut}(\overline{F}/F)$ be an extension of σ_j. We claim that $\bar{\sigma}_j|_{\tau_k(K)} \circ \tau_k$, $1 \le j \le r$, $1 \le k \le t$, are all the distinct F-embeddings of K in $\overline{F}$. If α is an F-embedding of K in $\overline{F}$, then $\alpha|_{F(u)} = \sigma_j$ for some $1 \le j \le r$. Since $\bar{\sigma}_j^{-1}|_{\alpha(K)} \circ \alpha$ is an $F(u)$-embedding of K in $\overline{F}$, $\bar{\sigma}_j^{-1}|_{\alpha(K)} \circ \alpha = \tau_k$ for some $1 \le k \le t$. Then $\alpha = \bar{\sigma}_j|_{\tau_k(K)} \circ \tau_k$. Since $[K:F]_s = rt$, the claim is proved.

Therefore

$$\begin{aligned} \mathrm{Tr}_{K/F}(u) &= [K:F]_i \sum_{j,k} \bar{\sigma}_j(\tau_k(u)) = [K:F]_i\, t \sum_j \sigma_j(u) \\ &= t\,[K:F(u)]_i\,[F(u):F]_i \sum_j \sigma_j(u) = -[K:F(u)]a_{n-1}. \end{aligned}$$

In the same way, we have $\mathrm{N}_{K/F}(u) = [(-1)^n a_0]^{[K:F(u)]}$.

(ii) A change of basis in (4.1) transforms A into a matrix similar to A. Thus $\mathrm{Tr}(A)$ and $\det A$ are independent of the basis. Let $\epsilon_1, \dots, \epsilon_k$ be a basis of $K/F(u)$. Then $\epsilon_i u^j$, $1 \le i \le k$, $0 \le j \le n-1$, form a basis of K/F. We have

$$u \begin{bmatrix} \epsilon_1 u^0 \\ \vdots \\ \epsilon_1 u^{n-1} \\ \vdots \\ \epsilon_k u^0 \\ \vdots \\ \epsilon_k u^{n-1} \end{bmatrix} = A \begin{bmatrix} \epsilon_1 u^0 \\ \vdots \\ \epsilon_1 u^{n-1} \\ \vdots \\ \epsilon_k u^0 \\ \vdots \\ \epsilon_k u^{n-1} \end{bmatrix},$$

where

$$A = \begin{bmatrix} C & & \\ & \ddots & \\ & & C \end{bmatrix}, \qquad C = \begin{bmatrix} 0 & 1 & & \\ & & \ddots & \\ & & & 1 \\ -a_0 & -a_1 & \cdots & -a_{n-1} \end{bmatrix}.$$

Then $\mathrm{Tr}(A) = k(-a_{n-1}) = \mathrm{Tr}_{K/F}(u)$ and $\det A = [(-1)^n a_0]^k = \mathrm{N}_{K/F}(u)$. □

Discriminant. Let $F \subset K$ be fields with $[K : F] = n$ and let $\alpha_1, \ldots, \alpha_n \in K$. Define the *discriminant* of $\alpha_1, \ldots, \alpha_n$ as

$$\Delta_{K/F}(\alpha_1, \ldots, \alpha_n) = \det\bigl(\mathrm{Tr}_{K/F}(\alpha_i \alpha_j)\bigr).$$

If

$$\begin{bmatrix} \beta_1 \\ \vdots \\ \beta_n \end{bmatrix} = A \begin{bmatrix} \alpha_1 \\ \vdots \\ \alpha_n \end{bmatrix}, \qquad A \in \mathrm{M}_{n\times n}(F),$$

then $\Delta_{K/F}(\beta_1, \ldots, \beta_n) = (\det A)^2 \Delta_{K/F}(\alpha_1, \ldots, \alpha_n)$. To prove this claim, we allow $\mathrm{Tr}_{K/F}$ to be applied to a matrix over K entrywise. Then

$$\begin{aligned} \Delta_{K/F}(\beta_1, \ldots, \beta_n) &= \det\bigl[\mathrm{Tr}_{K/F}\bigl((\beta_1, \ldots, \beta_n)^T (\beta_1, \ldots, \beta_n)\bigr)\bigr] \\ &= \det\bigl[\mathrm{Tr}_{K/F}\bigl(A(\alpha_1, \ldots, \alpha_n)^T (\alpha_1, \ldots, \alpha_n)\bigr) A^T\bigr] \\ &= (\det A)^2 \det\bigl[\mathrm{Tr}_{K/F}\bigl((\alpha_1, \ldots, \alpha_n)^T (\alpha_1, \ldots, \alpha_n)\bigr)\bigr] \\ &= (\det A)^2 \Delta_{K/F}(\alpha_1, \ldots, \alpha_n). \end{aligned}$$

Proposition 4.7. *Assume that K/F is separable with $[K : F] = n$. Then $\alpha_1, \ldots, \alpha_n \in K$ form a basis of K/F if and only if $\Delta_{K/F}(\alpha_1, \ldots, \alpha_n) \neq 0$.*

Proof. Since K/F is a finite-degree separable extension, there exists $u \in K$ such that $\mathrm{Tr}_{K/F}(u) \neq 0$. It follows that the map $K \times K \to F$, $(x, y) \mapsto \mathrm{Tr}_{K/F}(xy)$, is a nondegenerate F-bilinear form. Therefore, the conclusion is a well-known fact from linear algebra. □

Proposition 4.8. *Let F be a number field and $\mathfrak{a}$ a nonzero ideal of $\mathfrak{o}_F$. Then $\mathfrak{a}$ contains a basis $\alpha_1, \ldots, \alpha_n$ of $F/\mathbb{Q}$. (Note that $\Delta_{F/\mathbb{Q}}(\alpha_1, \ldots, \alpha_n) \in \mathbb{Z} \setminus \{0\}$.) If $|\Delta_{F/\mathbb{Q}}(\alpha_1, \ldots, \alpha_n)|$ is minimal, then $\mathfrak{a} = \mathbb{Z}\alpha_1 + \cdots + \mathbb{Z}\alpha_n$.*

Proof. 1° We claim that for each $\alpha \in F$, there exists $0 \neq m \in \mathbb{Z}$ such that $m\alpha \in \mathfrak{o}_F$.

Assume that α is a root of $\mathtt{X}^k + a_{k-1}\mathtt{X}^{k-1} + \cdots + a_0 \in \mathbb{Q}[\mathtt{X}]$. Choose $0 \neq m \in \mathbb{Z}$ such that $ma_0, \ldots, ma_{k-1} \in \mathbb{Z}$. Note that $m\alpha$ is a root of $\mathtt{X}^k + ma_{k-1}\mathtt{X}^{k-1} + \cdots + m^k a_0 \in \mathbb{Z}[\mathtt{X}]$.

2° We claim that $\mathfrak{a}$ contains a basis of $F/\mathbb{Q}$.

Let $\epsilon_1, \ldots, \epsilon_n$ be a basis of $F/\mathbb{Q}$. By 1°, there exists $0 \neq m \in \mathbb{Z}$ such that $m\epsilon_1, \ldots, m\epsilon_n \in \mathfrak{o}_F$. Choose $0 \neq \alpha \in \mathfrak{a}$. Then $\alpha m\epsilon_1, \ldots, \alpha m\epsilon_n$ is a basis of $F/\mathbb{Q}$ contained in $\mathfrak{a}$.

3° Assume that $\alpha_1 \ldots, \alpha_n \in \mathfrak{a}$ is a basis of $F/\mathbb{Q}$ such that $|\Delta_{F/\mathbb{Q}}(\alpha_1, \ldots, \alpha_n)|$ is minimal, and assume to the contrary that $\mathfrak{a} \neq \mathbb{Z}\alpha_1 + \cdots + \mathbb{Z}\alpha_n$. Choose $\alpha \in \mathfrak{a} \setminus (\mathbb{Z}\alpha_1 + \cdots + \mathbb{Z}\alpha_n)$ and write $\alpha = a_1\alpha_1 + \cdots + a_n\alpha_n$, $a_i \in \mathbb{Q}$.

Without loss of generality, assume that $a_1 \notin \mathbb{Z}$, say $a_1 = b + r$ for some $b \in \mathbb{Z}$ and $0 < r < 1$. Let $\alpha_1' = \alpha - b\alpha_1 \in \mathfrak{a}$. Then

$$\begin{bmatrix} \alpha_1' \\ \alpha_2 \\ \vdots \\ \alpha_n \end{bmatrix} = \begin{bmatrix} r & a_2 & \cdots & a_n \\ & 1 & & \\ & & \ddots & \\ & & & 1 \end{bmatrix} \begin{bmatrix} \alpha_1 \\ \alpha_2 \\ \vdots \\ \alpha_n \end{bmatrix}.$$

Thus $|\Delta_{F/\mathbb{Q}}(\alpha_1', \alpha_2, \dots, \alpha_n)| = r^2 |\Delta_{F/\mathbb{Q}}(\alpha_1, \dots, \alpha_n)| < |\Delta_{F/\mathbb{Q}}(\alpha_1, \dots, \alpha_n)|$, which is a contradiction. □

Corollary 4.9. *Every nonzero ideal* $\mathfrak{a}$ *of* $\mathfrak{o}_F$ *is a free* $\mathbb{Z}$*-module of rank* $[F : \mathbb{Q}]$. *A* $\mathbb{Z}$*-basis of* $\mathfrak{a}$ *is called an* integral basis *of* $\mathfrak{a}$.

Definition 4.10. Let F be a number field and let $\alpha_1, \dots, \alpha_n$ be an integral basis of $\mathfrak{o}_F$ over $\mathbb{Z}$. The *discriminant* of F is defined to be $\delta_F = \Delta_{F/\mathbb{Q}}(\alpha_1, \dots, \alpha_n)$, which is independent of the choice of integral basis.

Proposition 4.11. *Let* $\mathfrak{a}$ *be a nonzero ideal of* $\mathfrak{o}_F$.

(i) $\mathfrak{a} \cap \mathbb{Z} \neq \{0\}$.

(ii) $|\mathfrak{o}_F / \mathfrak{a}| < \infty$.

Proof. (i) Let $0 \neq \alpha \in \mathfrak{a}$. Then α is a root of some $\mathtt{X}^n + a_{n-1}\mathtt{X}^{n-1} + \cdots + a_0 \in \mathbb{Z}[\mathtt{X}]$, where $a_0 \neq 0$. We have $a_0 = -\alpha^n - a_{n-1}\alpha^{n-1} - \cdots - a_1\alpha \in \mathfrak{a} \cap \mathbb{Z}$.

(ii) Let $\alpha_1, \dots, \alpha_n$ be an integral basis of $\mathfrak{o}_F$ over $\mathbb{Z}$. Let $0 \neq a \in \mathfrak{a} \cap \mathbb{Z}$. Then $\mathfrak{o}_F/\mathfrak{a}$ is a homomorphic image of $\mathfrak{o}_F/(a)$, where

$$|\mathfrak{o}_F/(a)| = \left| \frac{\mathbb{Z}\alpha_1 + \cdots + \mathbb{Z}\alpha_n}{a\mathbb{Z}\alpha_1 + \cdots + a\mathbb{Z}\alpha_n} \right| = |a|^n < \infty.$$

□

Proposition 4.12. *Let* F *be a number field.*

(i) $\mathfrak{o}_F$ *is Noetherian.*

(ii) *Every nonzero prime ideal of* $\mathfrak{o}_F$ *is maximal.*

Proof. (i) Every nonzero ideal of $\mathfrak{o}_F$ is a free $\mathbb{Z}$-module of rank $[F : \mathbb{Q}]$ and hence is finitely generated.

(ii) Let $\mathfrak{p}$ be a prime ideal of $\mathfrak{o}_F$. Then $\mathfrak{o}_F/\mathfrak{p}$ is a finite integral domain and hence is a field. □

Proposition 4.13. *Let* $\mathfrak{a}$ *and* $\mathfrak{b}$ *be nonzero ideals of* $\mathfrak{o}_F$.

(i) *If* $\alpha \in F$ *is such that* $\alpha\mathfrak{a} \subset \mathfrak{a}$, *then* $\alpha \in \mathfrak{o}_F$.

(ii) *If* $\mathfrak{a} = \mathfrak{b}\mathfrak{a}$, *then* $\mathfrak{b} = \mathfrak{o}_F$.

(iii) *If* $\omega \in \mathfrak{o}_F$ *is such that* $\omega\mathfrak{a} = \mathfrak{b}\mathfrak{a}$, *then* $(\omega) = \mathfrak{b}$.

Proof. (i) Let $\epsilon_1, \dots, \epsilon_n$ be an integral basis of $\mathfrak{a}$ over $\mathbb{Z}$. Then

$$\alpha \begin{bmatrix} \epsilon_1 \\ \vdots \\ \epsilon_n \end{bmatrix} = A \begin{bmatrix} \epsilon_1 \\ \vdots \\ \epsilon_n \end{bmatrix} \qquad \text{for some } A \in \mathrm{M}_{n\times n}(\mathbb{Z}).$$

Thus $\det(\alpha I - A) = 0$, where $\det(\mathtt{X}I - A) \in \mathbb{Z}[\mathtt{X}]$ is monic.

(ii) Let $\epsilon_1, \dots, \epsilon_n$ be an integral basis of $\mathfrak{a}$ over $\mathbb{Z}$. Since $\mathfrak{a} = \mathfrak{b}\mathfrak{a}$, we have

$$\begin{bmatrix} \epsilon_1 \\ \vdots \\ \epsilon_n \end{bmatrix} = B \begin{bmatrix} \epsilon_1 \\ \vdots \\ \epsilon_n \end{bmatrix} \qquad \text{for some } B \in \mathrm{M}_{n\times n}(\mathfrak{b}).$$

So $\det(I - B) = 0$. The determinant expansion gives $\det(I - B) = 1 +$ terms in $\mathfrak{b}$. Hence $1 \in \mathfrak{b}$.

(iii) For each $\beta \in \mathfrak{b}$, $\beta\mathfrak{a} \subset \mathfrak{b}\mathfrak{a} = \omega\mathfrak{a}$, so $(\beta/\omega)\mathfrak{a} \subset \mathfrak{a}$. By (i), $\beta/\omega \in \mathfrak{o}_F$, i.e., $\beta \in (\omega)$. Thus $\mathfrak{b} \subset (\omega)$. Now $\omega^{-1}\mathfrak{b}$ is an ideal of $\mathfrak{o}_F$ and $\omega^{-1}\mathfrak{b}\mathfrak{a} = \omega^{-1}\omega\mathfrak{a} = \mathfrak{a}$. By (ii), $\omega^{-1}\mathfrak{b} = \mathfrak{o}_F$, i.e., $\mathfrak{b} = (\omega)$. □

The class number. For nonzero ideals $\mathfrak{a}$ and $\mathfrak{b}$ of $\mathfrak{o}_F$, define $\mathfrak{a} \sim \mathfrak{b}$ if there exist $\alpha, \beta \in \mathfrak{o}_F \setminus \{0\}$ such that $\alpha\mathfrak{a} = \beta\mathfrak{b}$. Then $\sim$ is an equivalence relation for all nonzero ideals of $\mathfrak{o}_F$. The number of $\sim$ equivalence classes, called the *class number* of F and denoted by h_F, is finite. The proof of this important fact presented here is due to Hurwitz [**19**, pp. 236–243].

Theorem 4.14. $h_F < \infty$.

Lemma 4.15. *There exists an integer* $M > 0$, *depending only on* F, *with the following property. For each* $\gamma \in F$, *there exist an integer* t *with* $1 \le t \le M$ *and an element* $\omega \in \mathfrak{o}_F$ *such that* $|\mathrm{N}_{F/\mathbb{Q}}(t\gamma - \omega)| < 1$.

Proof. Let $\omega_1, \dots, \omega_n$ be an integral basis of $\mathfrak{o}_F$ over $\mathbb{Z}$. Let $\sigma_1, \dots, \sigma_n$ be the distinct embeddings of F in $\mathbb{C}$. Write $\gamma = \sum_{i=1}^n \gamma_i\omega_i$, $\gamma_i \in \mathbb{Q}$.

Let $m > 0$ be an integer to be determined later. For each $0 \le s \le m^n$, we have

$$s\gamma = \sum_{i=1}^n \lfloor s\gamma_i \rfloor \omega_i + \sum_{i=1}^n \{s\gamma_i\}\omega_i,$$

where $\{x\} = x - \lfloor x \rfloor$ for $x \in \mathbb{R}$. Partition $[0,1)^n$ into m^n cubes of side length $1/m$. By the pigeonhole principle, there exist $0 \le s < s' \le m^n$ such that

$(\{s\gamma_i\})_{1\le i\le n}$ and $(\{s'\gamma_i\})_{1\le i\le n}$ belong to the same cube. Then

$$(s'-s)\gamma - \omega = \sum_{i=1}^{n}(\{s'\gamma_i\} - \{s\gamma_i\})\omega_i,$$

where $1 \le s'-s \le m^n$ and $\omega = \sum_{i=1}^{n}(\lfloor s'\gamma_i\rfloor - \lfloor s\gamma_i\rfloor)\omega_i \in \mathfrak{o}_F$. We have

$$\begin{aligned}\left|\mathrm{N}_{F/\mathbb{Q}}((s'-s)\gamma-\omega)\right| &= \left|\prod_{j=1}^{n}\Bigl[\sum_{i=1}^{n}(\{s'\gamma_i\}-\{s\gamma_i\})\sigma_j(\omega_i)\Bigr]\right| \\ &\le \left(\frac{1}{m}\right)^n \prod_{j=1}^{n}\Bigl[\sum_{i=1}^{n}|\sigma_j(\omega_i)|\Bigr].\end{aligned}$$

Choose m such that $m^n > \prod_{j=1}^{n}\bigl[\sum_{i=1}^{n}|\sigma_j(\omega_i)|\bigr]$. Then $M = m^n$ has the desired property. □

Proof of Theorem 4.14. Let $\mathfrak{a}$ be a nonzero ideal of $\mathfrak{o}_F$. Choose $\beta \in \mathfrak{a}$ such that $|\mathrm{N}_{F/\mathbb{Q}}(\beta)| = \min\{|\mathrm{N}_{F/\mathbb{Q}}(\alpha)| : 0 \ne \alpha \in \mathfrak{a}\}$. (Note that $|\mathrm{N}_{F/\mathbb{Q}}(\alpha)|$ is a positive integer for $\alpha \in \mathfrak{o}_F \setminus \{0\}$.) For each $\alpha \in \mathfrak{a}$, by Lemma 4.15, there exist $1 \le t \le M$ and $\omega \in \mathfrak{o}_F$ such that $|\mathrm{N}_{F/\mathbb{Q}}(t\alpha/\beta - \omega)| < 1$, i.e., $|\mathrm{N}_{F/\mathbb{Q}}(t\alpha - \omega\beta)| < |\mathrm{N}_{F/\mathbb{Q}}(\beta)|$, so $t\alpha - \omega\beta = 0$. Thus $M!\mathfrak{a} \subset (\beta)$. Let $\mathfrak{b} = (M!/\beta)\mathfrak{a}$. Then $\mathfrak{b}$ is a nonzero ideal of $\mathfrak{o}_F$ and $\mathfrak{a} \sim \mathfrak{b}$. Note that $M!\beta \in M!\mathfrak{a} = \beta\mathfrak{b}$. Thus $M! \in \mathfrak{b}$, i.e., $(M!) \subset \mathfrak{b}$. Since $|\mathfrak{o}_F/(M!)| < \infty$, $\mathfrak{b}$ is one of finitely many ideals. Hence $h_F < \infty$. □

Corollary 4.16. *Let $\mathfrak{a}$ be a nonzero ideal of $\mathfrak{o}_F$. Then there exists k with $1 \le k \le h_F$ such that $\mathfrak{a}^k$ is principal.*

Proof. Let $[\mathfrak{a}]$ denote the $\sim$ equivalence class of $\mathfrak{a}$. By the pigeonhole principle, at least two of $[\mathfrak{a}^0], \dots, [\mathfrak{a}^{h_F}]$ are identical, say $[\mathfrak{a}^i] = [\mathfrak{a}^j]$ for some $0 \le i < j \le h_F$. Then there exist $\alpha, \beta \in \mathfrak{o}_F \setminus \{0\}$ such that $\alpha\mathfrak{a}^i = \beta\mathfrak{a}^j$. Since $(\alpha/\beta)\mathfrak{a}^i = \mathfrak{a}^j \subset \mathfrak{a}^i$, by Proposition 4.13 (i), $\alpha/\beta \in \mathfrak{o}_F$. Since $(\alpha/\beta)\mathfrak{a}^i = \mathfrak{a}^j = \mathfrak{a}^{j-i}\mathfrak{a}^i$, by Proposition 4.13 (iii), $\mathfrak{a}^{j-i} = (\alpha/\beta)$. □

Proposition 4.17. *Let $\mathfrak{a}, \mathfrak{b}$, and $\mathfrak{c}$ be nonzero ideals of $\mathfrak{o}_F$.*

(i) *If $\mathfrak{a}\mathfrak{b} = \mathfrak{a}\mathfrak{c}$, then $\mathfrak{b} = \mathfrak{c}$.*

(ii) *If $\mathfrak{a} \subset \mathfrak{b}$, then there exists a nonzero ideal $\mathfrak{d}$ of $\mathfrak{o}_F$ such that $\mathfrak{a} = \mathfrak{b}\mathfrak{d}$.*

Proof. (i) By Corollary 4.16, there exists $k > 0$ such that $\mathfrak{a}^k = (\alpha)$ for some $\alpha \in \mathfrak{o}_F \setminus \{0\}$. Then $\alpha\mathfrak{b} = \mathfrak{a}^k\mathfrak{b} = \mathfrak{a}^k\mathfrak{c} = \alpha\mathfrak{c}$, and hence $\mathfrak{b} = \mathfrak{c}$.

(ii) There exists $k > 0$ such that $\mathfrak{b}^k = (\beta)$ for some $\beta \in \mathfrak{o}_F \setminus \{0\}$. Then $\mathfrak{a} = \mathfrak{b}(1/\beta)\mathfrak{b}^{k-1}\mathfrak{a}$, where $(1/\beta)\mathfrak{b}^{k-1}\mathfrak{a}$ is an ideal of $\mathfrak{o}_F$ since $(1/\beta)\mathfrak{b}^{k-1}\mathfrak{a} \subset (1/\beta)\mathfrak{b}^k = \mathfrak{o}_F$. □

Proof of Theorem 4.5. Let $\mathfrak{a}$ ($\neq \mathfrak{o}_F$) be a nonzero ideal of $\mathfrak{o}_F$. Since $|\mathfrak{o}_F/\mathfrak{a}| < \infty$, $\mathfrak{a}$ is contained in a maximal ideal $\mathfrak{p}_1$ of $\mathfrak{o}_F$. By Proposition 4.17 (ii), $\mathfrak{a} = \mathfrak{p}_1\mathfrak{a}_1$, where $\mathfrak{a}_1$ is a nonzero ideal of $\mathfrak{o}_F$ and $\mathfrak{a} \subsetneq \mathfrak{a}_1$. If $\mathfrak{a}_1 \neq \mathfrak{o}_F$, then $\mathfrak{a}_1 = \mathfrak{p}_2\mathfrak{a}_2$, where $\mathfrak{p}_2$ and $\mathfrak{a}_2$ are ideals of $\mathfrak{o}_F$ such that $\mathfrak{p}_2$ is maximal and $\mathfrak{a}_1 \subsetneq \mathfrak{a}_2$. Continuing in this manner, since $\mathfrak{o}_F$ is Noetherian, we must have $\mathfrak{a}_n = \mathfrak{o}_F$ for some n. Then $\mathfrak{a} = \mathfrak{p}_1 \cdots \mathfrak{p}_n$. □

It follows from Proposition 4.17 that the factorization of a nonzero ideal of $\mathfrak{o}_F$ into prime ideals is *unique*. In fact, the statement is true for any Dedekind domain; see Theorem 4.20.

Fractional ideals. Let $\mathfrak{o}$ be an integral domain with quotient field F. A *fractional ideal* of $\mathfrak{o}$ is a nonzero $\mathfrak{o}$-module $\mathfrak{a} \subset F$ such that $\alpha\mathfrak{a} \subset \mathfrak{o}$ for some $0 \neq \alpha \in F$, i.e., $\mathfrak{a} = \beta\mathfrak{b}$ for some nonzero ideal $\mathfrak{b}$ of $\mathfrak{o}$ and some $0 \neq \beta \in F$. A fractional ideal $\mathfrak{a}$ is said to be *invertible* if there exists a fractional ideal $\mathfrak{a}'$ such that $\mathfrak{a}\mathfrak{a}' = \mathfrak{o}$. The fractional ideal $\mathfrak{a}'$, if it exists, is unique and is denoted by $\mathfrak{a}^{-1}$. (If $\mathfrak{a}''$ is another fractional ideal such that $\mathfrak{a}\mathfrak{a}'' = \mathfrak{o}$, then $\mathfrak{a}' = \mathfrak{a}'\mathfrak{o} = \mathfrak{a}'\mathfrak{a}\mathfrak{a}'' = \mathfrak{o}\mathfrak{a}'' = \mathfrak{a}''$.)

Proposition 4.18. *Let F be a number field. Every fractional ideal of $\mathfrak{o}_F$ is invertible.*

Proof. We only have to show that every nonzero prime ideal $\mathfrak{p}$ of $\mathfrak{o}$ is invertible. Let $\mathfrak{p}' = \{\alpha \in F : \alpha\mathfrak{p} \subset \mathfrak{o}_F\}$. Then $\mathfrak{p}'$ is a fractional ideal since $\beta\mathfrak{p}' \subset \mathfrak{o}_F$ for any $0 \neq \beta \in \mathfrak{p}$; moreover, $\mathfrak{o}_F \subset \mathfrak{p}'$. We have $\mathfrak{p} \subset \mathfrak{p}'\mathfrak{p} \subset \mathfrak{o}_F$. If $\mathfrak{p}'\mathfrak{p} = \mathfrak{o}_F$, then $\mathfrak{p}' = \mathfrak{p}^{-1}$. Assume that $\mathfrak{p}'\mathfrak{p} \neq \mathfrak{o}_F$. Then $\mathfrak{p}'\mathfrak{p} = \mathfrak{p}$ since $\mathfrak{p}$ is maximal. By Proposition 4.13 (i), $\mathfrak{p}' \subset \mathfrak{o}_F$. Choose $0 \neq a \in \mathfrak{p}$. Since $(a) \subset \mathfrak{p}$, we have $(a) = \mathfrak{p}\mathfrak{b}$ for some nonzero ideal $\mathfrak{b}$ of $\mathfrak{o}_F$. Then $(1/a)\mathfrak{b}\mathfrak{p} \subset \mathfrak{o}_F$, and hence $(1/a)\mathfrak{b} \subset \mathfrak{p}' \subset \mathfrak{o}_F$, i.e., $\mathfrak{b} \subset (a)$, which contradicts the fact that $(a) = \mathfrak{p}\mathfrak{b}$. □

The ideal class group. Let I_F denote the multiplicative group of all fractional ideals of $\mathfrak{o}_F$. Let $\mathrm{Spec}^*(\mathfrak{o}_F)$ be the set of all nonzero prime ideals of $\mathfrak{o}_F$. Then I_F is the free abelian group generated by $\mathrm{Spec}^*(\mathfrak{o}_F)$. Let $P_F = \{\alpha\mathfrak{o}_F : 0 \neq \alpha \in F\}$, the subgroup of I_F consisting of all principal fractional ideals of $\mathfrak{o}_F$. The *ideal class group* of F is defined as $\mathrm{Cl}(F) = I_F/P_F$.

The equivalence relation for nonzero ideals of $\mathfrak{o}_F$ can be extended to fractional ideals of $\mathfrak{o}_F$. For $\mathfrak{a}, \mathfrak{b} \in I_F$, define $\mathfrak{a} \sim \mathfrak{b}$ if there exist $\alpha, \beta \in \mathfrak{o}_F\backslash\{0\}$ such that $\alpha\mathfrak{a} = \beta\mathfrak{b}$, i.e., if $\mathfrak{a}P_F = \mathfrak{b}P_F$ in $\mathrm{Cl}(F)$. Each equivalence class of fractional ideals of $\mathfrak{o}_F$ contains a unique equivalence class of nonzero ideals of $\mathfrak{o}_F$. Hence $|\mathrm{Cl}(F)| = h_F$.

$\mathfrak{p}$-adic valuation. Let F be a number field and let $0 \neq \alpha \in F$. Then $(\alpha) = \alpha\mathfrak{o}_F \in I_F$. Therefore

$$(\alpha) = \prod_{\mathfrak{p}\in\mathrm{Spec}^*(\mathfrak{o}_F)} \mathfrak{p}^{\nu_\mathfrak{p}(\alpha)}, \tag{4.2}$$

where $\nu_\mathfrak{p}(\alpha) \in \mathbb{Z}$ and $\nu_\mathfrak{p}(\alpha) = 0$ for all but finitely many $\mathfrak{p} \in \mathrm{Spec}^*(\mathfrak{o}_F)$. The integer $\nu_\mathfrak{p}(\alpha)$ is called the $\mathfrak{p}$-*adic order* of α. For each $\mathfrak{p} \in \mathrm{Spec}^*(\mathfrak{o}_F)$, (4.2) defines a function $\nu_\mathfrak{p} : F \setminus \{0\} \to \mathbb{Z}$. We also define $\nu_\mathfrak{p}(0) = \infty$.

Fact 4.19. *The function $\nu_\mathfrak{p} : F \to \mathbb{Z} \cup \{\infty\}$, called the* $\mathfrak{p}$-adic valuation *of F, has the following properties.*

(i) *$\nu_\mathfrak{p} : F^* \to \mathbb{Z}$ is an onto group homomorphism.*

(ii) *$\nu_\mathfrak{p}(0) = \infty$.*

(iii) *$\nu_\mathfrak{p}(x + y) \geq \min(\nu_\mathfrak{p}(x), \nu_\mathfrak{p}(y))$ for all $x, y \in F$.*

(A function $\nu : F \to \mathbb{Z}\cup\{\infty\}$ satisfying (i)–(iii) is called a discrete valuation *of F.)*

Proof. (i) Clearly, $\nu_\mathfrak{p} : F^* \to \mathbb{Z}$ is a group homomorphism. Take $\alpha \in \mathfrak{p} \setminus \mathfrak{p}^2$. Then $\nu_\mathfrak{p}(\alpha) = 1$, and hence $\nu_\mathfrak{p}$ is onto.

(iii) We may assume that $x, y \in \mathfrak{o}_F \setminus \{0\}$. Then $x, y \in \mathfrak{p}^{\min(\nu_\mathfrak{p}(x),\nu_\mathfrak{p}(y))}$, and hence $x + y \in \mathfrak{p}^{\min(\nu_\mathfrak{p}(x),\nu_\mathfrak{p}(y))}$, i.e., $(x + y) \subset \mathfrak{p}^{\min(\nu_\mathfrak{p}(x),\nu_\mathfrak{p}(y))}$. Thus $\nu_\mathfrak{p}(x + y) \geq \min(\nu_\mathfrak{p}(x), \nu_\mathfrak{p}(y))$. □

Characterizations of a Dedekind domain. A Dedekind domain can be characterized in several ways. We include the following theorem, whose proof can be found in [**18**]. Most of the facts proved in this section for the ring $\mathfrak{o}_F$ of integers of a number field F also hold for a general Dedekind domain.

Theorem 4.20. [**18**, VIII, Theorem 6.10] *Let R be an integral domain. The following statements are equivalent.*

(i) *R is a Dedekind domain.*

(ii) *Every nonzero ideal of R is uniquely a product of finitely many prime ideals.*

(iii) *Every nonzero ideal of R is invertible.*

(iv) *Every fractional ideal of R is invertible.*

(v) *The set of all fractional ideals of R is a group under multiplication.*

(vi) *Every ideal of R is projective.*

(vii) *Every fractional ideal of R is projective.*

(viii) *R is Noetherian and integrally closed, and every nonzero prime ideal of R is maximal.*

(ix) *R is Noetherian, and for every nonzero prime ideal $\mathfrak{p}$ of R, the localization $R_{\mathfrak{p}}$ is a discrete valuation ring.*

4.2. Ramification and Degree

Let F be a number field and $\mathfrak{p}$ a nonzero prime ideal of $\mathfrak{o}_F$. Then $\mathfrak{p}\cap\mathbb{Z}$ is a nonzero prime ideal of $\mathbb{Z}$, i.e., $\mathfrak{p}\cap\mathbb{Z}=p\mathbb{Z}$ for some prime $p\in\mathbb{Z}$. We say that $\mathfrak{p}$ *lies above* p. The positive integer $e=\nu_{\mathfrak{p}}(p)$ is called the *ramification index* of $\mathfrak{p}$. The homomorphism $\mathbb{Z}\to\mathfrak{o}_F/\mathfrak{p}$ has kernel $\mathfrak{p}\cap\mathbb{Z}=p\mathbb{Z}$, so it induces an embedding $\mathbb{Z}/p\mathbb{Z}\hookrightarrow\mathfrak{o}_F/\mathfrak{p}$. The positive integer $f=[\mathfrak{o}_F/\mathfrak{p}:\mathbb{Z}/p\mathbb{Z}]$ is called the *degree* of $\mathfrak{p}$. (Note that $\mathfrak{o}_F/\mathfrak{p}=\mathbb{F}_{p^f}$.)

Theorem 4.21. *Let $p\in\mathbb{Z}$ be a prime and write*

$$p\mathfrak{o}_F=\mathfrak{p}_1^{e_1}\cdots\mathfrak{p}_k^{e_k},$$

where $\mathfrak{p}_1,\dots,\mathfrak{p}_k\in\operatorname{Spec}^*(\mathfrak{o}_F)$ are distinct and $e_i>0$. Let f_i be the degree of $\mathfrak{p}_i$. Then

$$\sum_{i=1}^{k}e_if_i=[F:\mathbb{Q}].$$

Lemma 4.22 (The Chinese remainder theorem). *Let R be a ring and $\mathfrak{a}_1,\dots,\mathfrak{a}_k$ ideals of R such that $\mathfrak{a}_i+\mathfrak{a}_j=R$ for all $i\neq j$. Then*

$$R/\mathfrak{a}_1\cdots\mathfrak{a}_k\cong(R/\mathfrak{a}_1)\times\cdots\times(R/\mathfrak{a}_k).$$

Proof. 1° Define

$$\begin{array}{rccl}\phi: & R & \longrightarrow & (R/\mathfrak{a}_1)\times\cdots\times(R/\mathfrak{a}_k)\\ & x & \longmapsto & (x+\mathfrak{a}_1,\dots,x+\mathfrak{a}_k).\end{array}$$

We claim that ϕ is onto. To prove this, we show that for each $1\le i\le k$,

$$(0+\mathfrak{a}_1,\ \dots,\ 0+\mathfrak{a}_{i-1},\ 1+\mathfrak{a}_i,\ 0+\mathfrak{a}_{i+1},\ \dots,\ 0+\mathfrak{a}_k)\in\phi(R). \tag{4.3}$$

Of course, we only have to prove (4.3) with $i=1$. We have

$$R=(\mathfrak{a}_1+\mathfrak{a}_2)(\mathfrak{a}_1+\mathfrak{a}_3)\cdots(\mathfrak{a}_1+\mathfrak{a}_k)\subset\mathfrak{a}_1+\mathfrak{a}_2\cdots\mathfrak{a}_k. \tag{4.4}$$

Thus $1=y+x$ for some $y\in\mathfrak{a}_1$ and $x\in\mathfrak{a}_2\cdots\mathfrak{a}_k$. Then $\phi(x)=(1+\mathfrak{a}_1,\ 0+\mathfrak{a}_2,\dots,0+\mathfrak{a}_k)$.

2° Clearly, $\ker\phi = \mathfrak{a}_1 \cap \cdots \cap \mathfrak{a}_k$. It remains to show that $\mathfrak{a}_1 \cap \cdots \cap \mathfrak{a}_k = \mathfrak{a}_1 \cdots \mathfrak{a}_k$. By (4.4), $R = \mathfrak{a}_i + \prod_{j\neq i} \mathfrak{a}_j$ for all i. Thus

$$\begin{aligned}\mathfrak{a}_1 \cap \cdots \cap \mathfrak{a}_k &= (\mathfrak{a}_1 \cap \cdots \cap \mathfrak{a}_k) \prod_i \Big(\mathfrak{a}_i + \prod_{j\neq i} \mathfrak{a}_j\Big) \\ &\subset (\mathfrak{a}_1 \cap \cdots \cap \mathfrak{a}_k) \sum_i \prod_{j\neq i} \mathfrak{a}_j \subset \mathfrak{a}_1 \cdots \mathfrak{a}_k. \qquad \square\end{aligned}$$

Proof of Theorem 4.21. 1° We claim that for $i \neq j$, $\mathfrak{p}_i^{e_i} + \mathfrak{p}_j^{e_j} = \mathfrak{o}_F$.

We have $\mathfrak{p}_i + \mathfrak{p}_j = \mathfrak{o}_F$ since $\mathfrak{p}_i$ and $\mathfrak{p}_j$ are distinct maximal ideals of $\mathfrak{o}_F$. Thus $\mathfrak{o}_F = (\mathfrak{p}_i + \mathfrak{p}_j)^{e_i+e_j} \subset \mathfrak{p}_i^{e_i} + \mathfrak{p}_j^{e_j}$.

2° By Lemma 4.22 we have a ring isomorphism

$$\mathfrak{o}_F/p\mathfrak{o}_F \cong \mathfrak{o}_F/\mathfrak{p}_1^{e_1} \oplus \cdots \oplus \mathfrak{o}_F/\mathfrak{p}_k^{e_k}, \tag{4.5}$$

which is also an isomorphism of vector spaces over $\mathbb{Z}/p\mathbb{Z} = \mathbb{F}_p$. Since $\mathfrak{o}_F$ is a free $\mathbb{Z}$-module of rank $[F : \mathbb{Q}]$, $\dim_{\mathbb{F}_p} \mathfrak{o}_F/p\mathfrak{o}_F = [F : \mathbb{Q}]$. It remains to show that $\dim_{\mathbb{F}_p} \mathfrak{o}_F/\mathfrak{p}_i^{e_i} = e_i \dim_{\mathbb{F}_p} \mathfrak{o}_F/\mathfrak{p}_i$, i.e., $\dim_{\mathfrak{o}_F/\mathfrak{p}_i} \mathfrak{o}_F/\mathfrak{p}_i^{e_i} = e_i$. This follows from 3°.

3° Let $\mathfrak{p} \in \operatorname{Spec}^*(\mathfrak{o}_F)$ and $e > 0$. We claim that $\dim_{\mathfrak{o}_F/\mathfrak{p}} \mathfrak{o}_F/\mathfrak{p}^e = e$.

We have a short exact sequence of $\mathfrak{o}_F/\mathfrak{p}$-modules

$$0 \longrightarrow \mathfrak{p}^{e-1}/\mathfrak{p}^e \longrightarrow \mathfrak{o}_F/\mathfrak{p}^e \longrightarrow \mathfrak{o}_F/\mathfrak{p}^{e-1} \longrightarrow 0.$$

Using induction, it suffices to show that $\mathfrak{p}^{e-1}/\mathfrak{p}^e \cong \mathfrak{o}_F/\mathfrak{p}$. Choose $\alpha \in \mathfrak{p}^{e-1} \setminus \mathfrak{p}^e$. Then $\mathfrak{p}^e \subsetneq (\alpha) + \mathfrak{p}^e \subset \mathfrak{p}^{e-1}$, so $(\alpha) + \mathfrak{p}^e = \mathfrak{p}^{e-1}$. Define

$$\begin{aligned}\phi:\ \mathfrak{o}_F &\longrightarrow \mathfrak{p}^{e-1}/\mathfrak{p}^e \\ x &\longmapsto \alpha x + \mathfrak{p}^e.\end{aligned}$$

Then ϕ is an onto $\mathfrak{o}_F$-map with $\mathfrak{p} \subset \ker\phi$. Since $\ker\phi \neq \mathfrak{o}_F$, we must have $\ker\phi = \mathfrak{p}$. Thus ϕ induces an $\mathfrak{o}_F/\mathfrak{p}$-isomorphism $\mathfrak{o}_F/\mathfrak{p} \to \mathfrak{p}^{e-1}/\mathfrak{p}^e$. □

Example 4.23. Let $F = \mathbb{Q}(\sqrt{-5})$. Then $[F : \mathbb{Q}] = 2$, and by Exercise 4.3, $\mathfrak{o}_F = \mathbb{Z}[\sqrt{-5}]$. Let $\mathfrak{p}_1 = (3,\, 1 + \sqrt{-5})$ and $\mathfrak{p}_2 = (3,\, 1 - \sqrt{-5})$ be ideals of $\mathfrak{o}_F$. We have $\mathfrak{p}_i \neq \mathfrak{o}_F$, $i = 1, 2$. (Since $(1 - \sqrt{-5})\mathfrak{p}_1 \subset 3\mathfrak{o}_F \subsetneq (1 - \sqrt{-5})\mathfrak{o}_F$, $\mathfrak{p}_1 \neq \mathfrak{o}_F$. Similarly, $\mathfrak{p}_2 \neq \mathfrak{o}_F$.) Clearly, $\mathfrak{p}_1 \neq \mathfrak{p}_2$. Since $3 = 3^2 - (1-\sqrt{-5})(1+\sqrt{-5}) \in \mathfrak{p}_1\mathfrak{p}_2$, we have $3\mathfrak{o}_F \subset \mathfrak{p}_1\mathfrak{p}_2$. By Theorem 4.21, $\mathfrak{p}_1$ and $\mathfrak{p}_2$ are prime ideals of $\mathfrak{o}_F$ and $3\mathfrak{o}_F = \mathfrak{p}_1\mathfrak{p}_2$. Let e_i and f_i be the ramification index and the degree of $\mathfrak{p}_i$, $i = 1, 2$, respectively. Then $e_1 = e_2 = 1$ and $f_1 = f_2 = 1$.

Similarly, we have $5\mathfrak{o}_F = \mathfrak{p}^2$, where $\mathfrak{p} = (\sqrt{-5})$ is a prime ideal of $\mathfrak{o}_F$. The ramification index and the degree of $\mathfrak{p}$ are $e = 2$ and $f = 1$, respectively.

Fact 4.24. *Let F be a number field. If $\mathfrak{p} \in \operatorname{Spec}^*(\mathfrak{o}_F)$, then $\sigma\mathfrak{p} \in \operatorname{Spec}^*(\mathfrak{o}_F)$ for all $\sigma \in \operatorname{Aut}(F)$.*

Proof. Clearly, $\sigma\mathfrak{p}$ is a nonzero ideal of $\mathfrak{o}_F$. Since $\mathfrak{o}_F/\sigma\mathfrak{p} = \sigma\mathfrak{o}_F/\sigma\mathfrak{p} \cong \mathfrak{o}_F/\mathfrak{p}$, which is a field, $\sigma\mathfrak{p}$ is prime. □

Galois extensions. When $F/\mathbb{Q}$ is a finite Galois extension, primes in $\mathbb{Z}$ split in $\mathfrak{o}_F$ in a more predictable way.

Proposition 4.25. *Let $F/\mathbb{Q}$ be a finite Galois extension and $p \in \mathbb{Z}$ a prime. Then* $\mathrm{Aut}(F)$ *acts transitively on the prime ideals of $\mathfrak{o}_F$ lying above p.*

Proof. Assume to the contrary that there exist prime ideals $\mathfrak{p}$ and $\mathfrak{q}$ lying above p such that $\mathfrak{q} \notin \{\sigma\mathfrak{p} : \sigma \in \mathrm{Aut}(F)\}$. By Lemma 4.22, there exists $\alpha \in \mathfrak{o}_F$ such that $\alpha \equiv 0 \pmod{\mathfrak{q}}$ and $\alpha \equiv 1 \pmod{\sigma\mathfrak{p}}$ for all $\sigma \in \mathrm{Aut}(F)$. Then $\mathrm{N}_{F/\mathbb{Q}}(\alpha) = \prod_{\sigma\in\mathrm{Aut}(F)} \sigma(\alpha) \in \mathfrak{q} \cap \mathbb{Z} = p\mathbb{Z} \subset \mathfrak{p}$. On the other hand, since $\sigma^{-1}(\alpha) \equiv 1 \pmod{\mathfrak{p}}$ for all $\sigma \in \mathrm{Aut}(F)$, we have $\mathrm{N}_{F/\mathbb{Q}}(\alpha) = \prod_{\sigma\in\mathrm{Aut}(F)} \sigma(\alpha) \equiv 1 \pmod{\mathfrak{p}}$, which is a contradiction. □

Theorem 4.26. *Let $F/\mathbb{Q}$ be a finite Galois extension and $p \in \mathbb{Z}$ a prime. Write*

$$p\mathfrak{o}_F = \mathfrak{p}_1^{e_1} \cdots \mathfrak{p}_k^{e_k},$$

where $\mathfrak{p}_1, \dots, \mathfrak{p}_k \in$ $\mathrm{Spec}^*(\mathfrak{o}_F)$ *are distinct and $e_i > 0$. Then $e_1 = \cdots = e_k = e$, $f_1 = \cdots = f_k = f$, and $kef = [F : \mathbb{Q}]$.*

Proof. For each $1 \le i \le k$, there exists $\sigma \in \mathrm{Aut}(F)$ such that $\sigma\mathfrak{p}_1 = \mathfrak{p}_i$. Then $\mathfrak{o}_F/\mathfrak{p}_i = \sigma\mathfrak{o}_F/\sigma\mathfrak{p}_1 \cong \mathfrak{o}_F/\mathfrak{p}_1$, and hence $f_i = f_1$. Also,

$$\mathfrak{p}_1^{e_1} \cdots \mathfrak{p}_k^{e_k} = p\mathfrak{o}_F = \sigma(p\mathfrak{o}_F) = \sigma(\mathfrak{p}_1)^{e_1} \cdots \sigma(\mathfrak{p}_k)^{e_k}.$$

Comparing the multiplicities of $\mathfrak{p}_i$ gives $e_i = e_1$. □

4.3. Extensions of Number Fields

Let $F \subset K$ be number fields. For $\mathfrak{P} \in \mathrm{Spec}^*(\mathfrak{o}_K)$, we have $\mathfrak{p} = \mathfrak{P} \cap \mathfrak{o}_F \in \mathrm{Spec}^*(\mathfrak{o}_F)$. We say that $\mathfrak{P}$ *lies above* $\mathfrak{p}$ and we write $\mathfrak{P} \mid \mathfrak{p}$. For $\mathfrak{p} \in \mathrm{Spec}^*(\mathfrak{o}_F)$, we have

$$\mathfrak{p}\mathfrak{o}_K = \mathfrak{P}_1^{e_1} \cdots \mathfrak{P}_k^{e_k},$$

where $\mathfrak{P}_1, \dots, \mathfrak{P}_k \in \mathrm{Spec}^*(\mathfrak{o}_K)$ are distinct and $e_i > 0$; $\mathfrak{P}_1, \dots, \mathfrak{P}_k$ are all the prime ideals of $\mathfrak{o}_K$ lying above $\mathfrak{p}$. The positive integer e_i is called the *ramification index* of $\mathfrak{P}_i$ over $\mathfrak{p}$ and is denoted by $e(\mathfrak{P}_i \mid \mathfrak{p})$. We have $\mathfrak{o}_F/\mathfrak{p} \subset \mathfrak{o}_K/\mathfrak{P}_i$, and $f(\mathfrak{P}_i|\mathfrak{p}) = [\mathfrak{o}_K/\mathfrak{P}_i : \mathfrak{o}_F/\mathfrak{p}]$ is called the *degree* of $\mathfrak{P}_i$ over $\mathfrak{p}$.

Fact 4.27. *Let $F \subset K \subset L$ be number fields, and let $\mathfrak{p} \in$* $\mathrm{Spec}^*(\mathfrak{o}_F)$, *$\mathfrak{P} \in$* $\mathrm{Spec}^*(\mathfrak{o}_K)$, *and $\mathcal{P} \in$* $\mathrm{Spec}^*(\mathfrak{o}_L)$ *be such that $\mathcal{P} \mid \mathfrak{P} \mid \mathfrak{p}$. Then*

$$e(\mathcal{P}|\mathfrak{p}) = e(\mathcal{P}|\mathfrak{P})e(\mathfrak{P}|\mathfrak{p}), \qquad f(\mathcal{P}|\mathfrak{p}) = f(\mathcal{P}|\mathfrak{P})f(\mathfrak{P}|\mathfrak{p}).$$

Theorem 4.28. *Let $F \subset K$ be number fields and $\mathfrak{p} \in \mathrm{Spec}^*(\mathfrak{o}_F)$, and let $\mathfrak{P}_1, \ldots, \mathfrak{P}_k$ be all the prime ideals of $\mathfrak{o}_K$ lying above $\mathfrak{p}$. Then*

$$\sum_{i=1}^{k} e(\mathfrak{P}_i|\mathfrak{p})f(\mathfrak{P}_i|\mathfrak{p}) = [K : F].$$

Proof. Identical to the proof of Theorem 4.21, we have

$$\sum_{i=1}^{k} e(\mathfrak{P}_i|\mathfrak{p})f(\mathfrak{P}_i|\mathfrak{p}) = \dim_{\mathfrak{o}_F/\mathfrak{p}} \mathfrak{o}_K/\mathfrak{p}\mathfrak{o}_K.$$

It remains to show that $\dim_{\mathfrak{o}_F/\mathfrak{p}} \mathfrak{o}_K/\mathfrak{p}\mathfrak{o}_K = [K : F]$.

1° Let $S_\mathfrak{p} = \mathfrak{o}_F \setminus \mathfrak{p}$. For each $\mathfrak{o}_F$-submodule $A \subset \mathfrak{o}_K$, define $S_\mathfrak{p}^{-1}A = \{a/b : a \in A,\ b \in S_\mathfrak{p}\}$, which is the *localization* of A at $\mathfrak{p}$ and is an $S_\mathfrak{p}^{-1}\mathfrak{o}_F$-module. Consider the $\mathfrak{o}_F$-map

$$\begin{array}{rccc} g: & A & \longrightarrow & S_\mathfrak{p}^{-1}A/S_\mathfrak{p}^{-1}(\mathfrak{p}A) \\ & a & \longmapsto & a + S_\mathfrak{p}^{-1}(\mathfrak{p}A). \end{array}$$

First, g is onto. (For each $a/b \in S_\mathfrak{p}^{-1}A$, where $a \in A$ and $b \in S_\mathfrak{p}$, there exists $c \in \mathfrak{o}_F$ such that $cb \equiv 1 \pmod{\mathfrak{p}}$. We have $ca \in A$ and $ca \equiv a/b \pmod{S_\mathfrak{p}^{-1}(\mathfrak{p}A)}$.) Second, $\ker g = \mathfrak{p}A$. (If $a \in A$ is such that $a \in S_\mathfrak{p}^{-1}(\mathfrak{p}A)$, then $ba \in \mathfrak{p}A$ for some $b \in S_\mathfrak{p}$. Let $c \in \mathfrak{o}_F$ be such that $cb \equiv 1 \pmod{\mathfrak{p}}$. Then $a \equiv cba \equiv 0 \pmod{\mathfrak{p}A}$. Hence $\ker g \subset \mathfrak{p}A$. The reverse inclusion $\mathfrak{p}A \subset \ker g$ is obvious.) Therefore g induces an $\mathfrak{o}_F$-isomorphism

$$\bar{g} : A/\mathfrak{p}A \longrightarrow S_\mathfrak{p}^{-1}A/S_\mathfrak{p}^{-1}(\mathfrak{p}A), \tag{4.6}$$

which is clearly also an $\mathfrak{o}_F/\mathfrak{p}$-isomorphism.

2° Choose $t \in \mathfrak{o}_F$ such that $\nu_\mathfrak{p}(t) = 1$. Then every nonzero ideal of $S_\mathfrak{p}^{-1}\mathfrak{o}_F$ is generated by t^v, where $v = \min\{\nu_\mathfrak{p}(x) : x \in S_\mathfrak{p}^{-1}\mathfrak{o}_F\}$. Thus $S_\mathfrak{p}^{-1}\mathfrak{o}_F$ is a principal ideal domain (PID). (In fact, $S_\mathfrak{p}^{-1}\mathfrak{o}_F$ is a discrete valuation ring.) Since $\mathfrak{o}_K$ is finitely generated over $\mathbb{Z}$, $S_\mathfrak{p}^{-1}\mathfrak{o}_K$ is finitely generated over $S_\mathfrak{p}^{-1}\mathfrak{o}_F$. Clearly, $S_\mathfrak{p}^{-1}\mathfrak{o}_K$ is a torsion free $S_\mathfrak{p}^{-1}\mathfrak{o}_F$-module. Therefore, $S_\mathfrak{p}^{-1}\mathfrak{o}_K$ is a free $S_\mathfrak{p}^{-1}\mathfrak{o}_F$-module of finite rank. Let $\epsilon_1, \ldots, \epsilon_n$ be a basis of $S_\mathfrak{p}^{-1}\mathfrak{o}_K$ over $S_\mathfrak{p}^{-1}\mathfrak{o}_F$. We claim that $\epsilon_1, \ldots, \epsilon_n$ is also a basis of K over F. First, we have $\mathfrak{o}_K \subset S_\mathfrak{p}^{-1}\mathfrak{o}_K = \sum_i (S_\mathfrak{p}^{-1}\mathfrak{o}_F)\epsilon_i \subset \sum_i F\epsilon_i$, and hence $K = F\mathfrak{o}_K \subset \sum_i F\epsilon_i$. On the other hand, assume that $a_1\epsilon_1 + \cdots + a_n\epsilon_n = 0$, where $a_i \in F$. There exists $0 \neq b \in \mathfrak{o}_F$ such that $ba_i \in \mathfrak{o}_F$ for all i. Then $\sum_i ba_i\epsilon_i = 0$, which forces $ba_i = 0$, i.e., $a_i = 0$, for all i. Thus the claim is proved. In particular, $n = [K : F]$.

3° Now we have $\mathfrak{o}_F/\mathfrak{p}$-isomorphisms

$$\mathfrak{o}_K/\mathfrak{p}\mathfrak{o}_K \cong S_\mathfrak{p}^{-1}\mathfrak{o}_K/S_\mathfrak{p}^{-1}(\mathfrak{p}\mathfrak{o}_K) \qquad \text{(by (4.6))}$$

$$
\begin{aligned}
&= S_{\mathfrak{p}}^{-1}\mathfrak{o}_K/\mathfrak{p}(S_{\mathfrak{p}}^{-1}\mathfrak{o}_K)\\
&= \frac{(S_{\mathfrak{p}}^{-1}\mathfrak{o}_F)\epsilon_1+\cdots+(S_{\mathfrak{p}}^{-1}\mathfrak{o}_F)\epsilon_n}{\mathfrak{p}\big[(S_{\mathfrak{p}}^{-1}\mathfrak{o}_F)\epsilon_1+\cdots+(S_{\mathfrak{p}}^{-1}\mathfrak{o}_F)\epsilon_n\big]}\\
&= \frac{(S_{\mathfrak{p}}^{-1}\mathfrak{o}_F)\epsilon_1+\cdots+(S_{\mathfrak{p}}^{-1}\mathfrak{o}_F)\epsilon_n}{(S_{\mathfrak{p}}^{-1}\mathfrak{p})\epsilon_1+\cdots+(S_{\mathfrak{p}}^{-1}\mathfrak{p})\epsilon_n}\\
&\cong \underbrace{\frac{S_{\mathfrak{p}}^{-1}\mathfrak{o}_F}{S_{\mathfrak{p}}^{-1}\mathfrak{p}}\times\cdots\times\frac{S_{\mathfrak{p}}^{-1}\mathfrak{o}_F}{S_{\mathfrak{p}}^{-1}\mathfrak{p}}}_{n}\\
&\cong \underbrace{\mathfrak{o}_F/\mathfrak{p}\times\cdots\times\mathfrak{o}_F/\mathfrak{p}}_{n} \qquad \text{(by (4.6))}.
\end{aligned}
$$

Hence $\dim_{\mathfrak{o}_F/\mathfrak{p}}\mathfrak{o}_K/\mathfrak{p}\mathfrak{o}_K = n = [K:F]$. □

Theorem 4.29. *Let $F\subset K$ be number fields such that K/F is Galois. Let $\mathfrak{p}\in\operatorname{Spec}^*(\mathfrak{o}_F)$.*

(i) *$\operatorname{Aut}(K/F)$ acts transitively on the prime ideals of $\mathfrak{o}_K$ lying above $\mathfrak{p}$.*

(ii) *Let $\mathfrak{P}_1,\dots,\mathfrak{P}_k\in\operatorname{Spec}^*(\mathfrak{o}_K)$ be all the prime ideals lying above $\mathfrak{p}$. Then $e(\mathfrak{P}_i|\mathfrak{p})=e$ and $f(\mathfrak{P}_i|\mathfrak{p})=f$ for all i and $kef=[K:F]$.*

Proof. Same as the proofs of Proposition 4.25 and Theorem 4.26 □

Galois extensions. Let $F\subset K$ be number fields such that K/F is Galois. Let $\mathfrak{p}\in\operatorname{Spec}^*(\mathfrak{o}_F)$ and $\mathfrak{P}\in\operatorname{Spec}^*(\mathfrak{o}_K)$ be such that $\mathfrak{P}\mid\mathfrak{p}$. Define

$$
\begin{aligned}
D(\mathfrak{P}|\mathfrak{p}) &= \{\sigma\in\operatorname{Aut}(K/F):\sigma\mathfrak{P}=\mathfrak{P}\},\\
E(\mathfrak{P}|\mathfrak{p}) &= \{\sigma\in\operatorname{Aut}(K/F):\sigma(\alpha)\equiv\alpha \pmod{\mathfrak{P}} \text{ for all } \alpha\in\mathfrak{o}_K\}.
\end{aligned}
$$

We have $E(\mathfrak{P}|\mathfrak{p}) < D(\mathfrak{P}|\mathfrak{p}) < \operatorname{Aut}(K/F)$. The groups $D(\mathfrak{P}|\mathfrak{p})$ and $E(\mathfrak{P}|\mathfrak{p})$ are called the *decomposition group* and the *inertia group* of $\mathfrak{P}\mid\mathfrak{p}$, respectively. (For the definition of $D(\mathfrak{P}|\mathfrak{p})$ and $E(\mathfrak{P}|\mathfrak{p})$, K/F does not need to be Galois. But these groups are most useful when K/F is Galois.)

For a number field F, nonzero prime ideals of $\mathfrak{o}_F$ are sometimes referred to as *primes* of $\mathfrak{o}_F$ or F.

Theorem 4.30. *Let $F\subset K$ be number fields such that K/F is Galois, and let $\mathfrak{p}\in\operatorname{Spec}^*(\mathfrak{o}_F)$ and $\mathfrak{P}\in\operatorname{Spec}^*(\mathfrak{o}_K)$ be such that $\mathfrak{P}\mid\mathfrak{p}$. Let $e=e(\mathfrak{P}|\mathfrak{p})$ and $f=f(\mathfrak{P}|\mathfrak{p})$. Table 4.1 gives the data about the groups $D=D(\mathfrak{P}|\mathfrak{p})$ and $E=E(\mathfrak{P}|\mathfrak{p})$, their corresponding fields D' and E' (in the Galois correspondence of K/F), and the primes of D' and E' lying under $\mathfrak{P}$.*

Table 4.1. $D(\mathfrak{P}|\mathfrak{p})$, $E(\mathfrak{P}|\mathfrak{p})$, and the corresponding fields and primes

degree/index	field	group	prime	$e(\cdot\mid\cdot)$	$f(\cdot\mid\cdot)$
	K	$\{\text{id}\}$	$\mathfrak{P}$		
e	$\vert$	$\vert$	$\vert$	e	1
	E'	$E = E(\mathfrak{P}\mid\mathfrak{p})$	$\mathfrak{P}_E = \mathfrak{P} \cap \mathfrak{o}_{E'}$		
f	$\vert$	$\vert$	$\vert$	1	f
	D'	$D = D(\mathfrak{P}\mid\mathfrak{p})$	$\mathfrak{P}_D = \mathfrak{P} \cap \mathfrak{o}_{D'}$		
$k = [K : F]/ef$	$\vert$	$\vert$	$\vert$	1	1
	F	$\text{Aut}(K/F)$	$\mathfrak{p}$		

Proof. 1° We claim that $[D' : F] = k$.

Recall that k is the number of primes of $\mathfrak{o}_K$ lying above $\mathfrak{p}$: $\mathfrak{p}\mathfrak{o}_K = (\mathfrak{P}_1 \cdots \mathfrak{P}_k)^e$, where $\mathfrak{P}_1, \ldots, \mathfrak{P}_k \in \text{Spec}^*(\mathfrak{o}_K)$ are distinct and $\mathfrak{P}_1 = \mathfrak{P}$. Since $\text{Aut}(K/F)$ acts transitively on $\{\mathfrak{P}_1, \ldots, \mathfrak{P}_k\}$ and D is the stabilizer of $\mathfrak{P}_1$, we have $k = [\text{Aut}(K/F) : D] = [D' : F]$.

2° We claim that $e(\mathfrak{P}_D|\mathfrak{p}) = f(\mathfrak{P}_D|\mathfrak{p}) = 1$.

Since K/D' is Galois, D acts transitively on the primes of $\mathfrak{o}_K$ lying above $\mathfrak{P}_D$. By the definition of D, $\{\mathfrak{P}\}$ is the D-orbit of $\mathfrak{P}$. Hence $\mathfrak{P}$ is the only prime of $\mathfrak{o}_K$ lying above $\mathfrak{P}_D$. Thus $e(\mathfrak{P}|\mathfrak{P}_D)f(\mathfrak{P}|\mathfrak{P}_D) = [K : D']$, i.e.,

$$\frac{ef}{e(\mathfrak{P}_D|\mathfrak{p})f(\mathfrak{P}_D|\mathfrak{p})} = \frac{[K : F]}{[D' : F]} = ef.$$

Therefore $e(\mathfrak{P}_D|\mathfrak{p}) = f(\mathfrak{P}_D|\mathfrak{p}) = 1$.

3° We claim that $f(\mathfrak{P}|\mathfrak{P}_E) = 1$.

Denote both canonical homomorphisms $\mathfrak{o}_K \to \mathfrak{o}_K/\mathfrak{P}$ and $\mathfrak{o}_{E'} \to \mathfrak{o}_{E'}/\mathfrak{P}_E$ by $\overline{(\ )}$. Let $\bar{\alpha} \in \mathfrak{o}_K/\mathfrak{P}$, where $\alpha \in \mathfrak{o}_K$. Put $g = \prod_{\sigma \in E}(\mathtt{X} - \sigma(\alpha))$. Then $g \in E'[\mathtt{X}]$, and hence $g \in \mathfrak{o}_{E'}[\mathtt{X}]$. It follows that $(\mathfrak{o}_{E'}/\mathfrak{P}_E)[\mathtt{X}] \ni \bar{g} = (\mathtt{X} - \bar{\alpha})^{|E|}$, so $\bar{\alpha}$ is purely inseparable over $\mathfrak{o}_{E'}/\mathfrak{P}_E$. Since $\mathfrak{o}_K/\mathfrak{P}$ is separable over $\mathfrak{o}_{E'}/\mathfrak{P}_E$, we have $\bar{\alpha} \in \mathfrak{o}_{E'}/\mathfrak{P}_E$. Thus we have proved that $\mathfrak{o}_K/\mathfrak{P} = \mathfrak{o}_{E'}/\mathfrak{P}_E$.

4° It follows from 2° and 3° that $f(\mathfrak{P}_E|\mathfrak{P}_D) = f$.

5° Define

$$\begin{aligned} \phi:\ D &\longrightarrow \text{Aut}\big((\mathfrak{o}_K/\mathfrak{P})/(\mathfrak{o}_F/\mathfrak{p})\big) \\ \sigma &\longmapsto \phi(\sigma), \end{aligned}$$

where

$$\begin{aligned} \phi(\sigma): \quad \mathfrak{o}_K/\mathfrak{P} &\longrightarrow \mathfrak{o}_K/\mathfrak{P} \\ \alpha + \mathfrak{P} &\longmapsto \sigma(\alpha) + \mathfrak{P}, \quad \alpha \in \mathfrak{o}_K. \end{aligned}$$

Then ϕ is a homomorphism with $\ker \phi = E$. Hence ϕ induces an embedding $D/E \hookrightarrow \operatorname{Aut}\big((\mathfrak{o}_K/\mathfrak{P})/(\mathfrak{o}_F/\mathfrak{p})\big)$. In particular, $[D : E] \le f(\mathfrak{P}|\mathfrak{p}) = f$.

6° We have

$$[E' : D'] = [D : E] \le f \qquad \text{(by 5°)}$$

and

$$\begin{aligned} [E' : D'] &\ge f(\mathfrak{P}_E|\mathfrak{P}_D) && \text{(by Theorem 4.28)} \\ &= f && \text{(by 4°)}. \end{aligned}$$

Hence $[E' : D'] = f$. The remaining claims in Table 4.1 follow immediately. □

Corollary 4.31. *In Theorem 4.30, $E(\mathfrak{P}|\mathfrak{p}) \lhd D(\mathfrak{P}|\mathfrak{p})$ and the embedding $D(\mathfrak{P}|\mathfrak{p})/E(\mathfrak{P}|\mathfrak{p}) \hookrightarrow \operatorname{Aut}\big((\mathfrak{o}_K/\mathfrak{P})/(\mathfrak{o}_F/\mathfrak{p})\big)$ is an isomorphism. Therefore $D(\mathfrak{P}|\mathfrak{p})/E(\mathfrak{P}|\mathfrak{p})$ is cyclic of order $f(\mathfrak{P}|\mathfrak{p})$.*

Proof. By 6° of the proof of Theorem 4.30, $|D(\mathfrak{P}|\mathfrak{p})/E(\mathfrak{P}|\mathfrak{p})| = f(\mathfrak{P}|\mathfrak{p})$. □

Corollary 4.32. *In Theorem 4.30, we have the following special situations.*

(i) *If $D(\mathfrak{P}|\mathfrak{p}) \lhd \operatorname{Aut}(K/F)$, then $\mathfrak{p}$ splits into k distinct primes in $\mathfrak{o}_{D'}$: $\mathfrak{p}\mathfrak{o}_{D'} = \mathfrak{q}_1 \cdots \mathfrak{q}_k$.*

(ii) *If $D(\mathfrak{P}|\mathfrak{p}) \lhd \operatorname{Aut}(K/F)$ and $E(\mathfrak{P}|\mathfrak{p}) \lhd \operatorname{Aut}(K/F)$, then for each $1 \le i \le k$, $\mathfrak{q}_i$ remains prime in $\mathfrak{o}_{E'}$, i.e., $\mathfrak{q}_i\mathfrak{o}_{E'} \in \operatorname{Spec}^*(\mathfrak{o}_{E'})$. Moreover, $\mathfrak{q}_i$ becomes an eth power in $\mathfrak{o}_K$, i.e., $\mathfrak{q}_i\mathfrak{o}_K = \mathfrak{P}_i^e$ for some $\mathfrak{P}_i \in \operatorname{Spec}^*(\mathfrak{o}_K)$.*

Proof. (i) By Table 4.1, $e(\mathfrak{P}_D|\mathfrak{p}) = f(\mathfrak{P}_D|\mathfrak{p}) = 1$. Since D'/F is Galois, by Theorem 4.29, every prime of $\mathfrak{o}_{D'}$ lying above $\mathfrak{p}$ has ramification index 1 and degree 1, and there are k primes of $\mathfrak{o}_{D'}$ lying above $\mathfrak{p}$.

(ii) We have the following picture.

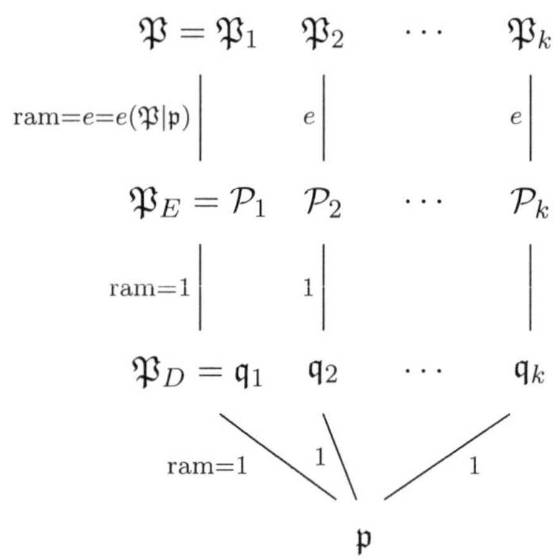

Since E'/F is Galois, for all $1 \le i \le k$, $e(\mathcal{P}_i|\mathfrak{p}) = 1$, and hence $e(\mathcal{P}_i|\mathfrak{q}_i) = 1$. Since K/F is Galois, for all $1 \le i \le k$, $e(\mathfrak{P}_i|\mathfrak{p}) = e$, and hence $e(\mathfrak{P}_i|\mathfrak{q}_i) = e$. ☐

Remark 4.33. In Table 4.1, D' and E' are called the *decomposition field* and the *inertia field* of $\mathfrak{P} \mid \mathfrak{p}$, respectively. Let $F \subset K$ be number fields such that K/F is Galois. If $F \subset L \subset M \subset K$ is a tower of fields with the data

field	prime	$e(\cdot \mid \cdot)$	$f(\cdot \mid \cdot)$
K	$\mathfrak{P}$		
e |	|	e	1
M	$\mathfrak{P}_M$		
f |	|	1	f
L	$\mathfrak{P}_L$		
|	|	1	1
F	$\mathfrak{p}$		

then L and M are the decomposition field and the inertia field of $\mathfrak{P} \mid \mathfrak{p}$, respectively.

Proof. Let $\sigma \in L' = \mathrm{Aut}(K/L)$. Since $e(\mathfrak{P}|\mathfrak{P}_L)f(\mathfrak{P}|\mathfrak{P}_L) = ef = [K : L]$, $\mathfrak{P}$ is the only prime of $\mathfrak{o}_K$ lying above $\mathfrak{P}_L$. Thus $\sigma(\mathfrak{P}) = \mathfrak{P}$, i.e., $\sigma \in D(\mathfrak{P}|\mathfrak{p}) = D$. Hence $L' \subset D$, i.e., $L \supset D'$. Since $[L : F] = [D' : F]$, we have $L = D'$. Let $\sigma \in M' = \mathrm{Aut}(K/M)$. Since $\mathfrak{o}_K/\mathfrak{P} = \mathfrak{o}_M/\mathfrak{P}_M$, σ induces the identity automorphism of $\mathfrak{o}_K/\mathfrak{P}$, i.e., $\sigma \in E(\mathfrak{P}|\mathfrak{p}) = E$. Thus $M' \subset E$, i.e., $M \supset E'$. Since $[K : M] = [K : E']$, we have $M = E'$. □

4.4. Factorization of Primes

Let $F \subset K$ be number fields. Since K/F is a finite separable extension, there exists $\alpha \in K$ such that $K = F(\alpha)$. We may assume that $\alpha \in \mathfrak{o}_K$. (There exists $0 \neq m \in \mathbb{Z}$ such that $m\alpha \in \mathfrak{o}_K$.) We have $|\mathfrak{o}_K/\mathfrak{o}_F[\alpha]| < \infty$. (In general, if $A < \mathbb{Z}^n$ and $A \cong \mathbb{Z}^n$, then $|\mathbb{Z}^n/A| < \infty$.) Let $\mathfrak{p} \in \operatorname{Spec}^*(\mathfrak{o}_F)$ be such that $\mathfrak{p} \mid p$, where p is a prime of $\mathbb{Z}$, and let $\overline{(\;)} : \mathfrak{o}_F[\mathtt{X}] \to (\mathfrak{o}_F/\mathfrak{p})[\mathtt{X}]$ be the reduction. Let $g \in \mathfrak{o}_F[\mathtt{X}]$ be the (monic) minimal polynomial of α over F, and let

$$\bar{g} = \bar{g}_1^{e_1} \cdots \bar{g}_k^{e_k}$$

be the factorization of $\bar{g}$ in $(\mathfrak{o}_F/\mathfrak{p})[\mathtt{X}]$, where each $g_i \in \mathfrak{o}_F[x]$ is monic and $\bar{g}_1, \dots, \bar{g}_k$ are distinct irreducibles in $(\mathfrak{o}_F/\mathfrak{p})[\mathtt{X}]$.

Theorem 4.34. *In the above notation, assume that $p \nmid |\mathfrak{o}_K/\mathfrak{o}_F[\alpha]|$. Then*

$$\mathfrak{p}\mathfrak{o}_K = \mathfrak{P}_1^{e_1} \cdots \mathfrak{P}_k^{e_k},$$

where $\mathfrak{P}_i = \mathfrak{p}\mathfrak{o}_K + (g_i(\alpha)) \in \operatorname{Spec}^*(\mathfrak{o}_K)$, $1 \le i \le k$, are distinct and $f(\mathfrak{P}_i|\mathfrak{p}) = \deg g_i$.

Proof. Let $f_i = \deg g_i = \deg \bar{g}_i$. We will prove the following.

(i) For each i, either $\mathfrak{P}_i = \mathfrak{o}_K$ or $\mathfrak{o}_K/\mathfrak{P}_i$ is a field of order $|\mathfrak{o}_F/\mathfrak{p}|^{f_i}$.

(ii) $\mathfrak{P}_i + \mathfrak{P}_j = \mathfrak{o}_K$ for all $i \neq j$.

(iii) $\mathfrak{p}\mathfrak{o}_K \supset \mathfrak{P}_1^{e_1} \cdots \mathfrak{P}_k^{e_k}$.

We first show that (i)–(iii) imply the theorem. Assume that $\mathfrak{P}_i \neq \mathfrak{o}_K$ for $1 \le i \le l$ and $\mathfrak{P}_i = \mathfrak{o}_K$ for $l < i \le k$. By (i), for $1 \le i \le l$, $\mathfrak{P}_i \in \operatorname{Spec}^*(\mathfrak{o}_K)$ with $f(\mathfrak{P}_i|\mathfrak{p}) = f_i$. By (ii), $\mathfrak{P}_1, \dots, \mathfrak{P}_l$ are distinct. By (iii), $\mathfrak{p}\mathfrak{o}_K = \mathfrak{P}_1^{d_1} \cdots \mathfrak{P}_l^{d_l}$, $0 \le d_i \le e_i$. We have

$$\begin{aligned} [K : F] &= d_1 f_1 + \cdots + d_l f_l \qquad \text{(by Theorem 4.28)} \\ &\le e_1 f_1 + \cdots + e_k f_k \\ &= \deg g = [K : F]. \end{aligned}$$

Thus we must have $l = k$ and $d_i = e_i$ for all i.

Proof of (i). Let $F_i = (\mathfrak{o}_F/\mathfrak{p})[\mathtt{X}]/(\bar{g}_i)$. The reduction homomorphism $\mathfrak{o}_F[\mathtt{X}] \to (\mathfrak{o}_F/\mathfrak{p})[\mathtt{X}]/(\bar{g}_i)$ is onto with kernel $(\mathfrak{p}, g_i)$. Hence we have an isomorphism $\mathfrak{o}_F[\mathtt{X}]/(\mathfrak{p}, g_i) \cong F_i$. Let $\psi : \mathfrak{o}_F[\mathtt{X}] \to \mathfrak{o}_K$ be the homomorphism that replaces $\mathtt{X}$ with α; then ψ induces a homomorphism $\bar{\psi} : \mathfrak{o}_F[\mathtt{X}] \to \mathfrak{o}_K/\mathfrak{P}_i$. We claim that $\bar{\psi}$ is onto. Note that $|\mathfrak{o}_K/(\mathfrak{o}_F[\alpha] + \mathfrak{p}\mathfrak{o}_K)|$ is a common divisor of $|\mathfrak{o}_K/\mathfrak{o}_F[\alpha]|$ and $|\mathfrak{o}_K/\mathfrak{p}\mathfrak{o}_K|$, where $|\mathfrak{o}_K/\mathfrak{p}\mathfrak{o}_K|$ is a power of p and $p \nmid |\mathfrak{o}_K/\mathfrak{o}_F[\alpha]|$.

Hence $|\mathfrak{o}_K/(\mathfrak{o}_F[\alpha]+\mathfrak{p}\mathfrak{o}_K)|=1$. Thus $\mathfrak{o}_K=\mathfrak{o}_F[\alpha]+\mathfrak{p}\mathfrak{o}_K\subset\mathfrak{o}_F[\alpha]+\mathfrak{P}_i$, which means that $\bar{\psi}$ is onto. Now we have $\mathfrak{o}_F[\mathtt{X}]/\ker\bar{\psi}\cong\mathfrak{o}_K/\mathfrak{P}_i$. Clearly, $(\mathfrak{p},g_i)\subset\ker\bar{\psi}$. We already know that $(\mathfrak{p},g_i)$ is a maximal ideal of $\mathfrak{o}_F[\mathtt{X}]$. Hence either $\ker\bar{\psi}=\mathfrak{o}_F[\mathtt{X}]$ or $\ker\bar{\psi}=(\mathfrak{p},g_i)$. If $\ker\bar{\psi}=\mathfrak{o}_F[\mathtt{X}]$, we have $\mathfrak{P}_i=\mathfrak{o}_K$. If $\ker\bar{\psi}=(\mathfrak{p},g_i)$, we have $\mathfrak{o}_K/\mathfrak{P}_i\cong\mathfrak{o}_F[\mathtt{X}]/(\mathfrak{p},g_i)\cong F_i$. Note that $|F_i|=|\mathfrak{o}_F/\mathfrak{p}|^{f_i}$.

Proof of (ii). Assume that $i\neq j$ $(1\le i,j\le k)$. Since $\gcd(\bar{g}_i,\bar{g}_j)=1$, there exist $u,v\in\mathfrak{o}_F[\mathtt{X}]$ such that $\bar{u}\bar{g}_i+\bar{v}\bar{g}_j=1$, i.e.,

$$ug_i+vg_j\equiv 1\pmod{\mathfrak{p}[\mathtt{X}]}.$$

Letting $\mathtt{X}=\alpha$, we have

$$u(\alpha)g_i(\alpha)+v(\alpha)g_j(\alpha)\equiv 1\pmod{\mathfrak{p}\mathfrak{o}_K}.$$

Hence $1\in\mathfrak{p}\mathfrak{o}_K+(g_i(\alpha))+(g_j(\alpha))=\mathfrak{P}_i+\mathfrak{P}_j$.

Proof of (iii). We have

$$\mathfrak{P}_1^{e_1}\cdots\mathfrak{P}_k^{e_k}=\prod_{i=1}^{k}\big(\mathfrak{p}\mathfrak{o}_K+(g_i(\alpha))\big)^{e_i}\subset\mathfrak{p}\mathfrak{o}_K+\Big(\prod_{i=1}^{k}g_i(\alpha)^{e_i}\Big).\tag{4.7}$$

Since $\prod_{i=1}^{k}\bar{g}_i^{e^i}=\bar{g}$, we have

$$\prod_{i=1}^{k}g_i^{e_i}\equiv g\pmod{\mathfrak{p}[\mathtt{X}]}.$$

Letting $\mathtt{X}=\alpha$, we have $\prod_{i=1}^{k}g_i(\alpha)^{e_i}\equiv g(\alpha)=0\pmod{\mathfrak{p}\mathfrak{o}_K}$. Thus by (4.7), $\mathfrak{P}_1^{e_1}\cdots\mathfrak{P}_k^{e_k}\subset\mathfrak{p}\mathfrak{o}_K$. □

4.5. Cyclotomic Fields

Let n be a positive integer and $\zeta_n=e^{2\pi i/n}$, and write $\mathbb{Q}(n)=\mathbb{Q}(\zeta_n)$, which is the nth *cyclotomic field.*

We first recall a few algebraic facts about the cyclotomic field. The extension $\mathbb{Q}(n)/\mathbb{Q}$ is Galois with $[\mathbb{Q}(n):\mathbb{Q}]=\phi(n)$, where ϕ is the Euler function. The minimal polynomial of ζ_n over $\mathbb{Q}$ is

$$\Phi_n(\mathtt{X})=\prod_{\substack{1\le i\le n\\ \gcd(i,n)=1}}(\mathtt{X}-\zeta_n^i).$$

The mapping

$$\begin{array}{ccc}\mathrm{Aut}(\mathbb{Q}(n)/\mathbb{Q}) & \longrightarrow & (\mathbb{Z}/n\mathbb{Z})^\times\\ \sigma & \longmapsto & i, \qquad \text{where } \sigma(\zeta_n)=\zeta_n^i,\end{array}$$

is an isomorphism.

Fact 4.35. *Let $m > 0$ and $s \geq 0$ be integers and $p \in \mathbb{Z}$ a prime with $p \nmid m$. Then*

$$\begin{array}{rccc} \psi : & \operatorname{Aut}(\mathbb{Q}(mp^s)/\mathbb{Q}) & \longrightarrow & \operatorname{Aut}(\mathbb{Q}(m)/\mathbb{Q}) \times \operatorname{Aut}(\mathbb{Q}(p^s)/\mathbb{Q}) \\ & \sigma & \longmapsto & (\sigma|_{\mathbb{Q}(m)}, \sigma|_{\mathbb{Q}(p^s)}) \end{array}$$

is an isomorphism.

Proof. Clearly, ψ is a one-to-one homomorphism. We also have

$$\begin{aligned} |\operatorname{Aut}(\mathbb{Q}(mp^s)/\mathbb{Q})| &= \phi(mp^s) = \phi(m)\phi(p^s) \\ &= |\operatorname{Aut}(\mathbb{Q}(m)/\mathbb{Q}) \times \operatorname{Aut}(\mathbb{Q}(p^s)/\mathbb{Q})|. \end{aligned}$$

□

Theorem 4.36. *Let $n > 0$ be an integer, $p \in \mathbb{Z}$ a prime, and $\mathfrak{P}$ a prime of $\mathbb{Q}(n)$ lying above p. Write $n = mp^s$, with $p \nmid m$. Let $\sigma_p \in \operatorname{Aut}(\mathbb{Q}(m)/\mathbb{Q})$ be defined by $\sigma_p(\zeta_m) = \zeta_m^p$ and let f be the order of p in $(\mathbb{Z}/m\mathbb{Z})^\times$. Let*

$$D = \langle \sigma_p \rangle \times \operatorname{Aut}(\mathbb{Q}(p^s)/\mathbb{Q}) \quad \text{and} \quad E = \{\mathrm{id}\} \times \operatorname{Aut}(\mathbb{Q}(p^s)/\mathbb{Q}),$$

treated as subgroups of $\operatorname{Aut}(\mathbb{Q}(n)/\mathbb{Q})$ through the isomorphism in Fact 4.35. Then D and E are the decomposition group and the inertia group of $\mathfrak{P} \mid p$, respectively. The relevant data about the primes $\mathfrak{P} \mid p$ are given in Table 4.2.

Table 4.2. Groups, fields, and primes of a cyclotomic extension

degree/index	field	group	prime	$e(\cdot\mid\cdot)$	$f(\cdot\mid\cdot)$
	$\mathbb{Q}(n)$	$\{\mathrm{id}\}$	$\mathfrak{P}$		
$\phi(p^s)$				$\phi(p^s)$	1
	$\mathbb{Q}(m)$	$E = \{\mathrm{id}\} \times \operatorname{Aut}(\mathbb{Q}(p^s)/\mathbb{Q})$	$\mathfrak{P}_E$		
f				1	f
	D'	$D = \langle \sigma_p \rangle \times \operatorname{Aut}(\mathbb{Q}(p^s)/\mathbb{Q})$	$\mathfrak{P}_D$		
$\phi(m)/f$				1	1
	$\mathbb{Q}$	$\operatorname{Aut}(\mathbb{Q}(n)/\mathbb{Q})$	$p\mathbb{Z}$		

We need a few facts for the proof of Theorem 4.36.

Fact 4.37. Let $F \subset K$ be fields such that K/F is separable and $[K : F] = n$. Let $\sigma_1, \ldots, \sigma_n$ be the F-embeddings of K into $\overline{K}$. Then for $\alpha_1, \ldots, \alpha_n \in K$,

$$\Delta_{K/F}(\alpha_1, \ldots, \alpha_n) = \left[\det(\sigma_j(\alpha_i))\right]^2.$$

Proof. We have

$$\mathrm{Tr}_{K/F}(\alpha_i\alpha_j) = \sum_{k=1}^{n} \sigma_k(\alpha_i)\sigma_k(\alpha_j),$$

i.e., $A = BB^T$, where $A = (\mathrm{Tr}_{K/F}(\alpha_i\alpha_j))$ and $B = (\sigma_j(\alpha_i))$. Thus $\Delta_{K/F}(\alpha_1, \ldots, \alpha_n) = \det A = (\det B)^2$. □

Lemma 4.38. *Assume that $K = F(\alpha)$, where α is separable of degree n over F with minimal polynomial $f \in F[\mathtt{X}]$. Then*

$$\Delta_{K/F}(\alpha^0, \ldots, \alpha^{n-1}) = (-1)^{\frac{1}{2}n(n-1)}\mathrm{N}_{K/F}(f'(\alpha)).$$

Proof. Let $\sigma_1, \ldots, \sigma_n$ be the F-embeddings of K into $\overline{K}$. Then

$$\begin{aligned}\Delta_{K/F}(\alpha^0, \ldots, \alpha^{n-1}) &= \left[\det(\sigma_j(\alpha^i))\right]^2 = \left[\det(\sigma_j(\alpha)^i)\right]^2 \\ &= \Big[\prod_{i<j}(\sigma_j(\alpha) - \sigma_i(\alpha))\Big]^2 \qquad \text{(Vandermonde)} \\ &= (-1)^{\frac{1}{2}n(n-1)} \prod_{i\neq j}(\sigma_j(\alpha) - \sigma_i(\alpha)).\end{aligned}$$

Since $f = \prod_i(\mathtt{X} - \sigma_i(\alpha))$, we have

$$f'(\sigma_j(\alpha)) = \prod_{\substack{i\\ i\neq j}}(\sigma_j(\alpha) - \sigma_i(\alpha)).$$

Hence

$$\begin{aligned}\mathrm{N}_{K/F}(f'(\alpha)) &= \prod_j \sigma_j(f'(\alpha)) = \prod_j f'(\sigma_j(\alpha)) \\ &= \prod_j \prod_{\substack{i\\ i\neq j}}(\sigma_j(\alpha) - \sigma_i(\alpha)) = \prod_{\substack{i,j\\ i\neq j}}(\sigma_j(\alpha) - \sigma_i(\alpha)).\end{aligned}$$

Therefore

$$\Delta_{K/F}(\alpha^0, \ldots, \alpha^{n-1}) = (-1)^{\frac{1}{2}n(n-1)}\mathrm{N}_{K/F}(f'(\alpha)).$$

□

Fact 4.39. Let n be a positive integer. Then $\Delta_{\mathbb{Q}(n)/\mathbb{Q}}(\zeta_n^0, \ldots, \zeta_n^{\phi(n)-1}) \mid n^{\phi(n)}$.

Proof. The minimal polynomial of ζ_n over $\mathbb{Q}$ is Φ_n. We have $\mathtt{X}^n - 1 = \Phi_n g$ for some $g \in \mathbb{Z}[\mathtt{X}]$. Thus $n\mathtt{X}^{n-1} = \Phi_n' g + \Phi_n g'$. Letting $\mathtt{X} = \zeta_n$, we have $n\zeta_n^{n-1} = \Phi_n'(\zeta_n)g(\zeta_n)$. Then

$$\mathrm{N}_{\mathbb{Q}(n)/\mathbb{Q}}\big(\Phi_n'(\zeta_n)\big)\mathrm{N}_{\mathbb{Q}(n)/\mathbb{Q}}\big(g(\zeta_n)\big) = \mathrm{N}_{\mathbb{Q}(n)/\mathbb{Q}}(n\zeta_n^{n-1}) = \pm n^{\phi(n)}.$$

Thus $\mathrm{N}_{\mathbb{Q}(n)/\mathbb{Q}}\big(\Phi_n'(\zeta_n)\big) \mid n^{\phi(n)}$. By Lemma 4.38, $\Delta_{\mathbb{Q}(n)/\mathbb{Q}}(\zeta_n^0, \ldots, \zeta_n^{\phi(n)-1}) \mid n^{\phi(n)}$. □

Fact 4.40. Let $F/\mathbb{Q}$ be a number field of degree n. Let $\alpha_1, \dots, \alpha_n \in \mathfrak{o}_F$ be linearly independent over $\mathbb{Q}$ and let $\Delta = \Delta_{F/\mathbb{Q}}(\alpha_1, \dots, \alpha_n)$. Then $\Delta \mathfrak{o}_F \subset \mathbb{Z}\alpha_1 + \cdots + \mathbb{Z}\alpha_n$.

Proof. Let $w \in \mathfrak{o}_F$ and write $w = \sum_i r_i \alpha_i$, $r_i \in \mathbb{Q}$. Let $A = (\mathrm{Tr}_{F/\mathbb{Q}}(\alpha_i \alpha_j))$. Then

$$A \begin{bmatrix} r_1 \\ \vdots \\ r_n \end{bmatrix} = \begin{bmatrix} \mathrm{Tr}_{F/\mathbb{Q}}(\alpha_1 w) \\ \vdots \\ \mathrm{Tr}_{F/\mathbb{Q}}(\alpha_n w) \end{bmatrix} \in \mathbb{Z}^n.$$

It follows that

$$\mathbb{Z}^n \ni \operatorname{adj}(A) A \begin{bmatrix} r_1 \\ \vdots \\ r_n \end{bmatrix} = (\det A) \begin{bmatrix} r_1 \\ \vdots \\ r_n \end{bmatrix} = \Delta \begin{bmatrix} r_1 \\ \vdots \\ r_n \end{bmatrix}.$$

Thus $\Delta w \in \mathbb{Z}\alpha_1 + \cdots + \mathbb{Z}\alpha_n$. □

Proof of Theorem 4.36. 1° Clearly, the fixed field of E in the Galois extension $\mathbb{Q}(n)/\mathbb{Q}$ is $\mathbb{Q}(m)$. Also obvious are the entries in the "degree/index" column of Table 4.2.

2° Let ϵ be a primitive mth root of unity in $\mathbb{F}_{p^f}$. (Recall that f is the order of p in $(\mathbb{Z}/m\mathbb{Z})^\times$.) For each $d \mid m$, let

$$h_d = \prod_{\substack{1 \le i \le d \\ \gcd(i,d)=1}} (\mathtt{X} - \epsilon^{\frac{m}{d} i}) \in \mathbb{F}_p[\mathtt{X}].$$

Let $\overline{(\ )} : \mathbb{Z}[\mathtt{X}] \to \mathbb{F}_p[\mathtt{X}]$ be the reduction. Then for all $d \mid m$,

$$\prod_{l \mid d} \bar{\Phi}_l = \overline{\prod_{l \mid d} \Phi_l} = \mathtt{X}^d - 1 = \prod_{l \mid d} h_l.$$

It follows that $\bar{\Phi}_d = h_d$ for all $d \mid m$.

Let $i_1, \dots, i_{\phi(m)/f} \in (\mathbb{Z}/m\mathbb{Z})^\times$ be the coset representatives of $\langle p \rangle$ in $(\mathbb{Z}/m\mathbb{Z})^\times$. Then

$$\bar{\Phi}_m = h_m = \prod_{j=1}^{\phi(m)/f} \gamma_j, \tag{4.8}$$

where $\gamma_j = \prod_{0 \le l \le f-1} (\mathtt{X} - \epsilon^{i_j p^l}) \in \mathbb{F}_p[\mathtt{X}]$ is the minimal polynomial of ϵ^{i_j} over $\mathbb{F}_p$. Let $g_j \in \mathbb{Z}[\mathtt{X}]$ be monic such that $\bar{g}_j = \gamma_j$.

3° Let $\Delta = \Delta_{\mathbb{Q}(m)/\mathbb{Q}}(\zeta_m^0, \dots, \zeta_m^{\phi(m)-1})$. Since $\Delta \mathfrak{o}_{\mathbb{Q}(m)} \subset \mathbb{Z}[\zeta_m]$ and $\Delta \mid m^{\phi(m)}$, we have $p \nmid |\mathfrak{o}_{\mathbb{Q}(m)}/\mathbb{Z}[\zeta_m]|$. Thus by (4.8) and Theorem 4.34, $e(\mathfrak{P}_E|p) = 1$ and $f(\mathfrak{P}_E|p) = f$. Moreover, $\mathfrak{P}_E = (p, g_j(\zeta_m))$. Since

$\sigma_p(g_j(\zeta_m)) = g_j(\zeta_m^p) \equiv (g_j(\zeta_m))^p \equiv 0 \pmod{\mathfrak{P}_E}$, we have $\sigma_p(\mathfrak{P}_E) = \mathfrak{P}_E$. Thus $\langle\sigma_p\rangle \subset D(\mathfrak{P}_E|p)$, and hence $D \subset D(\mathfrak{P}_E|p)$. Therefore $\mathfrak{P}_E$ is the only prime of $\mathbb{Q}(m)$ lying above $\mathfrak{P}_D$ and $f = [\mathbb{Q}(m) : D'] = f(\mathfrak{P}_E|\mathfrak{P}_D)$. It follows that $f(\mathfrak{P}_D|p) = 1$.

4° We claim that $e(\mathfrak{P}|\mathfrak{P}_E) = \phi(p^s)$. It suffices to show that $e(\mathfrak{P}|\mathfrak{P}_E) \geq \phi(p^s)$. We only have to consider the case $s > 0$. Let $\mathfrak{P}' = \mathfrak{P} \cap \mathfrak{o}_{\mathbb{Q}(p^s)}$. It suffices to show that $e(\mathfrak{P}'|p) \geq \phi(p^s)$. (Then $e(\mathfrak{P}|\mathfrak{P}_E) = e(\mathfrak{P}|p) \geq e(\mathfrak{P}'|p) \geq \phi(p^s)$.)

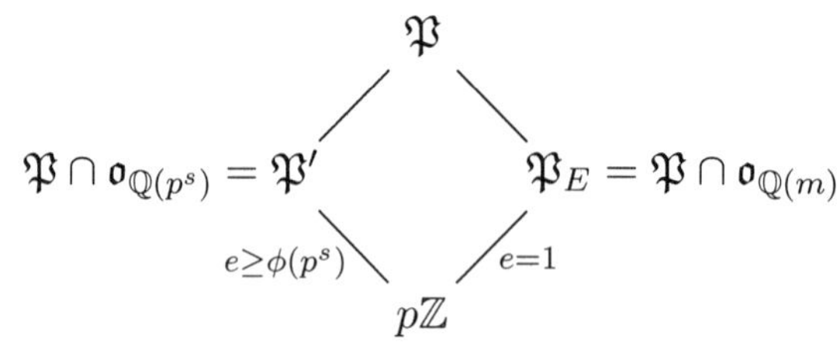

We have

$$1 + \mathtt{X}^{p^{s-1}} + \mathtt{X}^{p^{s-1}\cdot 2} + \cdots + \mathtt{X}^{p^{s-1}\cdot(p-1)} = \frac{\mathtt{X}^{p^s} - 1}{\mathtt{X}^{p^{s-1}} - 1} = \prod_{\substack{1\leq i\leq p^s \\ \gcd(i,p)=1}} (\mathtt{X} - \zeta_{p^s}^i).$$

Letting $\mathtt{X} = 1$, we have

$$p = \prod_{\substack{1\leq i\leq p^s \\ \gcd(i,p)=1}} (1 - \zeta_{p^s}^i) = (1 - \zeta_{p^s})^{\phi(p^s)} \prod_{\substack{1\leq i\leq p^s \\ \gcd(i,p)=1}} \frac{1 - \zeta_{p^s}^i}{1 - \zeta_{p^s}}.$$

For each $1 \leq i \leq p^s$ with $\gcd(i,p) = 1$, we have $(1 - \zeta_{p^s}^i)/(1 - \zeta_{p^s}) \in \mathfrak{o}_{\mathbb{Q}(p^s)}^\times$ since

$$\frac{1 - \zeta_{p^s}^i}{1 - \zeta_{p^s}} = 1 + \zeta_{p^s} + \cdots + \zeta_{p^s}^{i-1} \in \mathfrak{o}_{\mathbb{Q}(p^s)}$$

and

$$\frac{1 - \zeta_{p^s}}{1 - \zeta_{p^s}^i} = \frac{1 - \zeta_{p^s}^{ij}}{1 - \zeta_{p^s}^i} \in \mathfrak{o}_{\mathbb{Q}(p^s)},$$

where $ij \equiv 1 \pmod{p^s}$. Therefore

$$p\mathfrak{o}_{\mathbb{Q}(p^s)} = (1 - \zeta_{p^s})^{\phi(p^s)}\mathfrak{o}_{\mathbb{Q}(p^s)}. \tag{4.9}$$

Thus $e(\mathfrak{P}'|p) \geq \phi(p^s)$.

5° Now all the entries in Table 4.2 have been verified. By Remark 4.33, D and E are the decomposition group and the inertia group of $\mathfrak{P} \mid p$, respectively. □

Remark 4.41. Since $[\mathbb{Q}(p^s) : \mathbb{Q}] = \phi(p^s)$, (4.9) implies that $(1 - \zeta_{p^s})$ is the only prime of $\mathbb{Q}(p^s)$ lying above p and $e((1 - \zeta_{p^s})|p) = \phi(p^s)$.

Theorem 4.42. $\mathfrak{o}_{\mathbb{Q}(n)} = \mathbb{Z}[\zeta_n]$.

Theorem 4.42 will follow from a special case and a general fact.

Lemma 4.43. *Let p be a prime and $s > 0$. Then $\mathfrak{o}_{\mathbb{Q}(p^s)} = \mathbb{Z}[\zeta_{p^s}]$.*

Proof. We only have to show that $\mathfrak{o}_{\mathbb{Q}(p^s)} \subset \mathbb{Z}[\zeta_{p^s}]$. Let

$$\Delta = \Delta_{\mathbb{Q}(p^s)/\mathbb{Q}}((1-\zeta_{p^s})^0, \dots, (1-\zeta_{p^s})^{\phi(p^s)-1}).$$

Since $\mathbb{Z}[1-\zeta_{p^s}] = \mathbb{Z}[\zeta_{p^s}]$, we have $\Delta = \Delta_{\mathbb{Q}(p^s)/\mathbb{Q}}(\zeta_{p^s}^0, \dots, \zeta_{p^s}^{\phi(p^s)-1}) \mid p^{s\phi(p^s)}$ (by Fact 4.39). Let $w \in \mathfrak{o}_{\mathbb{Q}(p^s)}$ and write $w = a_0(1-\zeta_{p^s})^0 + \dots + a_{\phi(p^s)-1}(1-\zeta_{p^s})^{\phi(p^s)-1}$, $a_i \in \mathbb{Q}$. We want to show that $a_i \in \mathbb{Z}$ for all i. By Fact 4.40, $\Delta w \in \mathbb{Z}[1-\zeta_{p^s}]$. Thus $a_i = b_i/p^N$, $0 \le i \le \phi(p^s)-1$, where $N = s\phi(p^s)$ and $b_i \in \mathbb{Z}$. We have

$$\frac{1}{p^N}\left[b_0(1-\zeta_{p^s})^0 + \dots + b_{\phi(p^s)-1}(1-\zeta_{p^s})^{\phi(p^s)-1}\right] \in \mathfrak{o}_{\mathbb{Q}(p^s)}. \tag{4.10}$$

It remains to show that $p^N \mid b_i$ for all i. Assume the contrary. Let i $(0 \le i \le \phi(p^s)-1)$ be the smallest such that $p^N \nmid b_i$. Then from (4.10) we have

$$\frac{1}{p}\left[b_i'(1-\zeta_{p^s})^i + \dots + b_{\phi(p^s)-1}'(1-\zeta_{p^s})^{\phi(p^s)-1}\right] \in \mathfrak{o}_{\mathbb{Q}(p^s)}, \tag{4.11}$$

where $b_j' \in \mathbb{Z}$ $(i \le j \le \phi(p^s)-1)$ and $p \nmid b_i'$. There exists $a \in \mathbb{Z}$ such that $ab_i' \equiv 1 \pmod p$. Multiplying the element in (4.11) by a, we may assume that $b_i' = 1$. Thus, multiplying the element in (4.11) by $(1-\zeta_{p^s})^{\phi(p^s)-i-1}$ gives $\frac{1}{p}(1-\zeta_{p^s})^{\phi(p^s)-1} \in \mathfrak{o}_{\mathbb{Q}(p^s)}$. Then $(1-\zeta_{p^s})^{\phi(p^s)-1} \in p\mathfrak{o}_{\mathbb{Q}(p^s)} = (1-\zeta_{p^s})^{\phi(p^s)}\mathfrak{o}_{\mathbb{Q}(p^s)}$, which is impossible since $(1-\zeta_{p^s})$ is a prime of $\mathfrak{o}_{\mathbb{Q}(p^s)}$. □

Proposition 4.44. *Let F and K be number fields such that $[F:\mathbb{Q}] = m$, $[K:\mathbb{Q}] = n$ and $[FK:\mathbb{Q}] = mn$. Let $d = \gcd(\delta_F, \delta_K)$, where δ_F denotes the discriminant of F. Then $d\mathfrak{o}_{FK} \subset \mathfrak{o}_F\mathfrak{o}_K$.*

Proof. 1° Let $\mathrm{Emb}(F)$ be the set of all embeddings of F into $\mathbb{C}$. Define

$$\begin{array}{rccc} \psi: & \mathrm{Emb}(FK) & \longrightarrow & \mathrm{Emb}(F) \times \mathrm{Emb}(K) \\ & \sigma & \longmapsto & (\sigma|_F, \sigma|_K), \end{array}$$

which is clearly one-to-one. Since $|\mathrm{Emb}(FK)| = mn = |\mathrm{Emb}(F) \times \mathrm{Emb}(K)|$, ψ is onto. In particular, every $\tau \in \mathrm{Emb}(F)$ can be extended to some $\widetilde{\tau} \in \mathrm{Emb}(FK)$ such that $\widetilde{\tau}|_K = \mathrm{id}$.

2° Let $\alpha_1, \dots, \alpha_m$ be an integral basis of $\mathfrak{o}_F$ over $\mathbb{Z}$ and $\beta_1, \dots, \beta_n$ an integral basis of $\mathfrak{o}_K$ over $\mathbb{Z}$. Clearly, $\alpha_i\beta_j$, $1 \le i \le m$, $1 \le j \le n$, form a basis of FK over $\mathbb{Q}$. Let $\alpha = \sum_{i,j} a_{ij}\alpha_i\beta_j \in \mathfrak{o}_{FK}$, where $a_{ij} \in \mathbb{Q}$. It suffices to show that $\delta_F a_{ij} \in \mathbb{Z}$ for all i, j. (By symmetry, $\delta_K a_{ij} \in \mathbb{Z}$ and hence $da_{ij} \in \mathbb{Z}$ for all i, j.)

Let $\operatorname{Emb}(F) = \{\tau_1, \ldots, \tau_m\}$. We have

$$\widetilde{\tau_l}(\alpha) = \sum_{i,j} a_{ij}\tau_l(\alpha_i)\beta_j = \sum_i \tau_l(\alpha_i)x_i,$$

where $x_i = \sum_j a_{ij}\beta_j$. Let $A = (\tau_l(\alpha_i))_{l,i}$. Then

$$A\begin{bmatrix} x_1 \\ \vdots \\ x_n \end{bmatrix} = \begin{bmatrix} \widetilde{\tau_1}(\alpha) \\ \vdots \\ \widetilde{\tau_n}(\alpha) \end{bmatrix} \in \mathrm{M}_{n\times 1}(\mathbb{A}),$$

where $\mathbb{A} = \{x \in \mathbb{C} : x \text{ is integral over } \mathbb{Z}\}$. It follows that

$$\mathrm{M}_{n\times 1}(\mathbb{A}) \ni \operatorname{adj}(A)A\begin{bmatrix} x_1 \\ \vdots \\ x_n \end{bmatrix} = (\det A)\begin{bmatrix} x_1 \\ \vdots \\ x_n \end{bmatrix}.$$

Since $\delta_F = (\det A)^2$ we have $\delta_F x_i \in \mathbb{A}$. Thus

$$\sum_j \delta_F a_{ij}\beta_j = \delta_F x_i \in \mathbb{A} \cap K = \mathfrak{o}_K.$$

Since $\beta_1, \ldots, \beta_n$ is an integral basis of $\mathfrak{o}_K$ over $\mathbb{Z}$, we must have $\delta_F a_{ij} \in \mathbb{Z}$ for all i, j. □

Corollary 4.45. *In Proposition* 4.44, *if* $\gcd(\delta_F, \delta_K) = 1$, *then* $\mathfrak{o}_{FK} = \mathfrak{o}_F\mathfrak{o}_K$. *In particular, if* $\gcd(m, n) = 1$, *then* $\mathfrak{o}_{\mathbb{Q}(mn)} = \mathfrak{o}_{\mathbb{Q}(m)}\mathfrak{o}_{\mathbb{Q}(n)}$.

Proof. The first claim follows immediately from Proposition 4.44. For the second claim, note that $\phi(mn) = \phi(m)\phi(n)$ and that $\gcd(\delta_{\mathbb{Q}(m)}, \delta_{\mathbb{Q}(n)}) = 1$ since $\delta_{\mathbb{Q}(m)} \mid m^{\phi(m)}$ and $\delta_{\mathbb{Q}(n)} \mid n^{\phi(n)}$. Hence the first claim applies. □

Proof of Theorem 4.42. Write $n = p_1^{e_1} \cdots p_k^{e_k}$, where $p_1, \ldots, p_k$ are distinct primes and $e_i > 0$. By Corollary 4.45 and Lemma 4.43,

$$\mathfrak{o}_{\mathbb{Q}(n)} = \mathfrak{o}_{\mathbb{Q}(p_1^{e_1})} \cdots \mathfrak{o}_{\mathbb{Q}(p_k^{e_k})} = \mathbb{Z}[\zeta_{p_1^{e_1}}] \cdots \mathbb{Z}[\zeta_{p_k^{e_k}}] = \mathbb{Z}[\zeta_n].$$

□

4.6. Stickelberger's Congruence

Let p be a prime, $n > 0$ and $q = p^n$. Let $\mathfrak{p}$ be a prime of $\mathbb{Q}(q-1)$ lying above p. Then $f(\mathfrak{p}|p) = n$, so $\mathfrak{o}_{\mathbb{Q}(q-1)}/\mathfrak{p} = \mathbb{F}_q$. In $\mathfrak{o}_{\mathbb{Q}(q-1)}[\mathtt{X}]$, we have

$\mathtt{X}^q - \mathtt{X} = (\mathtt{X} - 0)(\mathtt{X} - \zeta_{q-1}^0) \cdots (\mathtt{X} - \zeta_{q-1}^{q-2})$. The image of this identity in $(\mathfrak{o}_{\mathbb{Q}(q-1)}/\mathfrak{p})[\mathtt{X}]$ gives

$$\mathfrak{o}_{\mathbb{Q}(q-1)}/\mathfrak{p} = \{0 + \mathfrak{p},\, \zeta_{q-1}^0 + \mathfrak{p},\, \ldots,\, \zeta_{q-1}^{q-2} + \mathfrak{p}\}.$$

Let $T = \{0, \zeta_{q-1}^0, \ldots, \zeta_{q-1}^{q-2}\} = \{0\} \cup \langle \zeta_{q-1} \rangle$. Define

$$\begin{aligned} \chi_{\mathfrak{p}} : \ \mathbb{F}_q = \mathfrak{o}_{\mathbb{Q}(q-1)}/\mathfrak{p} &\longrightarrow T \\ t + \mathfrak{p} &\longmapsto t, \quad t \in T. \end{aligned} \tag{4.12}$$

Clearly, $\chi_{\mathfrak{p}}$ is a multiplicative character of $\mathbb{F}_q$ of order $q-1$. Moreover,

$$\chi_{\mathfrak{p}}(x + \mathfrak{p}) \equiv x \pmod{\mathfrak{p}} \quad \text{for all } x \in \mathfrak{o}_{\mathbb{Q}(q-1)}.$$

(There exists $t \in T$ such that $x \equiv t \pmod{\mathfrak{p}}$. It follows that $\chi_{\mathfrak{p}}(x + \mathfrak{p}) = \chi_{\mathfrak{p}}(t + \mathfrak{p}) = t \equiv x \pmod{\mathfrak{p}}$.) There is a unique prime $\wp$ of $\mathfrak{o}_{\mathbb{Q}(p(q-1))}$ lying above $\mathfrak{p}$. Figure 4.1 depicts the relevant fields and primes to be used in the proof of Stickelberger's congruence. For any integer $a \geq 0$ with base-p expansion $a = a_0 + a_1 p + \cdots$, where $0 \leq a_i \leq p-1$, define $s(a) = a_0 + a_1 + \cdots$ and $\gamma(a) = a_0! a_1! \cdots$.

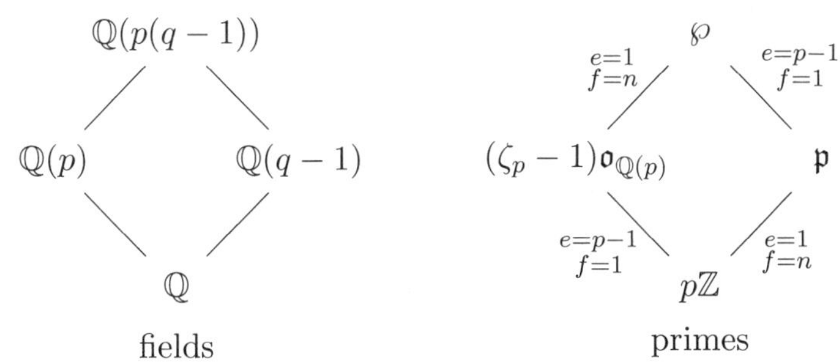

Figure 4.1. Fields and primes in Stickelberger's congruence

Theorem 4.46 (Stickelberger's congruence)**.** *For $1 \leq a \leq q-2$ we have*

$$\frac{g(\chi_{\mathfrak{p}}^{-a})}{(\zeta_p - 1)^{s(a)}} \equiv \frac{-1}{\gamma(a)} \pmod{\wp},$$

where $g(\chi_{\mathfrak{p}}^{-a})$ is the Gauss sum of $\chi_{\mathfrak{p}}^{-a}$.

Proof. We use induction on a. First assume that $a = 1$. We have

$$\begin{aligned} g(\chi_{\mathfrak{p}}^{-1}) &= \sum_{t\in\langle\zeta_{q-1}\rangle} \chi_{\mathfrak{p}}^{-1}(t+\mathfrak{p})\zeta_p^{\mathrm{Tr}_{q/p}(t+\mathfrak{p})} \\ &= \sum_{t\in\langle\zeta_{q-1}\rangle} \chi_{\mathfrak{p}}(t^{-1}+\mathfrak{p})\big(1+(\zeta_p-1)\big)^{\mathrm{Tr}_{q/p}(t+\mathfrak{p})} \\ &\equiv \sum_{t\in\langle\zeta_{q-1}\rangle} t^{-1}\big(1+(\zeta_p-1)\mathrm{Tr}_{q/p}(t+\mathfrak{p})\big) \pmod{(\zeta_p-1)^2} \\ &= (\zeta_p-1)\sum_{t\in\langle\zeta_{q-1}\rangle} t^{-1}\mathrm{Tr}_{q/p}(t+\mathfrak{p}) \\ &\equiv (\zeta_p-1)\sum_{t\in\langle\zeta_{q-1}\rangle} t^{-1}(t+t^p+\cdots+t^{p^{n-1}}) \pmod{(\zeta_p-1)\mathfrak{p}} \\ &= (\zeta_p-1)\sum_{t\in\langle\zeta_{q-1}\rangle} 1 = (q-1)(\zeta_p-1) \\ &\equiv -(\zeta_p-1) \pmod{(\zeta_p-1)^2}. \end{aligned}$$

Hence

$$\frac{g(\chi_{\mathfrak{p}}^{-1})}{\zeta_p-1} \equiv -1 \pmod{\wp}.$$

Now assume that $1 < a \le q-2$. If $p \mid a$, say $a = pb$, then $s(a) = s(b)$, $\gamma(a) = \gamma(b)$, and

$$g(\chi_{\mathfrak{p}}^{-a}) = \sum_{t\in\langle\zeta_{q-1}\rangle} \chi_{\mathfrak{p}}(t^{-pb})e_1(t) = \sum_{t\in\langle\zeta_{q-1}\rangle} \chi_{\mathfrak{p}}\big((t^p)^{-b}\big)e_1(t^p) = g(\chi_{\mathfrak{p}}^{-b}),$$

where $e_1(t) = \zeta_p^{\mathrm{Tr}_{q/p}(t)}$. Thus the induction hypothesis applies to $g(\chi_{\mathfrak{p}}^{-b})$. Therefore we may assume that $p \nmid a$. Let $a = a_0 + a_1p + \cdots + a_{n-1}p^{n-1}$ be the base-p expansion of a, where $a_0 > 0$. By Proposition 3.7 (ii),

$$\frac{g(\chi_{\mathfrak{p}}^{-a})}{(\zeta_p-1)^{s(a)}} = \frac{1}{(\zeta_p-1)^{s(a)}} \cdot \frac{g(\chi_{\mathfrak{p}}^{-1})g(\chi_{\mathfrak{p}}^{-(a-1)})}{J(\chi_{\mathfrak{p}}^{-1},\chi_{\mathfrak{p}}^{-(a-1)})},$$

where $J(\cdot\,,\cdot)$ is the Jacobi sum. In the above, by the induction hypothesis,

$$\frac{g(\chi_{\mathfrak{p}}^{-1})g(\chi_{\mathfrak{p}}^{-(a-1)})}{(\zeta_p-1)^{s(a)}} \equiv \frac{1}{(a_0-1)!a_1!\cdots a_{n-1}!} \pmod{\wp}.$$

Hence it suffices to prove that $J(\chi_{\mathfrak{p}}^{-1},\chi_{\mathfrak{p}}^{-(a-1)}) \equiv -a_0 \pmod{\mathfrak{p}}$. We have

$$\begin{aligned} &J(\chi_{\mathfrak{p}}^{-1},\chi_{\mathfrak{p}}^{-(a-1)}) \\ &= J(\chi_{\mathfrak{p}}^{-1},\chi_{\mathfrak{p}}^{q-a}) = \sum_{t\in\langle\zeta_{q-1}\rangle} \chi_{\mathfrak{p}}^{-1}(t+\mathfrak{p})\chi_{\mathfrak{p}}^{q-a}(1-t+\mathfrak{p}) \end{aligned}$$

$$\equiv \sum_{t\in\langle\zeta_{q-1}\rangle} t^{-1}(1-t)^{q-a} \pmod{\mathfrak{p}}$$

$$= \sum_{t\in\langle\zeta_{q-1}\rangle} t^{-1}\sum_{i=0}^{q-a}(-1)^i\binom{q-a}{i}t^i = \sum_{i=0}^{q-a}(-1)^i\binom{q-a}{i}\sum_{t\in\langle\zeta_{q-1}\rangle} t^{i-1}$$

$$= -\binom{q-a}{1}(q-1) \equiv -a \pmod{p}$$

$$\equiv -a_0 \pmod{p}.$$

□

Exercises

4.1. Let F be a number field. Prove that $\mathfrak{o}_F$ is a PID if and only if it is a UFD (unique factorization domain). (Note: The statement remains true when $\mathfrak{o}_F$ is replaced by a Dedekind domain.)

4.2. Let n be a positive integer and let A be a subgroup of $\mathbb{Z}^n$ such that $A \cong \mathbb{Z}^n$. Prove that $|\mathbb{Z}^n/A| < \infty$.

4.3. Let $F = \mathbb{Q}(\sqrt{d})$, where $d \in \mathbb{Z}\setminus\{0,1\}$ is square-free. Prove that

$$\mathfrak{o}_F = \begin{cases} \mathbb{Z}[\sqrt{d}] & \text{if } d \equiv 2,3 \pmod{4}, \\ \mathbb{Z}[(1+\sqrt{d})/2] & \text{if } d \equiv 1 \pmod{4}. \end{cases}$$

4.4. Let $d \in \mathbb{Z}\setminus\{0,\pm1\}$ be cube-free and let $F = \mathbb{Q}(\sqrt[3]{d})$.

(i) Let $\alpha = x + yd^{1/3} + zd^{2/3} \in F$, where $x, y, z \in \mathbb{Q}$. Prove that $\alpha \in \mathfrak{o}_F$ if and only if

$$\begin{cases} 3x, \\ 3x^2 - 3dyz, \\ x^3 + dy^3 - 3dxyz + d^2z^3 \end{cases}$$

are all integers.

(ii) Let $F = \mathbb{Q}(\sqrt[3]{2})$. Prove that $\mathfrak{o}_F = \mathbb{Z}[\sqrt[3]{2}]$.

(iii) Let $F = \mathbb{Q}(\sqrt[3]{2})$. Factor $2, 3, 5, 7, 11$, and 13 in $\mathfrak{o}_F$.

4.5. Let $F \subset L \subset K$ be number fields such that K/F is Galois. Let $\mathfrak{p} \in \mathrm{Spec}^*(\mathfrak{o}_F)$ and $\mathfrak{P} \in \mathrm{Spec}^*(\mathfrak{o}_K)$ be such that $\mathfrak{P} \mid \mathfrak{p}$, and let $\mathfrak{P}_L = \mathfrak{P}\cap L$. Let D' and E' be the decomposition field and the inertia field of $\mathfrak{P} \mid \mathfrak{p}$, respectively. Prove the following statements.

(i) $L \subset D'$ if and only if $e(\mathfrak{P}_L|\mathfrak{p}) = f(\mathfrak{P}_L|\mathfrak{p}) = 1$.

(ii) $L \supset D'$ if and only if $\mathfrak{P}$ is the only prime of K lying above $\mathfrak{P}_L$.

(iii) $L \subset E'$ if and only if $e(\mathfrak{P}_L|\mathfrak{p}) = 1$.

(iv) $L \supset E'$ if and only if $\mathfrak{P}_L$ is *totally ramified* in K, i.e., $\mathfrak{P}$ is the only prime of K lying above $\mathfrak{P}_L$ and $f(\mathfrak{P}|\mathfrak{P}_L) = 1$.

4.6. Let F be a number field and $\mathfrak{a}$ a nonzero ideal of $\mathfrak{o}_F$. The *norm* of $\mathfrak{a}$ is $\mathrm{N}(\mathfrak{a}) = |\mathfrak{o}_F/\mathfrak{a}|$, which is a positive integer. Prove the following statements.

(i) If $\mathfrak{a}$ and $\mathfrak{b}$ are nonzero ideals of $\mathfrak{o}_F$, then $\mathrm{N}(\mathfrak{ab}) = \mathrm{N}(\mathfrak{a})\mathrm{N}(\mathfrak{b})$.

(ii) If $0 \neq \alpha \in \mathfrak{o}_F$, then $\mathrm{N}(\alpha\mathfrak{o}_F) = |\mathrm{N}_{F/\mathbb{Q}}(\alpha)|$.

4.7. (Stickelberger) Let F be a number field of degree n over $\mathbb{Q}$ and let $a_1, \ldots, a_n \in \mathfrak{o}_F$. Prove that

$$\Delta_{F/\mathbb{Q}}(a_1, \ldots, a_n) \equiv 0 \text{ or } 1 \pmod 4.$$

Exercises 4.8–4.10 establish the following fact: A prime number p ramifies in a number field F, that is, there exists $\mathfrak{p} \in \mathrm{Spec}^*(\mathfrak{o}_F)$ such that $e(\mathfrak{p}|p) > 1$, if and only if $p \mid \delta_F$.

4.8. Let R be a commutative ring. For $a, b \in R$, define $a \sim b$ if $a = c^2 b$ for some $c \in R^\times$. Let $\pi_R : R \to R/\!\sim$ be the quotient map. Let S be an R-algebra which is a free R-module of finite rank with an R-basis $\epsilon_1, \ldots, \epsilon_n$. For $a \in S$, define

$$\mathrm{Tr}_{S/R}(a) = \mathrm{Tr}(A),$$

where $A \in \mathrm{M}_{n\times n}(R)$ is defined by

$$a \begin{bmatrix} \epsilon_1 \\ \vdots \\ \epsilon_n \end{bmatrix} = A \begin{bmatrix} \epsilon_1 \\ \vdots \\ \epsilon_n \end{bmatrix}.$$

Define the *discriminant* of S over R as

$$\delta_{S/R} = \pi_R\big(\det(\mathrm{Tr}_{S/R}(\epsilon_i\epsilon_j))\big).$$

(i) Prove that $\delta_{S/R}$ is independent of the choice of the basis $\epsilon_1, \ldots, \epsilon_n$ of S over R.

(ii) If S and T are two R-algebras which are free R-modules of finite rank, prove that

$$\delta_{(S\times T)/R} = \delta_{S/R}\,\delta_{T/R},$$

where the product in $R/\!\sim$ is defined in the obvious way.

4.9. Let p be a prime number and let F be a number field. Prove that in $\mathbb{F}_p/\!\sim$,

$$\overline{\delta_F} = \delta_{(\mathfrak{o}_F/p\mathfrak{o}_F)/\mathbb{F}_p},$$

where $\overline{(\ )} : \mathbb{Z} \to \mathbb{Z}/p\mathbb{Z} = \mathbb{F}_p$ is the reduction.

4.10. In Exercise 4.9, write

$$p\mathfrak{o}_F = \mathfrak{p}_1^{e_1} \cdots \mathfrak{p}_k^{e_k},$$

where $\mathfrak{p}_1, \ldots, \mathfrak{p}_k \in \operatorname{Spec}^*(\mathfrak{o}_F)$ are distinct and $e_1, \ldots, e_k > 0$. Prove that $p \mid \delta_F$ if and only if $e_i > 0$ for some $1 \leq i \leq k$.

4.11. (Stickelberger's congruence by the p-adic method) Let $q = p^n$. Let $\mathfrak{p}$ be a prime of $\mathbb{Q}(q-1)$ lying above p and let $\wp$ be the unique prime of $\mathbb{Q}(p(q-1))$ lying above $\mathfrak{p}$. We consider the Gauss sum $g(\chi_{\mathfrak{p}}^{-a})$, $1 \leq a \leq q-2$, in the completion of $\mathfrak{o}_{\mathbb{Q}(p(q-1))}$ at $\wp$.

(i) Show that

$$g(\chi_{\mathfrak{p}}^{-a}) = \sum_{i_0,\ldots,i_{n-1}} (\zeta_p - 1)^{i_0+\cdots+i_{n-1}} \sum_{t \in \langle \zeta_{q-1} \rangle} \binom{t}{i_0}\binom{t^p}{i_1} \cdots \binom{t^{p^{n-1}}}{i_{n-1}} t^{-a}.$$

(ii) Let $a = a_0 + a_1 p + \cdots + a_{n-1}p^{n-1}$, where $0 \leq a_i \leq p-1$, and let $s(a) = a_0 + \cdots + a_{n-1}$ and $\gamma(a) = a_0! \cdots a_{n-1}!$. Prove that

$$\frac{g(\chi_{\mathfrak{p}}^{-a})}{(\zeta_p - 1)^{s(a)}} \equiv \frac{-1}{\gamma(a)} \pmod{\wp}.$$

Exercises 4.12–4.14 lead to an explicit formula for the discriminant of the cyclotomic field.

4.12. Let p be a prime and e a positive integer such that $p^e > 2$. Prove that

$$\delta_{\mathbb{Q}(p^e)} = (-1)^{\phi(p^e)/2} p^{p^{e-1}(pe-e-1)}.$$

(Hint: Use Lemma 4.38.)

4.13. Let m and n be positive integers such that $\gcd(m,n) = 1$. By Corollary 4.45, $\{\zeta_m^i \zeta_n^j : 0 \leq i \leq \phi(m)-1,\ 0 \leq j \leq \phi(n)-1\}$ is an integral basis of $\mathfrak{o}_{\mathbb{Q}(mn)}$. Prove that

$$\begin{aligned}&\left(\operatorname{Tr}_{\mathbb{Q}(mn)/\mathbb{Q}}(\zeta_m^i \zeta_n^j \zeta_m^{i'} \zeta_n^{j'})\right)_{(i,j),\,(i',j')} \\ &= \left(\operatorname{Tr}_{\mathbb{Q}(m)/\mathbb{Q}}(\zeta_m^i \zeta_m^{i'})\right)_{i,i'} \otimes \left(\operatorname{Tr}_{\mathbb{Q}(n)/\mathbb{Q}}(\zeta_n^j \zeta_n^{j'})\right)_{j,j'}.\end{aligned}$$

Conclude that

$$\delta_{\mathbb{Q}(mn)} = \delta_{\mathbb{Q}(m)}^{\phi(n)} \delta_{\mathbb{Q}(n)}^{\phi(m)}.$$

4.14. Use Exercises 4.12 and 4.13 to show that for all $n > 2$,

$$\delta_{\mathbb{Q}(n)} = (-1)^{\phi(n)/2} \frac{n^{\phi(n)}}{\prod_{p \mid n} p^{\phi(n)/(p-1)}},$$

where the product is over all prime divisors p of n.

4.15. The sign of the Gauss quadratic sum can be determined using Stickelberger's congruence. Let p be an odd prime and η the quadratic character of $\mathbb{F}_p$.

(i) Let $\wp$ be a prime of $\mathfrak{o}_{\mathbb{Q}(p(p-1))}$ lying above p. Use Stickelberger's congruence to show that

$$g(\eta) \equiv \prod_{\substack{1 \le k \le p-1 \\ k \text{ odd}}} (\zeta_p^k - \zeta_p^{-k}) \pmod{\wp^{(p+1)/2}}.$$

(ii) Show that

$$\prod_{\substack{1 \le k \le p-1 \\ k \text{ odd}}} (\zeta_p^k - \zeta_p^{-k})^2 = (-1)^{(p-1)/2} p,$$

i.e.,

$$\prod_{\substack{1 \le k \le p-1 \\ k \text{ odd}}} (\zeta_p^k - \zeta_p^{-k}) = \begin{cases} \pm p^{1/2} & \text{if } p \equiv 1 \pmod 4, \\ \pm i p^{1/2} & \text{if } p \equiv -1 \pmod 4. \end{cases}$$

(iii) Show that

$$\prod_{\substack{1 \le k \le p-1 \\ k \text{ odd}}} (\zeta_p^k - \zeta_p^{-k}) = i^{(p-1)/2} \prod_{k=1}^{(p-1)/2} 2 \sin \frac{(4k-2)\pi}{p}$$

$$= \begin{cases} A & \text{if } p \equiv 1 \pmod 4, \\ iB & \text{if } p \equiv -1 \pmod 4, \end{cases}$$

where $A, B > 0$.

(iv) Conclude from (i)–(iii) that

$$g(\eta) \equiv \begin{cases} p^{1/2} \pmod{\wp^{(p+1)/2}} & \text{if } p \equiv 1 \pmod 4, \\ i p^{1/2} \pmod{\wp^{(p+1)/2}} & \text{if } p \equiv -1 \pmod 4. \end{cases}$$

The purpose of Exercises 4.16 and 4.17 is to derive the Stickelberger relation from Stickelberger's congruence.

4.16. Let $q = p^f$, where p is a prime and f is a positive integer. For $0 \le a < q - 1$, prove that

$$s(a) = (p-1) \sum_{j=0}^{f-1} \left\{ \frac{p^j a}{q-1} \right\},$$

where $s(a)$ is the sum of the base-p digits of a and $\{x\} = x - \lfloor x \rfloor$ for $x \in \mathbb{R}$.

4.17. Let $m > 1$ be an integer and let $\mathfrak{p} \in \operatorname{Spec}^*(\mathfrak{o}_{\mathbb{Q}(m)})$ be such that $m \notin \mathfrak{p}$. Let $\mathfrak{p} \cap \mathbb{Z} = p\mathbb{Z}$ and $q = p^f$, where f is the order of p in $(\mathbb{Z}/m\mathbb{Z})^\times$. Let $\mathfrak{P} \in \operatorname{Spec}^*(\mathfrak{o}_{\mathbb{Q}(q-1)})$ be such that $\mathfrak{P} \mid \mathfrak{p}$.

$$\begin{array}{cc} \mathbb{Q}(q-1) & \mathfrak{P} \\ | & | \\ \mathbb{Q}(m) & \mathfrak{p} \\ | & | \\ \mathbb{Q} & p\mathbb{Z} \end{array}$$

(i) Use Theorem 4.36 to show that $f(\mathfrak{p}|p) = f$ and $f(\mathfrak{P}|\mathfrak{p}) = 1$.

(ii) Let $\chi_{\mathfrak{P}}$ be the multiplicative character of $\mathbb{F}_q = \mathfrak{o}_{\mathbb{Q}(q-1)}/\mathfrak{P}$ defined in (4.12). Prove that

$$g(\chi_{\mathfrak{P}}^{-(q-1)/m})^m \in \mathfrak{o}_{\mathbb{Q}(m)}.$$

(iii) For each $1 \le t < m$ with $\gcd(t, m) = 1$, let $\sigma_t \in \operatorname{Aut}(\mathbb{Q}(m)/\mathbb{Q})$ be such that $\sigma_t(\zeta_m) = \zeta_m^t$. Prove that

$$\nu_{\sigma_t^{-1}\mathfrak{p}}\left(g(\chi_{\mathfrak{P}}^{-(q-1)/m})^m\right) = \frac{m}{p-1}\, s\left(t\frac{q-1}{m}\right).$$

(Hint: Use Stickelberger's congruence.)

(iv) Let $G = \operatorname{Aut}(\mathbb{Q}(m)/\mathbb{Q})$ and let $I_{\mathbb{Q}(m)}$ be the multiplicative group of all fractional ideals of $\mathbb{Q}(m)$. For $\mathfrak{a} \in I_{\mathbb{Q}(m)}$ and $\sum_{\sigma\in G} a_\sigma \sigma \in \mathbb{Z}[G]$, define

$$\Big(\sum_{\sigma\in G} a_\sigma \sigma\Big)\mathfrak{a} = \prod_{\sigma\in G} (\sigma\mathfrak{a})^{a_\sigma}.$$

This makes $I_{\mathbb{Q}(m)}$ a $\mathbb{Z}[G]$-module. Prove the *Stickelberger relation*

$$\Bigg(\sum_{\substack{1\le t<m\\ \gcd(t,m)=1}} t\sigma_t^{-1}\Bigg)\mathfrak{p} = g(\chi_{\mathfrak{P}}^{-(q-1)/m})^m \mathfrak{o}_{\mathbb{Q}(m)}.$$

(Hint: First note that since $g(\chi_{\mathfrak{P}}^{-(q-1)/m})\overline{g(\chi_{\mathfrak{P}}^{-(q-1)/m})} = q$, all primes of $\mathbb{Q}(m)$ dividing $g(\chi_{\mathfrak{P}}^{-(q-1)/m})^m \mathfrak{o}_{\mathbb{Q}(m)}$ lie above p. Then use (iii) and Exercise 4.16.)

Chapter 5

Zeros of Polynomials over Finite Fields

5.1. Ax's Theorem

Let $q = p^l$, where p is a prime and l is a positive integer. For each $f \in \mathbb{F}_q[\mathtt{X}_1, \dots, \mathtt{X}_n]$, let

$$Z(f) = \{(x_1, \dots, x_n) \in \mathbb{F}_q^n : f(x_1, \dots, x_n) = 0\}.$$

In 1935, Artin conjectured that if $f \in \mathbb{F}_q[\mathtt{X}_1, \dots, \mathtt{X}_n]$ is homogeneous with $0 < \deg f < n$, then $|Z(f)| > 1$. Almost immediately, Chevalley confirmed Artin's conjecture by proving the following result in [**7**]: If $f_1, \dots, f_r \in \mathbb{F}_q[\mathtt{X}_1, \dots, \mathtt{X}_n]$ are such that $\sum_{i=1}^r \deg f_i < n$, then $|Z(f_1) \cap \dots \cap Z(f_r)| > 0$ implies $|Z(f_1) \cap \dots \cap Z(f_r)| > 1$. Soon thereafter, Warning [**39**] strengthened the conclusion to $|Z(f_1) \cap \dots \cap Z(f_r)| \equiv 0 \pmod{p}$ under the same assumption. In 1964, Ax [**2**] gave a lower bound for the p-adic order of $|Z(f)|$ which is a much stronger version of Warning's result for $r = 1$. In 1971, Katz [**22**] generalized Ax's theorem to an arbitrary number of polynomials.

The current section is devoted to Ax's theorem. We follow the notation of section 4.6: $\mathfrak{p}$ is a prime of $\mathbb{Q}(q-1)$ lying above $p\mathbb{Z}$ and $\wp$ is the unique prime of $\mathbb{Q}(p(q-1))$ lying above $\mathfrak{p}$ (see Figure 4.1), $\mathbb{F}_q = \mathfrak{o}_{\mathbb{Q}(q-1)}/\mathfrak{p}$, $T = \{0\} \cup \langle \zeta_{q-1} \rangle$, and $\chi_{\mathfrak{p}} \in \widehat{\mathbb{F}_q^*}$ is defined by $\chi_{\mathfrak{p}}(t + \mathfrak{p}) = t$ for $t \in T$. For an integer i, $0 \le i \le q-1$, with base-p expansion $i = i^{(0)} + i^{(1)}p + \dots + i^{(l-1)}p^{l-1}$, where $0 \le i^{(j)} \le p-1$, recall that $s(i) = i^{(0)} + i^{(1)} + \dots + i^{(l-1)}$. We define

$$\tau(i) = i^{(l-1)} + i^{(0)}p + \dots + i^{(l-2)}p^{l-1}.$$

(The action of τ on i is to cyclically shift its base-p digits.) Clearly $\tau(i) \equiv pi \pmod{q-1}$ and

$$\sum_{j=0}^{l-1} \tau^j(i) = \frac{q-1}{p-1} s(i). \tag{5.1}$$

Lemma 5.1. *We have*

$$\zeta_p^{\mathrm{Tr}_{q/p}(t+\mathfrak{p})} = \sum_{i=0}^{q-1} c_i t^i, \qquad t \in T, \tag{5.2}$$

where

$$c_i = \begin{cases} 1 & \text{if } i = 0, \\ -\dfrac{q}{q-1} & \text{if } i = q-1, \\ \dfrac{1}{q-1} g(\chi_{\mathfrak{p}}^{-i}) & \text{if } 0 < i < q-1. \end{cases} \tag{5.3}$$

In particular,

$$\nu_{\wp}(c_i) = s(i), \quad 0 \le i \le q-1. \tag{5.4}$$

Proof. Let $\mathbb{C}^{\mathbb{F}_q^*}$ be the unitary space of all functions from $\mathbb{F}_q^*$ to $\mathbb{C}$ equipped with the inner product $\langle f, g\rangle = \frac{1}{q-1}\sum_{x\in\mathbb{F}_q^*} f(x)\overline{g(x)}$. Then $\{\chi_{\mathfrak{p}}^i|_{\mathbb{F}_q^*} : 0 \le i \le q-2\}$ is an orthonormal basis of $\mathbb{C}^{\mathbb{F}_q^*}$. Denote by $\zeta_p^{\mathrm{Tr}_{q/p}(\;)}$ the function $\mathbb{F}_q \to \mathbb{C}$, $x \mapsto \zeta_p^{\mathrm{Tr}_{q/p}(x)}$. Then

$$\begin{aligned} \zeta_p^{\mathrm{Tr}_{q/p}(\;)}|_{\mathbb{F}_q^*} &= \sum_{i=0}^{q-2} \langle \zeta_p^{\mathrm{Tr}_{q/p}(\;)}|_{\mathbb{F}_q^*}, \chi_{\mathfrak{p}}^i|_{\mathbb{F}_q^*}\rangle\, \chi_{\mathfrak{p}}^i|_{\mathbb{F}_q^*} \\ &= -\frac{1}{q-1}\chi_{\mathfrak{p}}^0|_{\mathbb{F}_q^*} + \frac{1}{q-1}\sum_{i=1}^{q-2} g(\chi_{\mathfrak{p}}^{-i})\chi_{\mathfrak{p}}^i|_{\mathbb{F}_q^*}. \end{aligned}$$

Hence for $t \in \langle \zeta_{q-1}\rangle$,

$$\begin{aligned} \zeta_p^{\mathrm{Tr}_{q/p}(t+\mathfrak{p})} &= -\frac{1}{q-1} + \frac{1}{q-1}\sum_{i=1}^{q-2} g(\chi_{\mathfrak{p}}^{-i})t^i \\ &= 1 + \frac{1}{q-1}\sum_{i=1}^{q-2} g(\chi_{\mathfrak{p}}^{-i})t^i - \frac{q}{q-1}t^{q-1} = \sum_{i=0}^{q-1} c_i t^i, \end{aligned}$$

where c_i is given by (5.3). Clearly, (5.2) also holds for $t = 0$.

For $0 < i < q-1$, (5.4) follows from (5.3) and Stickelberger's congruence. For $i = 0$ or $q-1$, (5.4) is obviously true. □

For $\boldsymbol{u} = (u_1, \dots, u_n) \in \mathbb{N}^n$, let $|\boldsymbol{u}| = u_1 + \cdots + u_n$. If $\boldsymbol{x} = (x_1, \dots, x_n)$ is an n-tuple of elements from a commutative ring, we define $\boldsymbol{x}^{\boldsymbol{u}} = x_1^{u_1} \cdots x_n^{u_n}$.

Theorem 5.2 (Ax [2]). *Let $f \in \mathbb{F}_q[\mathtt{X}_1, \dots, \mathtt{X}_n]$ be such that* $\deg f = d > 0$. *Then*

$$|Z(f)| \equiv 0 \pmod{q^{\lceil \frac{n}{d} \rceil - 1}}.$$

Proof. Let $\mathtt{X} = (\mathtt{X}_1, \dots, \mathtt{X}_n)$ and write

$$f = \sum_{j=1}^{m} a_j \mathtt{X}^{\boldsymbol{u}_j},$$

where $a_j \in \mathbb{F}_q$ and $\boldsymbol{u}_j \in \mathbb{N}^n$ with $|\boldsymbol{u}_j| \le d$. Since $\mathbb{F}_q = \mathfrak{o}_{\mathbb{Q}(q-1)}/\mathfrak{p}$, we have $a_j = \alpha_j + \mathfrak{p}$ for some $\alpha_j \in T$. We have

(5.5)

$$\begin{aligned}
q|Z(f)| &= \sum_{x_0 \in \mathbb{F}_q} \sum_{\boldsymbol{x} \in \mathbb{F}_q^n} \zeta_p^{\mathrm{Tr}_{q/p}(x_0 f(\boldsymbol{x}))} = \sum_{(x_0, \boldsymbol{x}) \in \mathbb{F}_q^{n+1}} \zeta_p^{\mathrm{Tr}_{q/p}(x_0 \sum_{j=1}^m a_j \boldsymbol{x}^{\boldsymbol{u}_j})} \\
&= \sum_{\boldsymbol{z} \in \mathbb{F}_q^{n+1}} \prod_{j=1}^{m} \zeta_p^{\mathrm{Tr}_{q/p}(a_j \boldsymbol{z}^{(1, \boldsymbol{u}_j)})} \qquad (\boldsymbol{z} = (x_0, \boldsymbol{x})) \\
&= \sum_{\boldsymbol{t} \in T^{n+1}} \prod_{j=1}^{m} \Big(\sum_{i=0}^{q-1} c_i \alpha_j^i \boldsymbol{t}^{i(1, \boldsymbol{u}_j)} \Big) \qquad \text{(by (5.2))} \\
&= \sum_{\boldsymbol{t} \in T^{n+1}} \sum_{0 \le i_1, \dots, i_m \le q-1} c_{i_1} \cdots c_{i_m} \alpha_1^{i_1} \cdots \alpha_m^{i_m} \boldsymbol{t}^{i_1(1, \boldsymbol{u}_1) + \cdots + i_m(1, \boldsymbol{u}_m)} \\
&= \sum_{0 \le i_1, \dots, i_m \le q-1} \alpha_1^{i_1} \cdots \alpha_m^{i_m} c_{i_1} \cdots c_{i_m} \sum_{\boldsymbol{t} \in T^{n+1}} \boldsymbol{t}^{i_1(1, \boldsymbol{u}_1) + \cdots + i_m(1, \boldsymbol{u}_m)}.
\end{aligned}$$

We claim that for all $(i_1, \dots, i_m) \in \{0, \dots, q-1\}^m$,

$$\nu_{\wp}\Big(c_{i_1} \cdots c_{i_m} \sum_{\boldsymbol{t} \in T^{n+1}} \boldsymbol{t}^{i_1(1, \boldsymbol{u}_1) + \cdots + i_m(1, \boldsymbol{u}_m)} \Big) \ge l(p-1) \Big\lceil \frac{n}{d} \Big\rceil. \tag{5.6}$$

Note that the conclusion of the theorem follows from (5.5) and (5.6). In fact, by (5.5) and (5.6),

$$\nu_{\wp}\big(|Z(f)|\big) + l(p-1) = \nu_{\wp}\big(q|Z(f)|\big) \ge l(p-1) \Big\lceil \frac{n}{d} \Big\rceil.$$

Thus $\nu_{\wp}(|Z(f)|) \ge l(p-1)(\lceil n/d \rceil - 1)$, which is the theorem.

To prove (5.6), first note that for $i \geq 0$,

$$\sum_{t\in T} t^i = \begin{cases} q & \text{if } i = 0, \\ q-1 & \text{if } i \neq 0 \text{ and } i \equiv 0 \pmod{q-1}, \\ 0 & \text{if } i \not\equiv 0 \pmod{q-1}. \end{cases} \tag{5.7}$$

We now proceed with three cases.

Case 1. Assume that $i_1(1, \boldsymbol{u}_1)+\cdots+i_m(1, \boldsymbol{u}_m) \not\equiv (0, \ldots, 0) \pmod{q-1}$. By (5.7),

$$\sum_{\boldsymbol{t}\in T^{n+1}} \boldsymbol{t}^{i_1(1,\boldsymbol{u}_1)+\cdots+i_m(1,\boldsymbol{u}_m)} = 0. \tag{5.8}$$

Case 2. Assume that $(i_1, \ldots, i_m) = (0, \ldots, 0)$. Then

$$\sum_{\boldsymbol{t}\in T^{n+1}} \boldsymbol{t}^{i_1(1,\boldsymbol{u}_1)+\cdots+i_m(1,\boldsymbol{u}_m)} = q^{n+1},$$

and the left side of (5.6) is $\geq \nu_\wp(q^{n+1}) = l(p-1)(n+1) > l(p-1)\lceil n/d \rceil$.

Case 3. Assume that $i_1(1, \boldsymbol{u}_1)+\cdots+i_m(1, \boldsymbol{u}_m) \equiv (0, \ldots, 0) \pmod{q-1}$ but $(i_1, \ldots, i_m) \neq (0, \ldots, 0)$. Then

$$\sum_{\boldsymbol{t}\in T^{n+1}} \boldsymbol{t}^{i_1(1,\boldsymbol{u}_1)+\cdots+i_m(1,\boldsymbol{u}_m)} = (q-1)^{k+1} q^{n-k}, \tag{5.9}$$

where k is the number of nonzero components of $i_1\boldsymbol{u}_1+\cdots+i_m\boldsymbol{u}_m$. Clearly,

$$(i_1 + \cdots + i_m)d \geq i_1|\boldsymbol{u}_1| + \cdots + i_m|\boldsymbol{u}_m| \geq k(q-1).$$

Since $i_1 + \cdots + i_m \equiv 0 \pmod{q-1}$, we get

$$i_1 + \cdots + i_m \geq (q-1)\left\lceil \frac{k}{d} \right\rceil. \tag{5.10}$$

Note that $\tau(i_1)(1, \boldsymbol{u}_1)+\cdots+\tau(i_m)(1, \boldsymbol{u}_m) \equiv p(i_1(1, \boldsymbol{u}_1)+\cdots+i_m(1, \boldsymbol{u}_m)) \equiv (0, \ldots, 0) \pmod{q-1}$ since $\tau(i_j) \equiv pi_j \pmod{q-1}$. The number of nonzero components of $\tau(i_1)\boldsymbol{u}_1 + \cdots + \tau(i_m)\boldsymbol{u}_m$ is also k since $i_j > 0$ if and only if $\tau(i_j) > 0$. Hence (5.10) holds with $(i_1, \ldots, i_m)$ replaced by $(\tau(i_1), \ldots, \tau(i_m))$. Thus

$$\tau^h(i_1) + \cdots + \tau^h(i_m) \geq (q-1)\left\lceil \frac{k}{d} \right\rceil, \qquad 0 \leq h \leq l-1. \tag{5.11}$$

By (5.11) and (5.1),

$$l(q-1)\left\lceil \frac{k}{d} \right\rceil \leq \sum_{h=0}^{l-1}\sum_{j=1}^{m} \tau^h(i_j) = \frac{q-1}{p-1}\sum_{j=1}^{m} s(i_j),$$

i.e.,

$$\sum_{j=1}^{m} s(i_j) \geq l(p-1)\left\lceil \frac{k}{d} \right\rceil.$$

Therefore,

$$\begin{aligned}\nu_{\wp}\big(c_{i_1}\cdots c_{i_m}(q-1)^{k+1}q^{n-k}\big) &= s(i_1)+\cdots+s(i_m)+l(p-1)(n-k)\\ &\geq l(p-1)\Big(\left\lceil \frac{k}{d}\right\rceil+n-k\Big) \geq l(p-1)\left\lceil \frac{n}{d}\right\rceil,\end{aligned}$$

which is (5.6). (In the last step, we used the fact that $\min\{\lceil k/d\rceil + n - k : 0 \leq k \leq n\} = \lceil n/d \rceil$.) □

Corollary 5.3. *Let $f \in \mathbb{F}_q[\mathtt{X}_1, \dots, \mathtt{X}_n]$ be such that $\deg f = d > 0$. Then either $|Z(f)| = 0$ or $|Z(f)| \geq q^{\lceil n/d \rceil - 1}$.*

Ax's theorem is equivalent to the inequality

$$\nu_p\big(|Z(f)|\big) \geq l\Big(\left\lceil \frac{n}{d}\right\rceil - 1\Big). \tag{5.12}$$

This lower bound is best possible as shown in Example 5.5.

Lemma 5.4. *Let $I_1, \dots, I_k$ be a partition of $\{1, \dots, n\}$, where $I_i \neq \emptyset$, $1 \leq i \leq n$. Let*

$$f = \sum_{i=1}^{k} \prod_{j \in I_i} \mathtt{X}_j \in \mathbb{F}_q[\mathtt{X}_1, \dots, \mathtt{X}_n].$$

Then

$$|Z(f)| = q^{n-1} + q^{k-1}(q-1)\prod_{i=1}^{k}\big(q^{|I_i|-1} - (q-1)^{|I_i|-1}\big).$$

In particular,

$$\nu_p\big(|Z(f)|\big) = l(k-1).$$

Proof. We have

$$\begin{aligned} q|Z(f)| &= \sum_{x_0 \in \mathbb{F}_q} \sum_{(x_1,\dots,x_n)\in\mathbb{F}_q^n} \zeta_p^{\mathrm{Tr}_{q/p}(x_0 f(x_1,\dots,x_n))}\\ &= q^n + \sum_{x_0\in\mathbb{F}_q^*}\sum_{(x_1,\dots,x_n)\in\mathbb{F}_q^n} \zeta_p^{\sum_{i=1}^k \mathrm{Tr}_{q/p}(x_0\prod_{j\in I_i} x_j)}\\ &= q^n + (q-1)\sum_{(x_1,\dots,x_n)\in\mathbb{F}_q^n}\prod_{i=1}^{k}\zeta_p^{\mathrm{Tr}_{q/p}(\prod_{j\in I_i} x_j)}\\ &= q^n + (q-1)\prod_{i=1}^{k}\Big(\sum_{x_j\in\mathbb{F}_q,\, j\in I_i}\zeta_p^{\mathrm{Tr}_{q/p}(\prod_{j\in I_i} x_j)}\Big)\end{aligned}$$

$$= q^n + (q-1)\prod_{i=1}^{k} q\big(q^{|I_i|-1} - (q-1)^{|I_i|-1}\big).$$

In the last step, we used the fact that

$$\sum_{(x_1,\dots,x_t)\in\mathbb{F}_q^t} \zeta_p^{\mathrm{Tr}_{q/p}(x_1\cdots x_t)} = q\big|\{(x_1,\dots,x_{t-1})\in\mathbb{F}_q^{t-1} : x_1\cdots x_{t-1}=0\}\big|$$
$$= q\big(q^{t-1} - (q-1)^{t-1}\big).$$

Thus the lemma is proved. □

Example 5.5. For each $0 < d \le n$, we choose a partition $I_1,\dots,I_{\lceil n/d\rceil}$ of $\{1,\dots,n\}$ such that $0 < |I_i| \le d$ for all i and $|I_1| = d$. Let $f = \sum_{i=1}^{\lceil n/d\rceil}\prod_{j\in I_i}\mathtt{X}_j \in \mathbb{F}_q[\mathtt{X}_1,\dots,\mathtt{X}_n]$. Then $\deg f = d$ and, by Lemma 5.4, $\nu_p(|Z(f)|) = l(\lceil n/d\rceil - 1)$. Hence the lower bound in (5.12) is attained.

If $d > n$, let $f = \mathtt{X}_1^{d-n+1}\mathtt{X}_2\cdots\mathtt{X}_n \in \mathbb{F}_q[\mathtt{X}_1,\dots,\mathtt{X}_n]$. Then $\deg f = d$, $|Z(f)| = q^n - (q-1)^n$, and hence the lower bound in (5.12) is also attained.

5.2. Katz's Theorem

In this section, we prove a generalization of Ax's theorem due to Katz [**22**]. It is a lower bound for the p-adic order of the number of common zeros of several polynomials in $\mathbb{F}_q[\mathtt{X}_1,\dots,\mathtt{X}_n]$. Katz's original proof uses sophisticated methods. Two simpler proofs, one based on Ax's method and the other using p-adic fields, were later found by Wan in [**37, 38**]. The proof given here is based on [**16**] by the author. For $a, b \in \mathbb{Q}$ and $k \in \mathbb{Z}$, congruence $a \equiv b \pmod{q^k}$ means that $\nu_p(a-b) \ge \nu_p(q^k)$.

Theorem 5.6 (Katz). *Let $f_i \in \mathbb{F}_q[\mathtt{X}_1,\dots,\mathtt{X}_n]$ be such that* $\deg f_i = d_i$*, $1 \le i \le r$, where $d_1 \ge \cdots \ge d_r \ge 1$. Then*

$$|Z(f_1)\cap\cdots\cap Z(f_r)| \equiv 0 \pmod{q^{\lceil \frac{n-d_1-\cdots-d_r}{d_1}\rceil}}. \tag{5.13}$$

Lemma 5.7. *Let $f_i \in \mathbb{F}_q[\mathtt{X}_1,\dots,\mathtt{X}_n]$, $1 \le i \le r$. Then*

$$|Z(f_1)\cap\cdots\cap Z(f_r)| = \frac{q^{1-r}}{q-1}\sum_{(a_1,\dots,a_r)\in\mathbb{F}_q^r}|Z(a_1f_1+\cdots+a_rf_r)| - \frac{q^n}{q-1}. \tag{5.14}$$

Proof. We have

$$\sum_{(a_1,\dots,a_r)\in\mathbb{F}_q^r}|Z(a_1f_1+\cdots+a_rf_r)|$$
$$= \sum_{(a_1,\dots,a_r)\in\mathbb{F}_q^r}\ \sum_{\substack{x\in\mathbb{F}_q^n\\ a_1f_1(x)+\cdots+a_rf_r(x)=0}} 1 = \sum_{x\in\mathbb{F}_q^n}\ \sum_{\substack{(a_1,\dots,a_r)\in\mathbb{F}_q^r\\ a_1f_1(x)+\cdots+a_rf_r(x)=0}} 1$$

$$= \Big(\sum_{x \in Z(f_1)\cap\cdots\cap Z(f_r)} + \sum_{x \in \mathbb{F}_q^n \setminus (Z(f_1)\cap\cdots\cap Z(f_r))} \Big) \sum_{\substack{(a_1,\dots,a_r)\in\mathbb{F}_q^r \\ a_1 f_1(x)+\cdots+a_r f_r(x)=0}} 1$$

$$= |Z(f_1)\cap\cdots\cap Z(f_r)|\, q^r + \big(q^n - |Z(f_1)\cap\cdots\cap Z(f_r)|\big)\, q^{r-1}.$$

Thus

$$|Z(f_1)\cap\cdots\cap Z(f_r)| = \frac{q^{1-r}}{q-1} \sum_{(a_1,\dots,a_r)\in\mathbb{F}_q^r} |Z(a_1 f_1 + \cdots + a_r f_r)| - \frac{q^n}{q-1}.$$

□

Proof of Theorem 5.6. We use induction on r. When $r = 1$, Theorem 5.6 is Ax's theorem.

Assume that $r > 1$. To prove that Theorem 5.6 holds for r, we use another induction on $\sum_{i=1}^r (d_1 - d_i)$.

If $\sum_{i=1}^r (d_1 - d_i) = 0$, then $d_1 = \cdots = d_r$. By Lemma 5.7, we have

$$|Z(f_1)\cap\cdots\cap Z(f_r)| \equiv \frac{q^{1-r}}{q-1} \sum_{(a_1,\dots,a_r)\in\mathbb{F}_q^r} |Z(a_1 f_1 + \cdots + a_r f_r)| \pmod{q^n}$$

$$\equiv 0 \pmod{q^{\lceil \frac{n}{d_1} \rceil - 1 + 1 - r}} \qquad \text{(Ax's theorem)}.$$

Since $\lceil (n - d_1 - \cdots - d_r)/d_1 \rceil = \lceil n/d_1 \rceil - r$, (5.13) holds.

Now assume that $\sum_{i=1}^r (d_1 - d_i) > 0$. Then $d_r < d_1$. By Lemma 5.7, we have

$$|Z(f_1)\cap\cdots\cap Z(f_r)| \equiv \frac{q^{1-r}}{q-1} \sum_{(a_1,\dots,a_r)\in\mathbb{F}_q^r} |Z(a_1 f_1 + \cdots + a_r f_r)| \pmod{q^n}$$

$$= \frac{q^{1-r}}{q-1} \sum_{(a_1,\dots,a_{r-1})\in\mathbb{F}_q^{r-1}} |Z(a_1 g_1 + \cdots + a_{r-1} g_{r-1} + g_r)|,$$

where $g_i = f_i(\mathtt{X}_1, \dots, \mathtt{X}_n) \in \mathbb{F}_q[\mathtt{X}_1, \dots, \mathtt{X}_n, \mathtt{X}_{n+1}]$, $1 \le i \le r-1$, and $g_r = \mathtt{X}_{n+1} f_r(\mathtt{X}_1, \dots, \mathtt{X}_n) \in \mathbb{F}_q[\mathtt{X}_1, \dots, \mathtt{X}_n, \mathtt{X}_{n+1}]$. Thus,

(5.15)

$$|Z(f_1)\cap\cdots\cap Z(f_r)|$$

$$\equiv \frac{q^{1-r}}{(q-1)^2} \sum_{\substack{(a_1,\dots,a_r)\in\mathbb{F}_q^r \\ a_r \ne 0}} |Z(a_1 g_1 + \cdots + a_r g_r)| \pmod{q^n}$$

$$= \frac{q^{1-r}}{(q-1)^2} \Big[\sum_{(a_1,\dots,a_r)\in\mathbb{F}_q^r} |Z(a_1 g_1 + \cdots + a_r g_r)|$$

$$- \sum_{(a_1,\dots,a_{r-1})\in\mathbb{F}_q^{r-1}} |Z(a_1g_1+\cdots+a_{r-1}g_{r-1})|\Big]$$

$$= \frac{1}{q-1}\Big[\frac{q^{1-r}}{q-1}\sum_{(a_1,\dots,a_r)\in\mathbb{F}_q^r} |Z(a_1g_1+\cdots+a_rg_r)|$$

$$- \frac{q^{1-(r-1)}}{q-1}\sum_{(a_1,\dots,a_{r-1})\in\mathbb{F}_q^{r-1}} |Z(a_1f_1+\cdots+a_{r-1}f_{r-1})|\Big]$$

$$\equiv \frac{1}{q-1}\big[|Z(g_1)\cap\cdots\cap Z(g_r)| - |Z(f_1)\cap\cdots\cap Z(f_{r-1})|\big] \pmod{q^n},$$

where the last step follows from Lemma 5.7. By the induction hypothesis on r, we have

$$(5.16)\qquad |Z(f_1)\cap\cdots\cap Z(f_{r-1})| \equiv 0 \pmod{q^{\lceil\frac{n-d_1-\cdots-d_{r-1}}{d_1}\rceil}}.$$

Note that $\deg g_i = d_i$, $1\le i\le r-1$, and $\deg g_r = d_r+1$. Hence, by the induction hypothesis on $\sum_{i=1}^r(d_1-d_i)$, we have

$$(5.17)\qquad |Z(g_1)\cap\cdots\cap Z(g_r)| \equiv 0 \pmod{q^{\lceil\frac{(n+1)-d_1-\cdots-d_{r-1}-(d_r+1)}{d_1}\rceil}}.$$

It follows from (5.15)–(5.17) that

$$|Z(f_1)\cap\cdots\cap Z(f_r)| \equiv 0 \pmod{q^{\lceil\frac{n-d_1-\cdots-d_r}{d_1}\rceil}}.$$

□

The p-adic bound in Theorem 5.6 is best possible as shown by the following proposition also due to Katz.

Proposition 5.8. *Let $q=p^l$ and let $d_1\ge\cdots\ge d_r\ge 1$ be integers. Then there exist $f_1,\dots,f_r\in\mathbb{F}_q[\mathtt{X}_1,\dots,\mathtt{X}_n]$ with $\deg f_i = d_i$, $1\le i\le r$, such that*

$$(5.18)\qquad \nu_p\big(|Z(f_1)\cap\cdots\cap Z(f_r)|\big) = \max\Big\{0,\ l\Big\lceil\frac{n-d_1-\cdots-d_r}{d_1}\Big\rceil\Big\}.$$

Proof. 1° We first prove a general fact: For any integers $d\ge m>0$, there exists $g\in\mathbb{F}_q[\mathtt{X}_1,\dots,\mathtt{X}_m]$ with $\deg g = d$ such that $Z(g)=\{(0,\dots,0)\}$.

Let $\epsilon_1,\dots,\epsilon_d$ be a basis of $\mathbb{F}_{q^d}$ over $\mathbb{F}_q$, and define

$$G = \prod_{\gamma\in\mathrm{Aut}(\mathbb{F}_{q^d}/\mathbb{F}_q)} \Big(\sum_{i=1}^d \gamma(\epsilon_i)\mathtt{X}_i\Big) \in \mathbb{F}_q[\mathtt{X}_1,\dots,\mathtt{X}_d].$$

Then $\deg G = d$. For $(x_1,\dots,x_d)\in\mathbb{F}_q^d$, $G(x_1,\dots,x_d)=0$ if and only if $\sum_{i=1}^d \gamma(\epsilon_i)x_i = 0$ for some $\gamma\in\mathrm{Aut}(\mathbb{F}_{q^d}/\mathbb{F}_q)$. This happens if and only if

$x_i = 0$ for all $1 \le i \le d$ since $\gamma(\epsilon_i)$, $1 \le i \le d$, are linearly independent over $\mathbb{F}_q$. Thus we have proved that $Z(G) = \{(0, \dots, 0)\}$. Let

$$g = G(\mathtt{X}_1, \dots, \mathtt{X}_{m-1}, \underbrace{\mathtt{X}_m, \dots, \mathtt{X}_m}_{d-m+1}) \in \mathbb{F}_q[\mathtt{X}_1, \dots, \mathtt{X}_m].$$

Then $\deg g = d$ and $Z(g) = \{(0, \dots, 0)\}$.

2° Now we construct the polynomials $f_1, \dots, f_r$.

First assume that $d_2 + \cdots + d_r \ge n$. Choose subsets $I_1, \dots, I_r$ of $\{1, \dots, n\}$ such that $\bigcup_{i=1}^r I_i = \{1, \dots, n\}$ and $|I_i| \le d_i$, $1 \le i \le r$. By 1°, for each $1 \le i \le r$, there exists $f_i \in \mathbb{F}_q[\mathtt{X}_j : j \in I_i] \subset \mathbb{F}_q[\mathtt{X}_1, \dots, \mathtt{X}_n]$ with $\deg f_i = d_i$ such that for $(x_1, \dots, x_n) \in \mathbb{F}_q^n$, $f_i(x_1, \dots, x_n) = 0$ if and only if $x_j = 0$ for all $j \in I_i$. Then $Z(f_1) \cap \cdots \cap Z(f_r) = \{(0, \dots, 0)\}$ and hence (5.18) holds.

Now assume that $d_2 + \cdots + d_r < n$. Let $I_2, \dots, I_r$ be a partition of $\{j \in \mathbb{Z} : n - (d_2 + \cdots + d_r) < j \le n\}$ with $|I_i| = d_i$, $2 \le i \le n$. By 1° again, for each $2 \le i \le r$, there exists $f_i \in \mathbb{F}_q[\mathtt{X}_1, \dots, \mathtt{X}_n]$ with $\deg f_i = d_i$ such that for $(x_1, \dots, x_n) \in \mathbb{F}_q^n$, $f_i(x_1, \dots, x_n) = 0$ if and only if $x_j = 0$ for all $j \in I_i$. By Example 5.5, there exists $h \in \mathbb{F}_q[\mathtt{X}_1, \dots, \mathtt{X}_{n-(d_2+\cdots+d_r)}]$ with $\deg h = d_1$ such that

$$\nu_p\big(|Z(h)|\big) = l\Big(\Big\lceil \frac{n - (d_2 + \cdots + d_r)}{d_1} \Big\rceil - 1\Big) = l\Big\lceil \frac{n - d_1 - \cdots - d_r}{d_1} \Big\rceil.$$

Let f_1 be h treated as an element of $\mathbb{F}_q[\mathtt{X}_1, \dots, \mathtt{X}_n]$. Then

$$Z(f_1) \cap \cdots \cap Z(f_r) = Z(h) \times \{(0, \dots, 0)\},$$

and hence (5.18) also holds. □

Theorem 5.6 implies that for $d_1 + \cdots + d_r \le n$, $|Z(f_1) \cap \cdots \cap Z(f_r)|$ is either 0 or at least $q^{\lceil (n-d_1-\cdots-d_r)/d_1 \rceil}$. We will see in the next section that $q^{\lceil (n-d_1-\cdots-d_r)/d_1 \rceil}$ can be replaced by $q^{n-d_1-\cdots-d_r}$.

5.3. Bounds on the Number of Zeros of Polynomials

We begin with a lower bound on the number of common zeros of several polynomials in $\mathbb{F}_q[\mathtt{X}_1, \dots, \mathtt{X}_n]$.

Theorem 5.9. *Let $f_1, \dots, f_r \in \mathbb{F}_q[\mathtt{X}_1, \dots, \mathtt{X}_n]$ be such that $\deg f_i > 0$, $1 \le i \le r$, and $d = \deg f_1 + \cdots + \deg f_r \le n$. Then either*

$$|Z(f_1) \cap \cdots \cap Z(f_r)| = 0 \tag{5.19}$$

or

$$|Z(f_1) \cap \cdots \cap Z(f_r)| \ge q^{n-d}. \tag{5.20}$$

Theorem 5.9 with $r = 1$ is due to Warning [**39**]; the general form appeared in Lidl and Niederreiter [**27**, §6.1]. We will follow the proof in [**27**]. First, we observe a few facts.

If $f \in \mathbb{F}_q[\mathtt{X}_1, \dots, \mathtt{X}_n]$ has $\deg f < n(q-1)$, then

$$\sum_{(x_1,\dots,x_n)\in\mathbb{F}_q^n} f(x_1, \dots, x_n) = 0.$$

To see this fact, it suffices to consider $f = \mathtt{X}_1^{i_1} \cdots \mathtt{X}_n^{i_n}$, where $i_1 + \cdots + i_n < n(q-1)$. We have $i_j < q-1$ for some $1 \le j \le n$. Therefore, by Lemma 2.21,

$$\sum_{(x_1,\dots,x_n)\in\mathbb{F}_q^n} x_1^{i_1} \cdots x_n^{i_n} = \Big(\sum_{x_1\in\mathbb{F}_q} x_1^{i_1}\Big) \cdots \Big(\sum_{x_n\in\mathbb{F}_q} x_n^{i_n}\Big) = 0.$$

If A_1 and A_2 are two distinct parallel d-dimensional affine subspaces of $\mathbb{F}_q^n$, i.e., two distinct cosets of a d-dimensional subspace of $\mathbb{F}_q^n$, then there exists $g \in \mathbb{F}_q[\mathtt{X}_1, \dots, \mathtt{X}_n]$ with $\deg g = (n-d)(q-1) - 1$ such that

$$g(x_1, \dots, x_n) = \begin{cases} 1 & \text{if } (x_1, \dots, x_n) \in A_1, \\ -1 & \text{if } (x_1, \dots, x_n) \in A_2, \\ 0 & \text{if } (x_1, \dots, x_n) \in \mathbb{F}_q^n \setminus (A_1 \cup A_2). \end{cases} \tag{5.21}$$

To see this fact, we may assume, through a suitable affine transformation of $\mathbb{F}_q^n$, that

$$\begin{aligned} A_1 &= \{(x_1, \dots, x_n) \in \mathbb{F}_q^n : x_1 = \cdots = x_{n-d} = 0\}, \\ A_2 &= \{(x_1, \dots, x_n) \in \mathbb{F}_q^n : x_1 = 1,\ x_2 = \cdots = x_{n-d} = 0\}. \end{aligned}$$

Then $g = (1+\mathtt{X}_1+\cdots+\mathtt{X}_1^{q-2})(1-\mathtt{X}_2^{q-1}) \cdots (1-\mathtt{X}_{n-d}^{q-1})$ has the desired property. A d-dimensional affine subspace of $\mathbb{F}_q^n$ is called a *d-flat* of $\mathbb{F}_q^n$.

Proof of Theorem 5.9. Let $Z = Z(f_1) \cap \cdots \cap Z(f_r)$ and assume that $|Z| > 0$.

1° We claim that for any two parallel d-flats A_1 and A_2 of $\mathbb{F}_q^n$,

$$|Z \cap A_1| \equiv |Z \cap A_2| \pmod p,$$

where $p = \operatorname{char} \mathbb{F}_q$. Assume that $A_1 \ne A_2$. Let $g \in \mathbb{F}_q[\mathtt{X}_1, \dots, \mathtt{X}_n]$ be a polynomial of degree $(n-d)(q-1) - 1$ satisfying (5.21). Then $(1 - f_1^{q-1}) \cdots (1 - f_r^{q-1})g$ has degree $d(q-1) + (n-d)(q-1) - 1 = n(q-1) - 1$ and takes the value 1 on $Z \cap A_1$, -1 on $Z \cap A_2$ and 0 elsewhere. Therefore in $\mathbb{F}_q$,

$$0 = \sum_{x\in\mathbb{F}_q^n} \big[(1 - f_1^{q-1}) \cdots (1 - f_r^{q-1})g\big](x) = |Z \cap A_1| - |Z \cap A_2|.$$

2° Assume that $|Z \cap A| \not\equiv 0 \pmod{p}$ for some d-flat A of $\mathbb{F}_q^n$. Let $A_1, \dots, A_{q^{n-d}}$ be all the d-flats of $\mathbb{F}_q^n$ parallel to A. By 1°, $|Z \cap A_i| \not\equiv 0 \pmod{p}$ for all $1 \le i \le q^{n-d}$. In particular, $|Z \cap A_i| \ge 1$, $1 \le i \le q^{n-d}$. Since $A_1, \dots, A_{q^{n-d}}$ form a partition of $\mathbb{F}_q^n$, we have

$$|Z| = \sum_{i=1}^{q^{n-d}} |Z \cap A_i| \ge q^{n-d}.$$

3° Assume that $|Z \cap A| \equiv 0 \pmod{p}$ for all d-flats A of $\mathbb{F}_q^n$. Then there is an integer $0 \le k < d$ such that $|Z \cap B| \equiv 0 \pmod{p}$ for all $(k+1)$-flats B of $\mathbb{F}_q^n$ and $|Z \cap C| \not\equiv 0 \pmod{p}$ for some k-flat C of $\mathbb{F}_q^n$. The number of $(k+1)$-flats of $\mathbb{F}_q^n$ containing C is $(q^{n-k}-1)/(q-1)$. Let B_i, $1 \le i \le (q^{n-k}-1)/(q-1)$, be all the $(k+1)$-flats of $\mathbb{F}_q^n$ containing C. Then C and $B_i \setminus C$, $1 \le i \le (q^{n-k}-1)/(q-1)$, form a partition of $\mathbb{F}_q^n$. Since

$$|Z \cap C| + |Z \cap (B_i \setminus C)| = |Z \cap B_i| \equiv 0 \pmod{p},$$

we have $|Z \cap (B_i \setminus C)| \not\equiv 0 \pmod{p}$. Thus,

$$\begin{aligned} |Z| &= |Z \cap C| + \sum_{i=1}^{(q^{n-k}-1)/(q-1)} |Z \cap (B_i \setminus C)| \\ &\ge 1 + \frac{q^{n-k}-1}{q-1} = 2 + q + \cdots + q^{n-k-1} > q^{n-d}. \end{aligned}$$

□

Example 5.10. In Theorem 5.9, let $d_i = \deg f_i$, $1 \le i \le r$. If $r = 1$ and $d_1 = 1$, (5.19) does not occur. For all other values of r and $d_1, \dots, d_r$, (5.19) can always be realized. If $d_1 \ge 2$, simply let $f_1 \in \mathbb{F}_q[\mathtt{X}_1]$ be irreducible of degree d_1. If $d_1 = d_2 = 1$, let $f_1 = \mathtt{X}_1$ and $f_2 = \mathtt{X}_1 + 1$.

The equality in (5.20) can be realized for all $d_i > 0$, $1 \le i \le r$, such that $d_1 + \cdots + d_r \le n$. By 1° in the proof of Proposition 5.8, for each $1 \le i \le r$, there exists $f_i \in \mathbb{F}_q[\mathtt{X}_1, \dots, \mathtt{X}_n]$ with $\deg f_i = d_i$ such that

$$Z(f_i) = \{(x_1, \dots, x_n) \in \mathbb{F}_q^n : x_j = 0 \text{ for } d_1 + \cdots + d_{i-1} + 1 \le j \le d_1 + \cdots + d_i\}.$$

Then $|Z(f_1) \cap \cdots \cap Z(f_r)| = |\{(0, \dots, 0)\} \times \mathbb{F}_q^{n-d}| = q^{n-d}$.

Next, we prove an upper bound for the number of zeros of polynomials in $\mathbb{F}_q[\mathtt{X}_1, \dots, \mathtt{X}_n]$ with a given degree. This result is conveniently stated in terms of the q-ary Reed-Muller code $R_q(d, n)$. Recall that $R_q(d, n)$ is the set of all functions from $\mathbb{F}_q^n$ to $\mathbb{F}_q$ represented by polynomials in $\mathbb{F}_q[\mathtt{X}_1, \dots, \mathtt{X}_n]$ of degree $\le d$. For convenience, we do not distinguish between a polynomial in $\mathbb{F}_q[\mathtt{X}_1, \dots, \mathtt{X}_n]$ and the function in $\mathcal{F}(\mathbb{F}_q^n, \mathbb{F}_q)$ it represents.

Theorem 5.11. *Let* $0 \le d \le n(q-1)$ *and write* $d = r(q-1)+s$, *where* $0 \le s < q-1$. *Then*

$$\max_{f \in R_q(d,n)\setminus\{0\}} |Z(f)| = q^{n-r-1}(q^{r+1}-q+s). \tag{5.22}$$

Proof. 1° We first show that for every $f \in R_q(d,n)\setminus\{0\}$, $|Z(f)| \le q^{n-r-1}\cdot(q^{r+1}-q+s)$. Identify $\mathbb{F}_q^n$ with $\mathbb{F}_{q^n}$. It suffices to show that f, as a function from $\mathbb{F}_{q^n}$ to $\mathbb{F}_q$, is represented by a polynomial in $\mathbb{F}_{q^n}[\mathtt{X}]$ of degree $\le q^{n-r-1}(q^{r+1}-q+s)$.

Let $E_{\le d} = \{(e_0,\dots,e_{n-1}) \in [0,q-1]^n : e_0+\cdots+e_{n-1} \le d\}$. By Theorem 2.17, the function $f(x)$ ($x \in \mathbb{F}_{q^n}$) is an $\mathbb{F}_q$-linear combination of expressions of the form $\mathrm{Tr}_{q^n/q}\big(ax^{\epsilon(q^0,\dots,q^{n-1})^T}\big)$, where $a \in \mathbb{F}_{q^n}$ and $\epsilon \in E_{\le d}$. Let τ be the Frobenius map of $\mathbb{F}_{q^n}/\mathbb{F}_q$. By (2.21),

$$\mathrm{Tr}_{q^n/q}\Big(ax^{\epsilon(q^0,\dots,q^{n-1})^T}\Big) = \sum_{m=0}^{n-1} \tau^m(a)\, x^{\tau^m(\epsilon)\cdot(q^0,\dots,q^{n-1})^T},$$

where $\tau^m(\epsilon)$ is the cyclic shift of ϵ m positions to the right. For each $0 \le m \le n-1$,

$$\begin{aligned}\tau^m(\epsilon)\cdot(q^0,\dots,q^{n-1})^T &\le (0,\dots,0,s,\overbrace{q-1,\dots,q-1}^{r})\cdot(q^0,\dots,q^{n-1})^T\\ &= sq^{n-r-1}+(q-1)(q^{n-r}+\cdots+q^{n-1})\\ &= q^{n-r-1}(q^{r+1}-q+s).\end{aligned}$$

Hence $\mathrm{Tr}_{q^n/q}\big(ax^{\epsilon(q^0,\dots,q^{n-1})^T}\big)$ is a polynomial in x of degree $\le q^{n-r-1}(q^{r+1}-q+s)$. The same is true for $f(x)$.

2° It remains to show that there exists $f \in R_q(d,n)\setminus\{0\}$ such that $|Z(f)| = q^{n-r-1}(q^{r+1}-q+s)$. Let S be an s-element subset of $\mathbb{F}_q$ and let

$$f_S = (\mathtt{X}_1^{q-1}-1)\cdots(\mathtt{X}_r^{q-1}-1)\prod_{a\in S}(\mathtt{X}_{r+1}-a) \in \mathbb{F}_q[\mathtt{X}_1,\dots,\mathtt{X}_n]. \tag{5.23}$$

Then $\deg f_S = r(q-1)+s = d$. Moreover, $f_S(x_1,\dots,x_n) \ne 0$ if and only if $x_1=\cdots=x_r=0$ and $x_{r+1}\notin S$. Hence the number of nonzeros of f_S is $(q-s)q^{n-r-1}$. Therefore $|Z(f_S)| = q^n-(q-s)q^{n-r-1} = q^{n-r-1}(q^{r+1}-q+s)$. □

For each function $f : \mathbb{F}_q^n \to \mathbb{F}_q$, its *support* is

$$\mathrm{supp}(f) = \{x \in \mathbb{F}_q^n : f(x) \ne 0\}.$$

The *Hamming weight* of f, or simply the *weight* of f, denoted by $|f|$, is

$$|f| = |\mathrm{supp}(f)| = q^n - |Z(f)|.$$

Corollary 5.12. *In the notation of Theorem* 5.11,

$$\min_{f\in R_q(d,n)\setminus\{0\}} |f| = (q-s)q^{n-r-1}. \tag{5.24}$$

Proof. In (5.22), observe that $q^{n-r-1}(q^{r+1}-q+s) = q^n-(q-s)q^{n-r-1}$. □

The function f_S in (5.23) attains the minimum value in (5.24), which is called the *minimum weight* of $R_q(d,n)$. In the next theorem, we determine all functions in $R_q(d,n)\setminus\{0\}$ which have the minimum weight. To state and prove this theorem, we need the affine linear group $\mathrm{AGL}(n,\mathbb{F}_q)$, which can be defined as

$$\mathrm{AGL}(n,\mathbb{F}_q) = \left\{\begin{bmatrix} A & 0\\ a & 1\end{bmatrix} : A\in \mathrm{GL}(n,\mathbb{F}_q),\ a\in\mathbb{F}_q^n\right\} < \mathrm{GL}(n+1,\mathbb{F}_q).$$

There is a right action of $\mathrm{AGL}(n,\mathbb{F}_q)$ on $\mathbb{F}_q^n$. For $\sigma = \left[\begin{smallmatrix} A & 0\\ a & 1\end{smallmatrix}\right] \in \mathrm{AGL}(n,\mathbb{F}_q)$ and $x\in\mathbb{F}_q^n$,

$$x^\sigma = xA + a.$$

This action induces a left action of $\mathrm{AGL}(n,\mathbb{F}_q)$ on $\mathcal{F}(\mathbb{F}_q^n,\mathbb{F}_q)$, the set of all functions from $\mathbb{F}_q^n$ to $\mathbb{F}_q$. For $\sigma\in\mathrm{AGL}(n,\mathbb{F}_q)$ and $f\in\mathcal{F}(\mathbb{F}_q^n,\mathbb{F}_q)$,

$$\sigma(f) = f\circ\sigma.$$

Lemma 5.13. *Let $S\subset\mathbb{F}_q^n$ be such that $|S\cap H|$ is a constant for every $(n-1)$-flat H of $\mathbb{F}_q^n$. Then $S=\emptyset$ or $\mathbb{F}_q^n$.*

Proof. 1° We claim that for $0\le k\le n$ and every k-flat B of $\mathbb{F}_q^n$, $|S\cap B|$ depends only on k.

We use induction on k. The case $k=n$ needs no proof and the case $k=n-1$ is the assumption. Assume that $k\le n-2$. Let B be a k-flat of $\mathbb{F}_q^n$. Choose a $(k+2)$-flat A of $\mathbb{F}_q^n$ containing B and let $C_1,\dots,C_{q+1}$ be the $(k+1)$-flats between B and A. Then B and $C_i\setminus B$, $1\le i\le q+1$, form a partition of A. By the induction hypothesis, we have

$$\begin{aligned}\frac{1}{q^{n-k-2}}|S| &= |S\cap A| = |S\cap B| + \sum_{i=1}^{q+1}\bigl(|S\cap C_i| - |S\cap B|\bigr)\\ &= (q+1)\frac{1}{q^{n-k-1}}|S| - q|S\cap B|.\end{aligned}$$

Thus $|S\cap B| = |S|/q^{n-k}$.

2° If $S\ne\emptyset$ and $S\ne\mathbb{F}_q^n$, then there exist 0-flats B_1 and B_2 of $\mathbb{F}_q^n$ such that $|S\cap B_1|=0$ and $|S\cap B_2|=1$, which is a contradiction to 1°. □

Lemma 5.14. *Let $0 \le d \le n(q-1)$ and write $d = r(q-1)+s$, $0 \le s < q-1$. Assume that $f \in R_q(d,n)$ has the minimum weight $(q-s)q^{n-r-1}$. Let $H_1, \dots, H_q$ be $(n-1)$-flats of $\mathbb{F}_q^n$ that form a parallel class. Then exactly one of the following occurs.*

(i) $\operatorname{supp}(f) \subset H_i$ *for some* i.

(ii) $|\operatorname{supp}(f) \cap H_i| = (q-s)q^{n-r-2}$ *for all* $1 \le i \le q$. *In this case,* $f1_{H_i} \in R_q((r+1)(q-1)+s, n)$ *has the minimum weight, where* $1_{H_i} : \mathbb{F}_q^n \to \mathbb{F}_q$ *is the indicator function of* H_i*:*

$$1_{H_i}(x) = \begin{cases} 1 & \text{if } x \in H_i, \\ 0 & \text{if } x \notin H_i. \end{cases}$$

(iii) $0 < s < q-1$ *and* $H_1, \dots, H_q$ *can be ordered so that* $\operatorname{supp}(f) \cap H_i = \emptyset$ *for* $1 \le i \le s$. *For each* $s < i \le q$, *there exists* $g_i \in R_q(q-1-s, n)$ *such that* $\operatorname{supp}(f) \cap H_i = \operatorname{supp}(fg_i)$ *and* $fg_i \in R_q((r+1)(q-1), n)$ *has the minimum weight.*

Proof. Without loss of generality, assume that

$$\operatorname{supp}(f) \cap H_i \begin{cases} = \emptyset & \text{if } 1 \le i \le k, \\ \ne \emptyset & \text{if } k < i \le q. \end{cases}$$

We may assume that $k \le q-2$. (If $k = q-1$, we are in case (i).) We may further assume that $H_i = \{a_i\} \times \mathbb{F}_q^{n-1}$, where $\{a_1, \dots, a_q\} = \mathbb{F}_q$. Then for $k < i \le q$, $\operatorname{supp}(f) \cap H_i = \operatorname{supp}(fg_i)$, where

$$g_i = \prod_{\substack{k<j\le q \\ j \ne i}} (\mathtt{X}_1 - a_j).$$

Since $fg_i \in R_q(r(q-1)+s+q-k-1, n)$ and $fg_i \ne 0$, by Corollary 5.12 we have

$$|fg_i| \ge \begin{cases} (k-s+1)q^{n-r-1} & \text{if } s-k < 0, \\ (q-s+k)q^{n-r-2} & \text{if } s-k \ge 0. \end{cases} \tag{5.25}$$

Thus
(5.26)

$$(q-s)q^{n-r-1} = |f| = \sum_{k<i\le q} |fg_i| \ge \begin{cases} (q-k)(k-s+1)q^{n-r-1} & \text{if } s-k < 0, \\ (q-k)(q-s+k)q^{n-r-2} & \text{if } s-k \ge 0. \end{cases}$$

The above inequality is equivalent to

$$\begin{cases} (q-1-k)(k-s) \le 0 & \text{if } s-k < 0, \\ k(s-k) \le 0 & \text{if } s-k \ge 0. \end{cases} \tag{5.27}$$

Thus we must have $k = 0$ or s. Tracing back from (5.27) to (5.26) and (5.25), we conclude that $|fg_i| = (q-s+k)q^{n-r-2}$, i.e., fg_i has the minimum weight of $R_q(r(q-1)+s+q-k-1, n)$. When $k = 0$, g_i represents the function -1_{H_i} and we have the conclusion of (ii). When $k = s$, we have (iii). □

Theorem 5.15 (Delsarte-Goethals-MacWilliams). *Let $0 \le d \le n(q-1)$ and write $d = r(q-1)+s$, $0 \le s < q-1$. Then $f \in R_q(d,n)$ has the minimum weight $(q-s)q^{n-r-1}$ if and only if $f = c\,\sigma(h)$ for some $c \in \mathbb{F}_q^*$, $\sigma \in \mathrm{AGL}(n, \mathbb{F}_q)$, and*

$$h = (\mathtt{X}_1^{q-1}-1)\cdots(\mathtt{X}_r^{q-1}-1)\prod_{1\le i\le s}(\mathtt{X}_{r+1}-a_i) \in R_q(d,n), \tag{5.28}$$

where $a_1, \dots, a_s \in \mathbb{F}_q$ are distinct.

Proof. The "if" part is obvious and we only have to prove the "only if" part. Assume that $f \in R_q(d,n)$ is such that $|f| = (q-s)q^{n-r-1}$. It suffices to show that there exists h of the form (5.28) and $\sigma \in \mathrm{AGL}(n, \mathbb{F}_q)$ such that $\mathrm{supp}(f) \subset \mathrm{supp}(\sigma(h))$. (Then there exists $c \in \mathbb{F}_q^*$ such that $|f - c\,\sigma(h)| < |f|$, and hence $f - c\,\sigma(h) = 0$.)

Note that $\mathrm{supp}(h)$ is a union of $q-s$ parallel $(n-r-1)$-flats contained in an $(n-r)$-flat of $\mathbb{F}_q^n$. (For convenience, we interpret "a union of q parallel (-1)-flats" as a 0-flat.) Therefore, it suffices to show that $\mathrm{supp}(f)$ is contained in a union of $q-s$ parallel $(n-r-1)$-flats in an $(n-r)$-flat of $\mathbb{F}_q^n$.

1° First assume that $r = 0$. We only need to consider the case $d > 0$. We claim that there exists an $(n-1)$-flat H of $\mathbb{F}_q^n$ such that $\mathrm{supp}(f) \cap H = \emptyset$. Otherwise, for every $(n-1)$-flat H, $f1_H : H \to \mathbb{F}_q$ is a nonzero function represented by a polynomial of degree $\le d$ in the coordinates of H, that is, $f1_H \in R_q(d, n-1) \setminus \{0\}$. By Corollary 5.12, $|f1_H| \ge (q-s)q^{n-2} = |f|/q$. It follows from Lemma 5.13 that $\mathrm{supp}(f) = \emptyset$ or $\mathbb{F}_q^n$, which is a contradiction.

Since $|f| = (q-s)q^{n-1} > q^{n-1}$, $\mathrm{supp}(f) \not\subset a + H$ for all $a \in \mathbb{F}_q^n$. Thus case (iii) of Lemma 5.14 must occur. Therefore $\mathrm{supp}(f)$ is contained in a union of $q-s$ parallels of H.

2° For $1 \le r \le n$, we use (backward) induction on r. The case $r = n$ is trivial. Assume that $1 \le r \le n-1$. Since $|f| > q^{n-r-1}$, we can choose $v_0, \dots, v_{n-r} \in \mathrm{supp}(f)$ which are affinely independent. Let H_1 be an $(n-1)$-flat of $\mathbb{F}_q^n$ containing $v_0, \dots, v_{n-r}$. We claim that $\mathrm{supp}(f) \subset H$, where $H = a + H_1$ for some $a \in \mathbb{F}_q^n$. Assume the contrary. Then by Lemma 5.14 (ii) and (iii), $\mathrm{supp}(f) \cap H_1 \subset \mathrm{supp}(g)$ for some $g \in R_q((r+1)(q-1)+s, n)$ (in case (ii)) or some $g \in R_q((r+1)(q-1), n)$ (in case (iii)) with minimum weight. By the induction hypothesis, $\operatorname{supp}(g)$ is contained in an $(n-r-1)$-flat of $\mathbb{F}_q^n$. But this is impossible since $v_0, \dots, v_{n-r}$ are affinely independent. Hence the claim is proved. Without loss of generality, assume that $H = \mathbb{F}_q^{n-1} \times \{0\}$. Then $\operatorname{supp}(f) \subset H$ implies that $f = (\mathtt{X}_n^{q-1} - 1)f_1$ for some $f_1 \in R_q(d-(q-1), n-1)$; note that $d-(q-1) = (r-1)(q-1)+s$. Moreover, $|f_1| = |f|$ is the minimum weight of $R_q((r-1)(q-1)+s, n-1)$. Another induction on n allows us to assume that $\operatorname{supp}(f_1)$ is contained in a union of $q-s$ parallel $(n-1-(r-1)-1)$-flats in an $(n-1-(r-1))$-flat of $\mathbb{F}_q^{n-1}$. Therefore, $\operatorname{supp}(f)$ is contained in a union of $q-s$ parallel $(n-r-1)$-flats in an $(n-r)$-flat of $\mathbb{F}_q^n$. □

Theorem 5.15 is due to Delsarte, Goethals and MacWilliams [**9**]. The proof given above is based on Leducq [**26**].

Corollary 5.16. *Let $0 \le d \le n(q-1)$ and write $d = r(q-1)+s$, where $0 \le s < q-1$.*

(i) *A subset $A \subset \mathbb{F}_q^n$ is the support of some $f \in R_q(d,n)$ with $|f| = (q-s)q^{n-r-1}$ if and only if $|A| = 1$ when $r = n$, and A is a union of $q-s$ parallel $(n-r-1)$-flats in an $(n-r)$-flat of $\mathbb{F}_q^n$ when $r < n$.*

(ii) *If $f_1, f_2 \in R_q(d,n)$ are such that $|f_1| = |f_2| = (q-s)q^{n-r-1}$ and $\operatorname{supp}(f_1) = \operatorname{supp}(f_2)$, then $f_1 = cf_2$ for some $c \in \mathbb{F}_q^*$.*

(iii) *We have*

$$\big|\{f \in R_q(d,n) : |f| = (q-s)q^{n-r-1}\}\big| = \begin{cases} q^r(q-1)\displaystyle\prod_{i=0}^{r-1}\frac{q^{n-i}-1}{q^{r-i}-1} & \text{if } s = 0, \\ q^r(q^{n-r}-1)\displaystyle\binom{q}{s}\prod_{i=0}^{r-1}\frac{q^{n-i}-1}{q^{r-i}-1} & \text{if } 0 < s < q-1. \end{cases} \tag{5.29}$$

Proof. (i) Let $h_S \in R_q(d,n)$ denote the function in (5.28), where $S = \{a_1, \dots, a_s\} \subset \mathbb{F}_q$. Then

$$\operatorname{supp}(h_S) = \{(x_1, \dots, x_n) \in \mathbb{F}_q^n : x_1 = \cdots = x_r = 0,\ x_{r+1} \in \mathbb{F}_q \setminus S\},$$

which is a union of $q-s$ parallel $(n-r-1)$-flats in an $(n-r)$-flat of $\mathbb{F}_q^n$. (Recall that we interpret "a union of q parallel (-1)-flats" as a 0-flat.) By Theorem 5.15, supports of elements in $R_q(d,n) \setminus \{0\}$ of minimum weight are the images of $\operatorname{supp}(h_S)$, where S runs through all s-element subsets of $\mathbb{F}_q$, under the action of $\mathrm{AGL}(n, \mathbb{F}_q)$. These images are precisely unions of $q-s$ parallel $(n-r-1)$-flats in an $(n-r)$-flat of $\mathbb{F}_q^n$.

(ii) Choose $a \in \operatorname{supp}(f_1)$ and let $c = f_1(a)/f_2(a)$. Then $f_1 - cf_2 \in R_q(d,n)$ and $|f_1 - cf_2| < (q-s)q^{n-r-1}$. Thus $f_1 - cf_2 = 0$.

(iii) By (i) and (ii),

$$\big|\{f \in R_q(d,n) : |f| = (q-s)q^{n-r-1}\}\big| = (q-1)|\mathcal{A}|,$$

where $\mathcal{A}$ is the set of all subsets $A \subset \mathbb{F}_q^n$ described in (i). If $s = 0$, $\mathcal{A}$ is the set of all $(n-r)$-flats of $\mathbb{F}_q^n$. Thus

$$|\mathcal{A}| = q^r \prod_{i=0}^{r-1} \frac{q^{n-i}-1}{q^{r-i}-1},$$

and (5.29) holds.

Now assume that $0 < s < q-1$. For each $A \in \mathcal{A}$, since $q-s \geq 2$, the affine span B of A is an $(n-r)$-flat of $\mathbb{F}_q^n$. If $A = A_1 \cup \cdots \cup A_{q-s} = A'_1 \cup \cdots \cup A'_{q-s}$, where each of $\{A_1, \dots, A_{q-s}\}$ and $\{A'_1, \dots, A'_{q-s}\}$ is a family of $q-s$ distinct parallel $(n-r-1)$-flats in B, then A_1 must be parallel to A'_1. (Otherwise, $|A'_1 \cap A_i| \leq |A'_1|/q$ for all $1 \leq i \leq q-s$, and hence $|A'_1| = \sum_{i=1}^{q-s} |A'_1 \cap A_i| \leq |A'_1|(q-s)/q < |A'_1|$, which is a contradiction.) Therefore, each $A \in \mathcal{A}$ can be uniquely obtained by first choosing an $(n-r)$-flat B of $\mathbb{F}_q^n$ and then selecting $q-s$ parallel $(n-r-1)$-flats in B. Hence

$$|\mathcal{A}| = \binom{q}{s} \frac{q^{n-r}-1}{q-1} \cdot q^r \prod_{i=0}^{r-1} \frac{q^{n-i}-1}{q^{r-i}-1},$$

from which (5.29) follows. □

5.4. Bounds Derived from Function Fields

This section serves as an informal introduction to algebraic functions in one variable. The main objective here is to state the Riemann hypothesis for function fields (RHFF) proved by Weil and several bounds derived from the RHFF—the Weil bound for the number of degree-one places of a function field over a finite field, the Hasse-Weil bound for the number of zeros of an absolutely irreducible homogeneous polynomial in $\mathbb{F}_q[\mathtt{X}, \mathtt{Y}, \mathtt{Z}]$, and the Lang-Weil bound for the number of zeros of an absolutely irreducible homogeneous polynomial in $\mathbb{F}_q[\mathtt{X}_0, \dots, \mathtt{X}_n]$. Basic concepts will be developed in reasonable detail, but facts and theorems are stated without proof except for a few occasions. For a more formal introduction to this important topic, the reader is referred to [**8, 10, 34, 35**]. For an elementary approach to the Hasse-Weil bound, see [**32**].

5.4.1. Algebraic function fields in one variable.

Let $K \subset F$ be fields. If there exists $x \in F$ which is transcendental over K such that $[F : K(x)] < \infty$, F is called an *algebraic function field in one variable* over K. Since we only deal with algebraic function fields in one variable here, we simply call F/K a *function field*. When the base field K is specified, a function field F/K is often referred to as a function field F and an object or a property X of F/K as X of F. Elements of F are called *algebraic functions* (in one variable) over K; those that are algebraic over K are called *constants*. The algebraic closure $\widetilde{K}$ of K in F is called the *constant field* of F/K. It is easy to see that $[\widetilde{K} : K] < \infty$. Note that $F/\widetilde{K}$ is also a function field.

If F/K is a function field and K is perfect, then the following are true:

(i) For each $x \in F$ which is transcendental over K, F is a finite simple extension over $K(x)$.

(ii) There exists $x \in F$ which is transcendental over K such that F is a finite separable extension over $K(x)$. Hence $F = K(x, y)$, where y is a root of a separable irreducible polynomial over $K(x)$.

Let F/K be a function field. A *valuation ring* of F/K is a subring $\mathcal{O}$ of F such that (i) $K \subsetneq \mathcal{O} \subsetneq F$ and (ii) for each $z \in F$, either $z \in \mathcal{O}$ or $z^{-1} \in \mathcal{O}$. Valuation rings of F/K are precisely maximal subrings of F that contain K and are not fields.

Every valuation ring $\mathcal{O}$ of F/K is local, and the maximal ideal P of $\mathcal{O}$ is principal. A generator t of P (i.e., an element t such that $P = t\mathcal{O}$) is called a *prime* of $\mathcal{O}$; $\mathcal{O}$ is uniquely determined by P since $\mathcal{O} = \{z \in F : z^{-1} \notin P\}$. The maximal ideal of a valuation ring of F/K is called a *place* of F/K. The set of all places of F/K is denoted by $\mathbb{P}_F$.

A *discrete valuation* of a field F is an onto map $\nu : F \to \mathbb{Z} \cup \{\infty\}$ satisfying

(i) $\nu(x) = \infty$ if and only if $x = 0$;

(ii) $\nu(xy) = \nu(x) + \nu(y)$ for all $x, y \in F$;

(iii) $\nu(x + y) \geq \min\{\nu(x), \nu(y)\}$ for all $x, y \in F$.

Let P be a place of a function field F/K, and let t be a prime of the corresponding valuation ring $\mathcal{O}$. Then each element $z \in F \setminus \{0\}$ has a unique representation $z = ut^n$, where $u \in \mathcal{O}^\times$ and $n \in \mathbb{Z}$. Define

$$\begin{aligned} \nu_P : \ F \ &\longrightarrow \ \mathbb{Z} \cup \{\infty\} \\ z \ &\longmapsto \ \begin{cases} \infty & \text{if } z = 0, \\ n & \text{if } z = ut^n, \ u \in \mathcal{O}^\times, \ n \in \mathbb{Z}. \end{cases} \end{aligned}$$

Then ν_P is a discrete valuation of F such that $\nu_P(K^*) = \{0\}$. Let $\mathbb{R}_F$ be the set of all valuation rings of F/K and let $\mathbb{V}_F$ be the set of all discrete valuations of F such that $\nu(K^*) = \{0\}$. Then the following maps are bijections:

$$\begin{array}{ccccc} \mathbb{V}_F & \longleftrightarrow & \mathbb{P}_F & \longleftrightarrow & \mathbb{R}_F \\ \nu_P & \longleftrightarrow & P & \longleftrightarrow & \mathcal{O}_P, \end{array} \tag{5.30}$$

where $\mathcal{O}_P = \{z \in F : z^{-1} \notin P\}$ is the corresponding valuation ring of P.

The *residue field* of a place $P \in \mathbb{P}_F$ is defined to be $F_P = \mathcal{O}_P/P$. The canonical homomorphism $\mathcal{O}_P \to \mathcal{O}_P/P = F_P$ restricts to an embedding $K \hookrightarrow F_P$. The *degree* of P is defined to be $\deg P = [F_P : K]$, which is always finite.

A function field F/K has infinitely many places. The *divisor group* of F, denoted by $\mathcal{D}_F$, is the free abelian group generated by $\mathbb{P}_F$. Elements of $\mathcal{D}_F$, called *divisors* of F, are of the form

$$A = \sum_{P \in \mathbb{P}_F} n_P P,$$

where $n_P \in \mathbb{Z}$ for all $P \in \mathbb{P}_F$ and $n_P \neq 0$ for only finitely many $P \in \mathbb{P}_F$. The valuation of A at $P \in \mathbb{P}_F$ is defined to be $\nu_P(A) = n_P$. The degree of $A \in \mathcal{D}_F$ is

$$\deg A = \sum_{P \in \mathbb{P}_F} \nu_P(A) \deg P.$$

The group $\mathcal{D}_F$ is equipped with a partial order: for $A, B \in \mathcal{D}_F$, "$A \leq B$" means that $\nu_P(A) \leq \nu_P(B)$ for all $P \in \mathbb{P}_F$. For each $z \in F^*$, there are only finitely many $P \in \mathbb{P}_F$ such that $\nu_P(z) \neq 0$. The *divisor* of z is defined to be

$$(z) = \sum_{P \in \mathbb{P}_F} \nu_P(z) P.$$

Divisors of F of the form (z), where $z \in F^*$, are said to be *principal*. The set of principal divisors of F, denoted by $\mathcal{P}_F$, is a subgroup of $\mathcal{D}_F$, and the quotient group

$$\mathcal{C}_F = \mathcal{D}_F/\mathcal{P}_F$$

is called the *divisor class group* of F.

Every $z \in F$ can be treated as a function defined on $\mathbb{P}_F$ whose value at $P \in \mathbb{P}_F$ is

$$z(P) = \begin{cases} z + P \in F_P & \text{if } z \in \mathcal{O}_P, \\ \infty & \text{if } z \notin \mathcal{O}_P. \end{cases}$$

Let $z \in F^*$. We have $\deg(z) = 0$; moreover, $(z) = 0$ if and only if z is a nonzero constant. If $\nu_P(z) > 0$ (resp., < 0) at $P \in \mathbb{P}_F$, P is called a *zero* (resp., *pole*) of order $\nu_P(z)$ (resp., $-\nu_P(z)$) of z. We can write

$$(z) = (z)_0 - (z)_\infty,$$

where

$$(z)_0 = \sum_{\substack{P \in \mathbb{P}_F \\ \nu_P(z)>0}} \nu_P(z)P \quad \text{and} \quad (z)_\infty = -\sum_{\substack{P \in \mathbb{P}_F \\ \nu_P(z)<0}} \nu_P(z)P$$

are the *zero divisor* and the *pole divisor* of z, respectively. If z is not a constant,

$$\deg(z)_0 = \deg(z)_\infty = [F : K(z)].$$

5.4.2. The Riemann-Roch theorem.

Let F/K be a function field; in this subsection we always assume that K is the constant field of F/K, i.e., K is algebraically closed in F. For $A \in \mathcal{D}_F$, let

$$\mathcal{L}(A) = \{x \in F : (x) + A \geq 0\},$$

which is a finite-dimensional vector space over K, and define

$$\dim A = \dim_K \mathcal{L}(A).$$

Further define $s(A) = \deg A - \dim A + 1$, which is ≥ 0 whenever $A \geq 0$. The integers $\deg A$, $\dim A$, and $s(A)$ depend only on the class of A in $\mathcal{C}_F$. The function s is increasing in the sense that $s(A) \leq s(B)$ if $A, B \in \mathcal{D}_F$ are such that $A \leq B$. The *genus* of F is defined to be

$$g = \max\{s(A) : A \in \mathcal{D}_F\},$$

which is a nonnegative integer. An *adele* of F is a mapping $\alpha : \mathbb{P}_F \to F$ such that $\alpha(P) \in \mathcal{O}_P$ for all but finitely many $P \in \mathbb{P}_F$. Equivalently, an adele of F is a sequence $\alpha = (\alpha_P)_{P\in\mathbb{P}_F}$ such that $\alpha_P \in F$ for all $P \in \mathbb{P}_F$ and $\alpha_P \in \mathcal{O}_P$ for all but finitely many $P \in \mathbb{P}_F$. The valuation of α at $P \in \mathbb{P}_F$, denoted by $\nu_P(\alpha)$, is $\nu_P(\alpha_P)$. Let $\mathcal{A}_F$ denote the set of all adeles of F, and for each $A \in \mathcal{D}_F$, let $\mathcal{A}_F(A) = \{\alpha \in \mathcal{A}_F : \nu_P(\alpha) + \nu_P(A) \geq 0 \text{ for all } P \in \mathbb{P}_F\}$. A *Weil differential* of F/K is a K-linear map $\omega : \mathcal{A}_F \to K$ such that $\omega(\mathcal{A}_F(A) + F) = \{0\}$ for some $A \in \mathcal{D}_F$. Let Ω_F be the set of all Weil differentials of F/K. The product of an element $x \in F$ and a Weil differential $\omega \in \Omega_F$ is a Weil differential $x\omega \in \Omega_F$ defined by

$$\begin{array}{rccc} x\omega : & \mathcal{A}_F & \longrightarrow & K \\ & \alpha & \longmapsto & \omega(x\alpha). \end{array}$$

It turns out that Ω_F is a 1-dimensional vector space over F. For each $0 \neq \omega \in \Omega_F$, there is a unique divisor $(\omega) \in \mathcal{D}_F$, called the divisor of ω, with the following properties.

(i) ω vanishes on $\mathcal{A}_F((\omega)) + F$.

(ii) If ω vanishes on $\mathcal{A}_F(A) + F$ for some $A \in \mathcal{D}_F$, then $A \leq (\omega)$.

The divisor of a nonzero Weil differential of F/K is called a *canonical divisor* of F/K. The canonical divisors of F/K are precisely those with degree $2g-2$ and dimension g, where g is the genus of F/K. All canonical divisors form a class in $\mathcal{C}_F$.

The Riemann-Roch theorem is a fundamental equation frequently used throughout the theory of function fields.

Theorem 5.17 (Riemann-Roch). *Let W be a canonical divisor of F/K. Then for all $A \in \mathcal{D}_F$,*

$$\dim A = \deg A + 1 - g + \dim(W - A), \tag{5.31}$$

where g is the genus of F/K.

Theorem 5.18. *A function field F/K is rational, i.e., $F = K(x)$ for some $x \in F$ which is transcendental over K, if and only if F/K is of genus 0 and has a divisor of degree 1.*

Theorem 5.19 (Strong approximation). *Let $P_1, \dots, P_r$ be distinct places of F/K. For any given $x_1, \dots, x_r \in F$, $n_1, \dots, n_r \in \mathbb{Z}$, and $Q \in \mathbb{P}_F \setminus \{P_1, \dots, P_r\}$, there exists $x \in F$ such that*

$$\begin{cases} \nu_{P_i}(x - x_i) = n_i & \text{for } 1 \le i \le r, \\ \nu_P(x) \ge 0 & \text{for } P \in \mathbb{P}_F \setminus \{P_1, \dots, P_r, Q\}. \end{cases}$$

5.4.3. Algebraic extensions of function fields.

Let F/K be a function field and fix an algebraic closure Φ of F. An algebraic extension of F/K is a function field F'/K' such that $F \subset F' \subset \Phi$, F' is algebraic over F, and $K \subset K'$. It follows that K' is algebraic over K. Moreover, if $[F' : F] < \infty$, then $[K' : K] < \infty$.

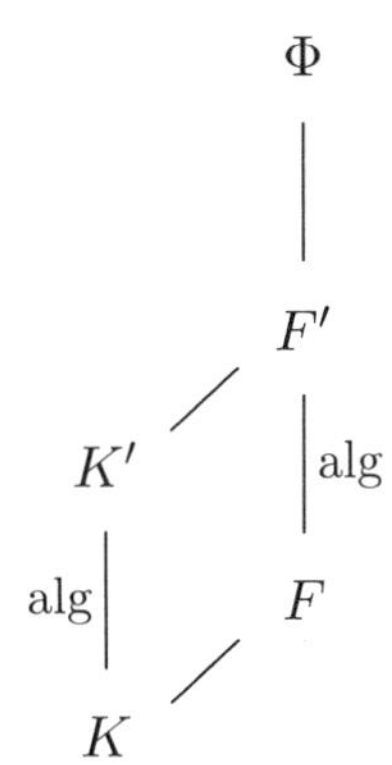

Let F'/K' be an algebraic extension of a function field F/K. For $P \in \mathbb{P}_F$ and $P' \in \mathbb{P}_{F'}$, if $P' \supset P$, we say that P' lies above P and write $P' \mid P$. For each $P' \in \mathbb{P}_{F'}$, $P = P' \cap F$ is the only place of F lying under P'. For each $P \in \mathbb{P}_F$, there is only a finite number (at least one) of places of F' lying above P. Let $P \in \mathbb{P}_F$ and $P' \in \mathbb{P}_{F'}$ be places such that $P' \mid P$, and let t be a prime of $\mathcal{O}_P$. The inclusion $\mathcal{O}_P \hookrightarrow \mathcal{O}_{P'}$ induces an embedding $F_P = \mathcal{O}_P/P \hookrightarrow \mathcal{O}_{P'}/P' = F_{P'}$. The *ramification index* and the *relative degree* of P' over P are defined as

$$e(P'|P) = \nu_{P'}(t)$$

and

$$f(P'|P) = [F_{P'} : F_P],$$

respectively.

Theorem 5.20. *Let F'/K' be a finite extension of a function field F/K and let $P \in \mathbb{P}_F$. Then*

$$\sum_{P' \in \mathbb{P}_{F'},\, P'|P} e(P'|P) f(P'|P) = [F' : F].$$

5.4.4. Bounds for the genus.

In this subsection, F/K is a function field where K is perfect and is algebraically closed in F. Let g denote the genus of F/K.

Theorem 5.21 (Castelnuovo)**.** *Let F_1 and F_2 be subfields of F such that F_1/K and F_2/K are function fields and $F = F_1F_2$. For $i = 1, 2$, let g_i denote the genus of F_i/K, and let $n_i = [F : F_i]$, which is finite since F is algebraic and finitely generated over F_i. Then*

$$g \leq n_1g_1 + n_2g_2 + (n_1 - 1)(n_2 - 1). \tag{5.32}$$

One can always write $F = K(x, y)$ for some $x, y \in F$ which are transcendental over K. Setting $F_1 = K(x)$ and $F_2 = K(y)$ in Theorem 5.21 gives

$$g \leq ([F : K(x)] - 1)([F : K(y)] - 1). \tag{5.33}$$

Theorem 5.22. *Assume that $F = K(x, y)$, where $x, y \in F$ are transcendental over K and $f(x, y) = 0$ for some irreducible $f \in K[\mathtt{X}, \mathtt{Y}]$ with $\deg f = d$. Then*

$$g \leq \frac{1}{2}(d - 1)(d - 2). \tag{5.34}$$

5.4.5. The zeta function of a function field over a finite field.

Throughout this subsection, $F/\mathbb{F}_q$ is a function field of genus g, and $\mathbb{F}_q$ is assumed to be algebraically closed in F. For each integer $n \geq 0$, let

$$A_n = |\{A \in \mathcal{D}_F : A \geq 0,\ \deg A = n\}|, \tag{5.35}$$

which is always positive and finite. Obviously, $A_0 = 1$. We have

$$A_n = \frac{h}{q-1}(q^{n+1-g} - 1) \quad \text{for } n > 2g-2, \tag{5.36}$$

where h is the *class number* of F, which is a positive integer defined by

$$h = |\{A \in \mathcal{D}_F : \deg A = 0\}/\mathcal{P}_F|. \tag{5.37}$$

For $1 \leq n \leq 2g-2$, A_n is more subtle and less predictable.

The *zeta function* of F is defined to be

$$Z_F(\mathtt{t}) = \sum_{n=0}^{\infty} A_n \mathtt{t}^n \in \mathbb{C}[[\mathtt{t}]]. \tag{5.38}$$

Equivalently,

$$Z_F(\mathtt{t}) = \prod_{P \in \mathbb{P}_F} (1 - \mathtt{t}^{\deg P})^{-1}. \tag{5.39}$$

For $A \in \mathcal{D}_F$, let $[A] = A + \mathcal{P}_F \in \mathcal{D}_F/\mathcal{P}_F = \mathcal{C}_F$, and define $\deg[A] = \deg A$ and $\dim[A] = \dim A$. (Recall that the degree and the dimension of a divisor depend only on its class in $\mathcal{C}_F$.)

Theorem 5.23. *We have*

(5.40)

$$Z_F(\mathtt{t}) = \begin{cases} \dfrac{1}{(1-\mathtt{t})(1-q\mathtt{t})} & \text{if } g = 0, \\ \dfrac{1}{q-1} \displaystyle\sum_{\substack{[A] \in \mathcal{C}_F \\ 0 \leq \deg[A] \leq 2g-2}} q^{\dim[A]} \mathtt{t}^{\deg[A]} + \dfrac{h}{q-1}\left[\dfrac{q^g \mathtt{t}^{2g-1}}{1-q\mathtt{t}} - \dfrac{1}{1-\mathtt{t}}\right] & \text{if } g > 0. \end{cases}$$

The zeta function satisfies the functional equation

$$Z_F(\mathtt{t}) = q^{g-1}\mathtt{t}^{2g-2} Z_F\Big(\frac{1}{q\mathtt{t}}\Big). \tag{5.41}$$

The L-polynomial. It follows from (5.40) that

$$Z_F(\mathtt{t}) = \frac{L_F(\mathtt{t})}{(1-\mathtt{t})(1-q\mathtt{t})},$$

where

(5.42) $$L_F(\mathtt{t}) = (1-\mathtt{t})(1-q\mathtt{t})Z_F(\mathtt{t}) \in \mathbb{Z}[\mathtt{t}]$$

is a polynomial of degree $2g$ and is called the *L-polynomial* of F. All information about the zeta function is encoded in the L-polynomial. The coefficients of $L_F(\mathtt{t})$ are known to the following extent: Let $L_F(\mathtt{t}) = a_0 + a_1\mathtt{t} + \cdots + a_{2g}\mathtt{t}^{2g}$. Then $a_0 = 1$, $a_{2g} = q^g$, $a_1 = A_1 - (q+1)$, where $A_1 = |\{P \in \mathbb{P}_F : \deg P = 1\}|$, and $a_{2g-i} = q^{g-i}a_i$ for all $0 \le i \le g$. Let $\alpha_1^{-1}, \ldots, \alpha_{2g}^{-1} \in \mathbb{C}$ be the roots of $L_F(\mathtt{t})$. Then

(5.43) $$L_F(\mathtt{t}) = \prod_{i=1}^{2g}(1-\alpha_i\mathtt{t}).$$

The complex numbers $\alpha_1, \ldots, \alpha_{2g}$, when suitably labeled, satisfy $\alpha_i\alpha_{g+i} = q$, $1 \le i \le g$. The following profound theorem, known as the *Riemann hypothesis for function fields* (RHFF), was proved by Weil [**41**].

Theorem 5.24 (RHFF). *Let $\alpha_1, \ldots, \alpha_{2g}$ be the reciprocals of the roots of the L-polynomial of a function field $F/\mathbb{F}_q$. Then*

(5.44) $$|\alpha_i| = q^{1/2} \quad \text{for all } 1 \le i \le 2g.$$

Corollary 5.25 (The Weil bound). *The number A_1 of degree-1 places of a function field $F/\mathbb{F}_q$ satisfies*

(5.45) $$|A_1 - (q+1)| \le 2gq^{1/2}.$$

Inequality (5.45) immediately follows from (5.44):

$$|A_1 - (q+1)| = \Big|-\sum_{i=1}^{2g}\alpha_i\Big| \le \sum_{i=1}^{2g}|\alpha_i| = 2gq^{1/2}.$$

The zeta function of $\mathbb{F}_{q^r}F/\mathbb{F}_{q^r}$. Let Ω be an extension of F that contains $\overline{\mathbb{F}}_q$. For each integer $r > 0$, let $F_r = \mathbb{F}_{q^r}F$. Then $F_r/\mathbb{F}_{q^r}$ is a function field where $\mathbb{F}_{q^r}$ is algebraically closed in F_r, and $F_r/\mathbb{F}_{q^r}$ has the same genus as $F/\mathbb{F}_q$. The zeta function and the L-polynomial of F_r are related to those of F. We have

$$Z_{F_r}(\mathtt{t}^r) = \prod_{\zeta\in\mathbb{C},\,\zeta^r=1} Z_F(\zeta\mathtt{t})$$

and

$$L_{F_r}(\mathtt{t}) = \prod_{i=1}^{2g}(1-\alpha_i^r\mathtt{t}),$$

where $L_F(\mathtt{t}) = \prod_{i=1}^{2g}(1-\alpha_i\mathtt{t})$.

5.4.6. The intersection number and Bézout's theorem.

Throughout this subsection, K is an algebraically closed field.

The affine case. Let $f \in K[\mathtt{X}, \mathtt{Y}] \setminus \{0\}$ and $P = (a, b) \in K^2$. Write

$$f = f_m(\mathtt{X} - a, \mathtt{Y} - b) + f_{m+1}(\mathtt{X} - a, \mathtt{Y} - b) + \cdots ,$$

where $f_i \in K[\mathtt{X}, \mathtt{Y}]$ is homogeneous of degree i (called an *i-form* in $\mathtt{X}, \mathtt{Y}$) and $f_m \neq 0$. The integer m is called the *multiplicity* of f at P and is denoted by $m_P(f)$. If $m_P(f) > 0$, the linear factors of f_m are called the *tangent lines* of f at P. If $m_P(f) = 1$, which happens if and only if $f(P) = 0$ and $((\partial f/\partial \mathtt{X})(P), (\partial f/\partial \mathtt{Y})(P)) \neq (0, 0)$, P is called a *simple point* of f; if $m_P(f) > 1$, P is called a *multiple point* of f. We define $m_P(0) = \infty$.

For $f, g \in K[\mathtt{X}, \mathtt{Y}]$ and $P \in K^2$, the *intersection number* of f and g at P is defined to be

$$I(P, f \cap g) = \dim_K(R/(f, g)R), \tag{5.46}$$

where

$$R = \left\{ \frac{u}{v} : u, v \in K[\mathtt{X}, \mathtt{Y}],\ v(P) \neq 0 \right\}$$

is the local ring of the affine plane K^2 at P.

The intersection number obeys the following axioms. Let $f, g, g_1, g_2 \in K[\mathtt{X}, \mathtt{Y}]$ and $P \in K^2$.

(i) $I(P, f \cap g) \in \mathbb{N} \cup \{\infty\}$; $I(P, f \cap g) = \infty$ if and only if $\gcd(f, g)(P) = 0$.

(ii) $I(P, f \cap g) = 0$ if and only if $(f(P), g(P)) \neq (0, 0)$.

(iii) Let T be an invertible affine transformation of K^2. Then $I(P, f \cap g) = I(T^{-1}(P), (f \circ T) \cap (g \circ T))$.

(iv) $I(P, f \cap g) = I(P, g \cap f)$.

(v) $I(P, f \cap g) \geq m_P(f) m_P(g)$, and the equality holds if and only if f and g have no common tangent lines at P.

(vi) $I(P, f \cap g_1 g_2) = I(P, f \cap g_1) + I(P, f \cap g_2)$.

(vii) If $g_1 \equiv g_2 \pmod{f}$, then $I(P, f \cap g_1) = I(P, f \cap g_2)$.

Moreover, $(f, g) \mapsto I(P, f \cap g)$ is the unique function from $K[\mathtt{X}, \mathtt{Y}]^2$ to $\mathbb{N} \cup \{\infty\}$ satisfying the above axioms.

An algorithm. There is an effective algorithm for computing the intersection number $I(P, f \cap g)$. Without loss of generality, assume that $P = (0, 0)$ and $f, g \in K[\mathtt{X}, \mathtt{Y}]$ are such that $f(P) = g(P) = 0$.

Case 1. Assume that one of $f(\mathtt{X},0)$ and $g(\mathtt{X},0)$ is 0, say $f(\mathtt{X},0)=0$. Then $f=\mathtt{Y}f_1$ for some $f_1\in K[\mathtt{X},\mathtt{Y}]$. By (iv) and (vi),

$$I(P,f\cap g)=I(P,\mathtt{Y}\cap g)+I(P,f_1\cap g). \tag{5.47}$$

Write $g=\mathtt{X}^m g_1(\mathtt{X})+\mathtt{Y}g_2(\mathtt{X},\mathtt{Y})$, where $g_1\in K[\mathtt{X}]$, $g_1(0)\neq 0$, and $g_2\in K[\mathtt{X},\mathtt{Y}]$. Then in (5.47),

$$\begin{aligned} I(P,\mathtt{Y}\cap g) &= I(P,\mathtt{Y}\cap \mathtt{X}^m g_1(\mathtt{X})) && \text{(by (vii))}\\ &= I(P,\mathtt{Y}\cap\mathtt{X}^m) && \text{(by (vi) and (ii))}\\ &= m && \text{(by (v))}. \end{aligned}$$

To find $I(P,f_1\cap g)$ in (5.47), replace f with f_1 and repeat the procedure.

Case 2. Assume that $f(\mathtt{X},0)\neq 0$ and $g(\mathtt{X},0)\neq 0$. Let $f(\mathtt{X},0)=a_r\mathtt{X}^r+\cdots+a_1\mathtt{X}$ and $g(\mathtt{X},0)=b_s\mathtt{X}^s+\cdots+b_1\mathtt{X}$, where $a_rb_s\neq 0$. Without loss of generality, assume that $r\le s$. Let $g_1=g-b_sa_r^{-1}\mathtt{X}^{s-r}f$. Then by (vii),

$$I(P,f\cap g)=I(P,f\cap g_1),$$

where $\deg g_1(\mathtt{X},0)<s=\deg g(\mathtt{X},0)$. Replace g with g_1 and repeat the procedure. Continue in this way until Case 1 applies.

In a finite number of steps, the algorithm either determines $I(P,f\cap g)$ or finds that $I(P,f\cap g)>\deg f\cdot\deg g$. In the latter case, $I(P,f\cap g)=\infty$ by Theorem 5.27.

Example 5.26. Let $f=\mathtt{X}+\mathtt{Y}^2+\mathtt{X}^4$, $g=\mathtt{X}+\mathtt{XY}+\mathtt{Y}^2-\mathtt{X}^3+\mathtt{X}^2\mathtt{Y}\in K[\mathtt{X},\mathtt{Y}]$, and $P=(0,0)$. Then

$$\begin{aligned} &I(P,f\cap g)\\ &=I\big(P,\ (\mathtt{X}+\mathtt{Y}^2+\mathtt{X}^4)\cap(\mathtt{X}+\mathtt{XY}+\mathtt{Y}^2-\mathtt{X}^3+\mathtt{X}^2\mathtt{Y})\big)\\ &=I\big(P,\ (\mathtt{X}+\mathtt{Y}^2+\mathtt{X}^4)\cap(\mathtt{XY}-\mathtt{X}^3+\mathtt{X}^2\mathtt{Y}-\mathtt{X}^4)\big)\\ &=I\big(P,\ (\mathtt{X}+\mathtt{Y}^2+\mathtt{X}^4)\cap\mathtt{X}(\mathtt{Y}-\mathtt{X}^2+\mathtt{XY}-\mathtt{X}^3)\big)\\ &=I\big(P,\ (\mathtt{X}+\mathtt{Y}^2+\mathtt{X}^4)\cap\mathtt{X}\big)+I\big(P,\ (\mathtt{X}+\mathtt{Y}^2+\mathtt{X}^4)\cap(\mathtt{Y}-\mathtt{X}^2+\mathtt{XY}-\mathtt{X}^3)\big)\\ &=2+1=3. \end{aligned}$$

The projective case. The n-dimensional projective space over K, denoted by $\mathbb{P}^n(K)$, is the quotient set $(K^{n+1}\setminus\{0\})/\sim$, where $\sim$ is an equivalence relation on $K^{n+1}\setminus\{0\}$ defined as follows: For $x,y\in K^{n+1}\setminus\{0\}$, $x\sim y$ if and only if $x=ky$ for some $k\in K^*$. The $\sim$ equivalence class of $(x_0,\dots,x_n)\in K^{n+1}\setminus\{0\}$ is denoted by $(x_0:\cdots:x_n)$.

Let $f\in K[\mathtt{X},\mathtt{Y},\mathtt{Z}]$ be a homogeneous polynomial and $P\in\mathbb{P}^2(K)$. Write P in such a way that one of its nonzero coordinates is 1, say $P=(a:b:1)$, and define

$$m_P(f)=m_{(a,b)}(f(\mathtt{X},\mathtt{Y},1))$$

as the *multiplicity* of f at P. The multiplicity so defined does not depend on the choice of the "special" coordinate of P that is scaled to 1. A point $P \in \mathbb{P}^2(K)$ is a multiple point of f if and only if $f(P) = (\partial f/\partial \mathtt{X})(P) = (\partial f/\partial \mathtt{Y})(P) = (\partial f/\partial \mathtt{Z})(P) = 0$.

Let $f, g \in K[\mathtt{X}, \mathtt{Y}, \mathtt{Z}]$ be two homogeneous polynomials with $\deg f = m$ and $\deg g = n$, and let $P \in \mathbb{P}^2(K)$. The *intersection number* of f and g at P is defined to be

$$I(P, f \cap g) = \dim_K(S/(f/l^m, g/l^n)S), \tag{5.48}$$

where $l \in K[\mathtt{X}, \mathtt{Y}, \mathtt{Z}]$ is any linear form such that $l(P) \neq 0$ and

$$S = \left\{\frac{u}{v} : u, v \in K[\mathtt{X}, \mathtt{Y}, \mathtt{Z}] \text{ are homogeneous of the same degree, } v(P) \neq 0\right\}$$

is the local ring of $\mathbb{P}^2(K)$ at P. The intersection number $I(P, f \cap g)$ is independent of the choice of l. In particular, if $P = (a : b : 1)$, we can choose $l = \mathtt{Z}$, and hence

$$I(P, f \cap g) = I((a, b), f(\mathtt{X}, \mathtt{Y}, 1) \cap g(\mathtt{X}, \mathtt{Y}, 1)), \tag{5.49}$$

where the right side of (5.49) is the "affine" intersection number discussed earlier.

Theorem 5.27 (Bézout). *Let $f, g \in K[\mathtt{X}, \mathtt{Y}, \mathtt{Z}]$ be homogeneous of degree m and n, respectively, such that $\gcd(f, g) = 1$. Then*

$$\sum_{P \in \mathbb{P}^2(K)} I(P, f \cap g) = mn. \tag{5.50}$$

5.4.7. The Hasse-Weil bound and the Lang-Weil bound.

A polynomial $f \in \mathbb{F}_q[\mathtt{X}_0, \dots, \mathtt{X}_n]$ is said to be *absolutely irreducible* if it is irreducible in $\overline{\mathbb{F}}_q[\mathtt{X}_0, \dots, \mathtt{X}_n]$. If $f \in \mathbb{F}_q[\mathtt{X}_0, \dots, \mathtt{X}_n]$ is homogeneous, define

$$V_{\mathbb{P}^n(\mathbb{F}_q)}(f) = \{(x_0 : \cdots : x_n) \in \mathbb{P}^n(\mathbb{F}_q) : f(x_0, \dots, x_n) = 0\}.$$

Theorem 5.28 (The Hasse-Weil bound). *Let $f \in \mathbb{F}_q[\mathtt{X}, \mathtt{Y}, \mathtt{Z}]$ be an absolutely irreducible homogeneous polynomial of degree d. Then*

$$\left||V_{\mathbb{P}^2(\mathbb{F}_q)}(f)| - q\right| \leq (d-1)(d-2)q^{1/2} + c(d), \tag{5.51}$$

where $c(d)$ is a constant depending only on d.

Theorem 5.28 is derived from the Weil bound (Corollary 5.25) and Theorem 5.22. Moreover, the constant $c(d)$ can be chosen as

$$c(d) = \frac{1}{2}d(d-1)^2 + 1. \tag{5.52}$$

Without loss of generality, we may assume that $\deg_{\mathtt{Y}}(\mathtt{X}, \mathtt{Y}, 1) > 0$. Let $F = \mathbb{F}_q(x, y)$, where x is transcendental over $\mathbb{F}_q$ and y is a root of $f(x, \mathtt{Y}, 1)$. Then

$F/\mathbb{F}_q$ is a function field. The absolute irreducibility of f implies that $\mathbb{F}_q$ is algebraically closed in F. (Let K be the algebraic closure of $\mathbb{F}_q$ in F. Then $\mathbb{F}_q(x) \subset K(x) \subset \mathbb{F}_q(x,y)$. Since $f(x,\mathtt{Y},1)$ is the minimal polynomial of y over $\overline{\mathbb{F}}_q(x)$, $[\mathbb{F}_q(x,y) : K(x)] = \deg_{\mathtt{Y}} f(x,\mathtt{Y},1) = [\mathbb{F}_q(x,y) : \mathbb{F}_q(x)]$. Thus $K(x) = \mathbb{F}_q(x)$, i.e., $K = \mathbb{F}_q$.) The genus g of $F/\mathbb{F}_q$ satisfies $g \le (d-1)(d-2)/2$ by Theorem 5.22. Let $\mathcal{A}_1 = \{P \in \mathbb{P}_F : \deg P = 1\}$ and $A_1 = |\mathcal{A}_1|$. By Corollary 5.25,

$$|A_1 - (q+1)| \le (d-1)(d-2)q^{1/2}. \tag{5.53}$$

For each $P \in \mathcal{A}_1$, let $z \in \{x, y, 1\}$ be such that $\nu_P(z) = \min\{\nu_P(x), \nu_P(y), \nu_P(1)\}$. Then

$$\Big(\frac{x}{z}(P) : \frac{y}{z}(P) : \frac{1}{z}(P)\Big) \in V_{\mathbb{P}^2(\mathbb{F}_q)}(f).$$

This defines a mapping

$$\begin{array}{rccl} \psi: & \mathcal{A}_1 & \longrightarrow & V_{\mathbb{P}^2(\mathbb{F}_q)}(f) \\ & P & \longmapsto & \Big(\frac{x}{z}(P) : \frac{y}{z}(P) : \frac{1}{z}(P)\Big). \end{array}$$

If $(x_0 : y_0 : z_0) \in V_{\mathbb{P}^2(\mathbb{F}_q)}(f)$ is a simple point, then $|\psi^{-1}(x_0 : y_0 : z_0)| = 1$. In general, given $(x_0 : y_0 : z_0) \in V_{\mathbb{P}^2(\mathbb{F}_q)}(f)$, $|\psi^{-1}(x_0 : y_0 : z_0)| \le d$. To see this, assume, without loss of generality, that $z_0 = 1$. Let P denote the place of the rational function field $\mathbb{F}_q(x)/\mathbb{F}_q$ that is the zero of $x - x_0$. There are at most $[F : \mathbb{F}_q(x)] \le d$ places of $F/\mathbb{F}_q$ lying above P. Since all places in $\psi^{-1}(x_0 : y_0 : 1)$ lie above P, we have $|\psi^{-1}(x_0 : y_0 : 1)| \le d$. Let S be the set of all multiple points on $V_{\mathbb{P}^2(\mathbb{F}_q)}(f)$. It follows that

$$|V_{\mathbb{P}^2(\mathbb{F}_q)}(f)| - |S| \le A_1 \le |V_{\mathbb{P}^2(\mathbb{F}_q)}(f)| - |S| + |S|d. \tag{5.54}$$

Without loss of generality, assume that $\partial f/\partial\mathtt{X} \ne 0$. Since $S \subset V_{\mathbb{P}^2(\mathbb{F}_q)}(f) \cap V_{\mathbb{P}^2(\mathbb{F}_q)}(\partial f/\partial\mathtt{X})$, by Bézout's theorem (Theorem 5.27),

$$d(d-1) \ge \deg f \cdot \deg \frac{\partial f}{\partial\mathtt{X}} \ge \sum_{P\in S} I\Big(P,\ f \cap \frac{\partial f}{\partial\mathtt{X}}\Big) \ge 2|S|.$$

Hence

$$|S| \le \frac{1}{2}d(d-1). \tag{5.55}$$

Combining (5.53)–(5.55) gives

$$\begin{aligned} |V_{\mathbb{P}^2(\mathbb{F}_q)}(f)| - q &\le A_1 + |S| - q = A_1 - (q+1) + |S| + 1 \\ &\le (d-1)(d-2)q^{1/2} + \frac{1}{2}d(d-1) + 1 \end{aligned}$$

and

$$\begin{aligned}|V_{\mathbb{P}^2(\mathbb{F}_q)}(f)| - q &\geq A_1 - (d-1)|S| - q = A_1 - (q+1) - (d-1)|S| + 1 \\ &\geq -(d-1)(d-2)q^{1/2} - \frac{1}{2}d(d-1)^2 + 1.\end{aligned}$$

Hence we may choose $c(d)$ as in (5.52).

The following generalization of Theorem 5.28 is known as the *Lang-Weil bound* [**25**].

Theorem 5.29 (Lang-Weil). *Let $f \in \mathbb{F}_q[\mathtt{X}_0, \ldots, \mathtt{X}_n]$ be an absolutely irreducible homogeneous polynomial of degree d. Then*

$$\big||V_{\mathbb{P}^n(\mathbb{F}_q)}(f)| - q^{n-1}\big| \leq (d-1)(d-2)q^{n-\frac{3}{2}} + c(n,d)q^{n-2},$$

where $c(n,d)$ is a constant depending only on n and d.

For an explicit estimate of the constant $c(n,d)$, see [**6**].

Exercises

5.1. Find all roots of $f(\mathtt{X},\mathtt{Y}) = \mathtt{Y}^2 - \mathtt{X}(\mathtt{X}-1)(\mathtt{X}-2) \in \mathbb{F}_5[\mathtt{X},\mathtt{Y}]$ in $\mathbb{F}_5^2$.

5.2. Use Corollary 5.12 to compute the minimum weight of the Reed-Muller code $R_{2^3}(d,3)$ for all $0 \leq d \leq 3(2^3-1)$.

5.3. Let n be a positive integer.
  (i) (Chevalley) Prove that every homogeneous polynomial $f \in \mathbb{F}_q[\mathtt{X}_1, \ldots, \mathtt{X}_n]$ with $0 < \deg f < n$ has a nontrivial zero, i.e., there exists $(0,\ldots,0) \neq (x_1,\ldots,x_n) \in \mathbb{F}_q^n$ such that $f(x_1,\ldots,x_n) = 0$.
  (ii) Let m be the largest integer such that there are m matrices $A_1,\ldots,A_m \in \mathrm{M}_{n\times n}(\mathbb{F}_q)$ having the property that $x_1A_1 + \cdots + x_mA_m \in \mathrm{GL}(n,\mathbb{F}_q)$ for all $(0,\ldots,0) \neq (x_1,\ldots,x_m) \in \mathbb{F}_q^m$. Prove that $m = n$.

5.4. Let $f \in \mathbb{F}_q[\mathtt{X}_0,\ldots,\mathtt{X}_n]$ be homogeneous of degree $d > 0$. Prove that

$$|V_{\mathbb{P}^n(\mathbb{F}_q)}(f)| \equiv 1 + q + \cdots + q^{\lceil (n+1)/d\rceil - 2} \pmod{q^{\lceil (n+1)/d\rceil - 1}}.$$

5.5. Prove that

$$\sum_{\substack{f\in\mathbb{F}_q[\mathtt{X}_1,\ldots,\mathtt{X}_n]\\ \deg f\leq d}} |Z(f)| = q^{n-1+\binom{n+d}{d}}.$$

5.6. Let $\mathbb{F}_q[\mathtt{X}_1,\ldots,\mathtt{X}_n]^{(d)}$ be the set of all homogeneous polynomials of degree d in $\mathbb{F}_q[\mathtt{X}_1,\ldots,\mathtt{X}_n]$, including 0. Prove that for $d > 0$,

$$\sum_{f\in\mathbb{F}_q[\mathtt{X}_1,\ldots,\mathtt{X}_n]^{(d)}} |Z(f)| = q^{\binom{n-1+d}{d}-1}(q^n + q - 1).$$

5.7. Let $f = a_1\mathtt{X}_1^{k_1} + \cdots + a_n\mathtt{X}_n^{k_n} \in \mathbb{F}_q[\mathtt{X}_1, \dots, \mathtt{X}_n]$, where $k_1, \dots, k_n > 0$ and $a_1, \dots, a_n \in \mathbb{F}_q^*$. Let $d_i = \gcd(k_i, q-1)$.

(i) Prove that for $b \in \mathbb{F}_q^*$,

$$|Z(f-b)| = q^{n-1} + \sum_{l_1=1}^{d_1-1} \cdots \sum_{l_n=1}^{d_n-1} \lambda\Big(b^{(q-1)\sum_i l_i/d_i} \prod_i a_i^{-l_i(q-1)/d_i}\Big) \cdot J(\lambda^{l_1(q-1)/d_1}, \dots, \lambda^{l_n(q-1)/d_n}),$$

where $\widehat{\mathbb{F}_q^*} = \langle\lambda\rangle$ and J is the Jacobi sum. (Hint: $|Z(f-b)| = \sum_{c_1+\cdots+c_n=b} \prod_i |Z(a_i\mathtt{X}_i^{k_i} - c_i)|$, where $|Z(a_i\mathtt{X}_i^{k_i} - c_i)| = \sum_{l_i=0}^{d_i-1} \lambda^{l_i(q-1)/d_i}(a_i^{-1}c_i)$.)

(ii) Prove that

$$|Z(f)| = q^{n-1} + \frac{q-1}{q} \sum_{\substack{1 \le l_i \le d_i - 1 \\ \sum_i l_i/d_i \in \mathbb{Z}}} \lambda\Big(\prod_i a_i^{-l_i(q-1)/d_i}\Big) \prod_i g(\lambda^{l_i(q-1)/d_i}).$$

5.8. (i) Let $A \in \mathrm{M}_{(q(q^n-1)/(q-1)) \times q^n}(\mathbb{C})$ be the incidence matrix whose columns are labeled by the elements of $\mathbb{F}_q^n$, whose rows are labeled by the $(n-1)$-flats of $\mathbb{F}_q^n$, and whose (i,j)-entry is

$$a_{ij} = \begin{cases} 1 & \text{if the } j\text{th element of } \mathbb{F}_q^n \text{ is contained} \\ & \text{in the } i\text{th } (n-1)\text{-flat of } \mathbb{F}_q^n, \\ 0 & \text{otherwise.} \end{cases}$$

Prove that $\operatorname{rank} A = q^n$.

(ii) Use (i) to give an alternate proof of Lemma 5.13: If $S \subset \mathbb{F}_q^n$ is such that $|S \cap H|$ is a constant for every $(n-1)$-flat H of $\mathbb{F}_q^n$, then $S = \emptyset$ or $\mathbb{F}_q^n$.

5.9. Let p be a prime. Prove that the Reed-Muller code $R_p(d, n)$, $0 \le d \le n(p-1)$, is generated over $\mathbb{F}_p$ by the elements of minimum (nonzero) weight.

5.10. Consider $f \in \mathbb{F}_q[\mathtt{X}]$ which is not of the form g^p, where $p = \operatorname{char} \mathbb{F}_q$ and $g \in \mathbb{F}_q[\mathtt{X}]$. Assume that $(f(\mathtt{X}) - f(\mathtt{Y}))/(\mathtt{X} - \mathtt{Y})$ has a factor $A \in \mathbb{F}_q[\mathtt{X}, \mathtt{Y}]$ which is absolutely irreducible with $\deg A = d$. Prove that f is not a permutation polynomial of $\mathbb{F}_q$ when

$$q^{1/2} > \frac{1}{2}\Big[(d-1)(d-2) + \sqrt{(d-1)^2(d-2)^2 + 8d + 4 + 4c(d)}\Big],$$

where $c(d)$ is the constant in the Hasse-Weil bound (Theorem 5.28). (Hint: Use the Hasse-Weil bound to show that $|V_{\mathbb{F}_q^2}(A)| > d$.)

5.11. Let F be a field and let $\mathtt{t}, \mathtt{X}$, and $\mathtt{Y}$ be algebraically independent over F. Let $A, B \in F[\mathtt{X}]$ be such that $B \neq 0$, $\gcd(A, B) = 1$, and A' and B' are not both 0. Prove that $[A(\mathtt{X})B(\mathtt{Y}) - A(\mathtt{Y})B(\mathtt{X})]/(\mathtt{X} - \mathtt{Y})$ is irreducible in $F[\mathtt{X}, \mathtt{Y}]$ if and only if $\mathrm{Gal}\,(A(\mathtt{X}) - B(\mathtt{X})\mathtt{t}/F(\mathtt{t}))$, the Galois group of the polynomial $A(\mathtt{X}) - B(\mathtt{X})\mathtt{t}$ over $F(\mathtt{t})$, acts 2-transitively on the roots of $A(\mathtt{X}) - B(\mathtt{X})\mathtt{t}$. (The Galois group $\mathrm{Gal}\,(A(\mathtt{X}) - B(\mathtt{X})\mathtt{t}/F(\mathtt{t}))$ is called the *arithmetic monodromy group* of $A(\mathtt{X})/B(\mathtt{X})$.)

5.12. Let

$$f = \mathtt{X}^3\mathtt{Z} + \mathtt{X}^2\mathtt{Y}^2 + \mathtt{Z}^4, \quad g = \mathtt{X}^2\mathtt{Z} + \mathtt{X}\mathtt{Y}^2 + \mathtt{X}\mathtt{Y}\mathtt{Z} + \mathtt{Y}\mathtt{Z}^2 - \mathtt{Z}^3 \in F[\mathtt{X}, \mathtt{Y}, \mathtt{Z}],$$

where F is an algebraically closed field.

(i) Find all common zeros of f and g in $\mathbb{P}^2(F)$.

(ii) Determine $m_P(f)$ and $m_P(g)$ for each common zero P of f and g.

(iii) Use the algorithm in subsection 5.4.6 to find all intersection numbers of f and g.

(Note: The solution is dependent on $\mathrm{char}\, F$.)

5.13. Let K be an algebraically closed field. Let $f \in K[\mathtt{X}, \mathtt{Y}, \mathtt{Z}] \setminus K$ be homogeneous without multiple points in $\mathbb{P}^2(K)$. Prove that f is irreducible in $K[\mathtt{X}, \mathtt{Y}, \mathtt{Z}]$. (Hint: Use Bézout's theorem.)

5.14. (Elliptic curve) Let K be an algebraically closed field with $\mathrm{char}\, K \neq 2, 3$. Let $f = \mathtt{Y}^2\mathtt{Z} - a_0\mathtt{Z}^3 - a_1\mathtt{Z}^2\mathtt{X} - a_2\mathtt{Z}\mathtt{X}^2 - a_3\mathtt{X}^3 \in K[\mathtt{X}, \mathtt{Y}, \mathtt{Z}]$, where $a_3 \neq 0$ and $a_0 + a_1\mathtt{X} + a_2\mathtt{X}^2 + a_3\mathtt{X}^3 \in K[\mathtt{X}]$ has no multiple roots. Prove that f has no multiple points in $\mathbb{P}^2(K)$.

5.15. Let $F/\mathbb{F}_q$ be a function field with genus g. ($\mathbb{F}_q$ is assumed to be algebraically closed in F.) Let $A_n = |\{A \in \mathcal{D}_F : A \geq 0,\ \deg A = n\}|$.

(i) Assume that $g = 1$. Express $L_F(\mathtt{t})$ in terms of A_1.

(ii) Assume that $g = 2$. Express $L_F(\mathtt{t})$ in terms of A_1 and A_2.

5.16. (Lambert series) Let R be a commutative ring. Prove the following identity in $R[[\mathtt{t}]]$ (the ring of formal power series over R):

$$\sum_{n \geq 1} a_n \mathtt{t}^n (1 - \mathtt{t}^n)^{-1} = \sum_{n \geq 1} \Big(\sum_{d | n} a_d \Big) \mathtt{t}^n.$$

5.17. Let $F/\mathbb{F}_q$ be a function field, where $\mathbb{F}_q$ is algebraically closed in F. For $n \geq 0$, let B_n denote the number of places of F of degree n.

(i) Prove that

$$\frac{Z_F'(\mathtt{t})}{Z_F(\mathtt{t})} = \sum_{n \geq 1} B_n n \mathtt{t}^{n-1} (1 - \mathtt{t}^n)^{-1}.$$

(Hint: Start with $Z_F(\mathtt{t}) = \prod_{n\geq 1}(1-\mathtt{t}^n)^{-B_n}$ and then differentiate $\log Z_F(\mathtt{t})$.)

(ii) Use (i) and Exercise 5.16 to prove that

$$\frac{Z_F'(\mathtt{t})}{Z_F(\mathtt{t})} = \sum_{n\geq 1}\Bigl(\sum_{d\mid n} d\,B_d\Bigr)\mathtt{t}^{n-1}.$$

Chapter 6

Classical Groups

The classical groups usually refer to the group of invertible linear transformations of an n-dimensional vector space V over a field F, its subgroups preserving certain forms defined on V, and other groups related to these groups. Equivalently, elements of a classical group can be treated as $n \times n$ invertible matrices over F with additional properties. Our approach to classical groups in this book is more matrix theoretic than geometric but not entirely the former. A practical reason for this preference is that matrices are easier to handle in computations. For a more geometric approach to the subject, the reader is referred to [**14**] and [**21**, Chapter 6].

In this chapter, it is not always necessary to assume that the field is finite. Many results are stated and proved for an arbitrary field.

In our notation, the center of a group G is denoted by $\mathrm{Z}(G)$ and the commutator subgroup of G is denoted by G'. For $H < G$, $X \subset G$ and $a \in G$, $\mathrm{N}_G(H)$ is the normalizer of H in G, $\mathrm{C}_G(X)$ is the centralizer of X in G, and $[a]_G$ is the conjugacy class of a in G. We let $\mathfrak{S}_n$ denote the symmetric group of degree n; more generally, $\mathfrak{S}_X$ is the symmetric group on a set X. The set of all $m \times n$ matrices over a field F is denoted by $\mathrm{M}_{m\times n}(F)$, and $\mathrm{M}_n(F)$ stands for $\mathrm{M}_{n\times n}(F)$. The identity matrix in $\mathrm{M}_n(F)$ is denoted by I_n or I. The $n \times n$ zero matrix is denoted by 0_n. When n is specified, E_{ij} is the $n \times n$ matrix whose (i, j)-entry is 1 and whose other entries are 0. The vector space F^n can be interpreted either as the row space $\mathrm{M}_{1\times n}(F)$ or as the column space $\mathrm{M}_{n\times 1}(F)$, depending on the context. For $A \in \mathrm{M}_{m\times n}(F)$, we define $\ker A = \{x \in F^n : Ax = 0\}$. For any two matrices A and B, we write

$$A \oplus B = \begin{bmatrix} A & \\ & B \end{bmatrix}.$$

6.1. The General Linear Group and Its Related Groups

Let F be a field. The *general linear group* of degree n over F, denoted by $\mathrm{GL}(n,F)$, is the group of all $n\times n$ invertible matrices over F. The *affine linear group* of degree n over F is

$$\mathrm{AGL}(n,F)=\left\{\begin{bmatrix}A&0\\a&1\end{bmatrix}:A\in\mathrm{GL}(n,F),\ a\in F^n\right\}<\mathrm{GL}(n+1,F).$$

Both $\mathrm{GL}(n,F)$ and $\mathrm{AGL}(n,F)$ act on F^n from the right. For $A\in\mathrm{GL}(n,F)$, $\sigma=\left[\begin{smallmatrix}A&0\\a&1\end{smallmatrix}\right]\in\mathrm{AGL}(n,F)$ and $x\in F^n$,

$$x^A=xA\qquad\text{and}\qquad x^\sigma=xA+a.$$

Through these actions, $\mathrm{GL}(n,F)$ is identified with the group of invertible linear transformations of F^n and $\mathrm{AGL}(n,F)$ with the group of invertible affine transformations of F^n.

The action of $\mathrm{GL}(n,F)$ on F^n defines a semidirect product $\mathrm{GL}(n,F)\ltimes F^n$: For $(A,a),(B,b)\in\mathrm{GL}(n,F)\times F^n$, their product is defined as

$$(A,a)\cdot(B,b)=(AB,\ a^B+b)=(AB,\ aB+b).$$

Clearly,

$$\begin{array}{rccc}\phi: & \mathrm{AGL}(n,F) & \longrightarrow & \mathrm{GL}(n,F)\ltimes F^n\\ & \begin{bmatrix}A&0\\a&1\end{bmatrix} & \longmapsto & (A,a)\end{array}$$

is an isomorphism.

The group

$$\mathrm{SL}(n,F)=\{A\in\mathrm{GL}(n,F):\det A=1\}\lhd\mathrm{GL}(n,F)$$

is called the *special linear group* of degree n over F. It is the kernel of the homomorphism $\det:\mathrm{GL}(n,F)\to F^*$.

Proposition 6.1. *We have*

$$\mathrm{C}_{\mathrm{GL}(n,F)}\big(\mathrm{SL}(n,F)\big)=\{aI_n:a\in F^*\},\tag{6.1}$$

$$\mathrm{Z}\big(\mathrm{GL}(n,F)\big)=\{aI_n:a\in F^*\},\tag{6.2}$$

$$\mathrm{Z}\big(\mathrm{SL}(n,F)\big)=\{aI_n:a\in F,\ a^n=1\}.\tag{6.3}$$

Proof. To prove (6.1), it suffices to show that the left side is contained in the right side. Let $A=(a_{ij})\in\mathrm{C}_{\mathrm{GL}(n,F)}\big(\mathrm{SL}(n,F)\big)$. Then for $i\neq j$, $I+E_{ij}\in\mathrm{SL}(n,F)$, so $A(I+E_{ij})=(I+E_{ij})A$, i.e.,

$$AE_{ij}=E_{ij}A.\tag{6.4}$$

The (j,j)-entry of AE_{ij} is a_{ji}, and the (j,j)-entry of $E_{ij}A$ is 0. Thus $a_{ji}=0$. The (i,j)-entry of AE_{ij} is a_{ii}, and the (i,j)-entry of $E_{ij}A$ is a_{jj}. Thus $a_{ii}=a_{jj}$. Therefore $A=aI_n$ for some $a\in F^*$.

Equations (6.2) and (6.3) immediately follow from (6.1). □

Define

$$\begin{aligned}\mathrm{PGL}(n,F)&=\mathrm{GL}(n,F)/\mathrm{Z}\big(\mathrm{GL}(n,F)\big),\\ \mathrm{PSL}(n,F)&=\mathrm{SL}(n,F)/\mathrm{Z}\big(\mathrm{SL}(n,F)\big).\end{aligned}$$

Respectively, they are called the *projective general linear group* and the *projective special linear group* of degree n over F. Since $\mathrm{Z}\big(\mathrm{SL}(n,F)\big)=\mathrm{Z}\big(\mathrm{GL}(n,F)\big)\cap\mathrm{SL}(n,F)$, the inclusion $\mathrm{SL}(n,F)\subset\mathrm{GL}(n,F)$ induces an embedding $\mathrm{PSL}(n,F)\hookrightarrow\mathrm{PGL}(n,F)$.

Let $\mathbb{P}^n(F)$ be the n-dimensional projective space over F. For $x=(x_0,\dots,x_n)\in F^{n+1}\setminus\{0\}$, let $\langle x\rangle$ denote the 1-dimensional subspace of F^{n+1} generated by x. The element $(x_0:\dots:x_n)\in\mathbb{P}^n(F)$ can be identified with $\langle x\rangle$. The action of $\mathrm{GL}(n+1,F)$ on F^{n+1} induces an action of $\mathrm{GL}(n+1,F)$ on $\mathbb{P}^n(F)$ with kernel $\mathrm{Z}\big(\mathrm{GL}(n+1,F)\big)$. Therefore, $\mathrm{PGL}(n+1,F)$ acts faithfully, i.e., with a trivial kernel, on $\mathbb{P}^n(F)$. More explicitly, for $x\in F^{n+1}\setminus\{0\}$ and $A\in\mathrm{GL}(n+1,F)$, with $\bar{A}$ denoting the image of A in $\mathrm{PGL}(n+1,F)$, the action of $\bar{A}$ on $\langle x\rangle$ is given by

$$\langle x\rangle^{\bar{A}}=\langle xA\rangle.$$

When $F=\mathbb{F}_q$, the orders of the groups introduced above are easily determined.

Proposition 6.2. *We have*

$$|\mathrm{GL}(n,\mathbb{F}_q)|=q^{\frac{1}{2}n(n-1)}\prod_{i=1}^{n}(q^i-1). \tag{6.5}$$

Proof. Let us construct a matrix $A\in\mathrm{GL}(n,\mathbb{F}_q)$ row by row. The first row of A can be any nonzero vector in $\mathbb{F}_q^n$, so there are q^n-1 choices. Assume that the first i rows $a_1,\dots,a_i$ of A have been chosen. Then the $(i+1)$th row can be any element of $\mathbb{F}_q^n$ which is not a linear combination of $a_1,\dots,a_i$, so there are q^n-q^i choices for the $(i+1)$th row. Thus,

$$|\mathrm{GL}(n,\mathbb{F}_q)|=(q^n-1)(q^n-q)\cdots(q^n-q^{n-1})=q^{\frac{1}{2}n(n-1)}\prod_{i=1}^{n}(q^i-1).$$

□

The orders of $\mathrm{AGL}(n, \mathbb{F}_q)$, $\mathrm{PGL}(n, \mathbb{F}_q)$, $\mathrm{SL}(n, \mathbb{F}_q)$ and $\mathrm{PSL}(n, \mathbb{F}_q)$ are given by the following obvious formulas:

$$|\mathrm{AGL}(n, \mathbb{F}_q)| = q^n |\mathrm{GL}(n, \mathbb{F}_q)|,$$

$$|\mathrm{PGL}(n, \mathbb{F}_q)| = |\mathrm{SL}(n, \mathbb{F}_q)| = \frac{1}{q-1} |\mathrm{GL}(n, \mathbb{F}_q)|,$$

$$|\mathrm{PSL}(n, \mathbb{F}_q)| = \frac{1}{(q-1)\gcd(n, q-1)} |\mathrm{GL}(n, \mathbb{F}_q)|.$$

In the last formula, $\gcd(n, q-1)$ is the number of nth roots of unity in $\mathbb{F}_q$.

Let $T(n, F) < \mathrm{GL}(n, F)$ be the subgroup of invertible lower triangular matrices and $T^*(n, F) < T(n, F)$ the subgroup of lower triangular matrices with 1's on the diagonal. Then $|T(n, \mathbb{F}_q)| = (q-1)^n q^{n(n-1)/2}$ and $|T^*(n, \mathbb{F}_q)| = q^{n(n-1)/2}$. By (6.5), $T^*(n, \mathbb{F}_q)$ is a Sylow p-subgroup of $\mathrm{GL}(n, \mathbb{F}_q)$, where $p = \operatorname{char} \mathbb{F}_q$.

Proposition 6.3. *We have*

$$\mathrm{N}_{\mathrm{GL}(n,F)}\big(T^*(n, F)\big) = T(n, F).$$

Consequently, the number of Sylow p-subgroups in $\mathrm{GL}(n, \mathbb{F}_q)$ *is*

$$\frac{|\mathrm{GL}(n, \mathbb{F}_q)|}{|T(n, \mathbb{F}_q)|} = (q-1)^{-n} \prod_{i=1}^{n} (q^i - 1).$$

Proof. It suffices to show that $\mathrm{N}_{\mathrm{GL}(n,F)}\big(T^*(n, F)\big) \subset T(n, F)$. Let $A \in \mathrm{N}_{\mathrm{GL}(n,F)}\big(T^*(n, F)\big)$ and let the last column of A be $\left[\begin{smallmatrix} \alpha \\ a \end{smallmatrix}\right]$, where $\alpha \in F^{n-1}$ and $a \in F$. Let

$$B = \begin{bmatrix} I_{n-1} & 0 \\ \beta & 1 \end{bmatrix} \in T^*(n, F)$$

and $C = A^{-1}BA \in T^*(n, F)$. Comparing the (n, n)-entries on both sides of $AC = BA$, we have $a = \beta\alpha + a$, i.e., $\beta\alpha = 0$. Since $\beta \in F^{n-1}$ is arbitrary, it follows that $\alpha = 0$. Now,

$$A = \begin{bmatrix} A_1 & 0 \\ * & a \end{bmatrix},$$

where $A_1 \in \mathrm{N}_{\mathrm{GL}(n-1,F)}\big(T^*(n-1, F)\big)$. By induction on n, $A_1 \in T(n-1, F)$, so $A \in T(n, F)$. □

6.2. Simplicity of PSL(n, F)

The main result of this section is the following theorem.

Theorem 6.4. *If* (i) $n \geq 3$, *or* (ii) $n = 2$ *and* $|F| \geq 4$, *then* $\mathrm{PSL}(n, F)$ *is simple.*

By Theorem 6.4, PSL$(n, \mathbb{F}_q)$ is a finite simple group for $n \geq 3$ and for $n = 2$ with $q \geq 4$. These groups form one of the sixteen infinite families of finite simple groups of Lie type in the classification of finite simple groups [**12, 13**].

We will prove a slightly stronger version of Theorem 6.4.

Theorem 6.5. *Assume that either $n \geq 3$ or $n = 2$ and $|F| \geq 4$. Let $\mathcal{H}$ be a subgroup of* GL(n, F) *such that $\mathcal{H}$ is invariant under conjugation by* SL(n, F) *and $\mathcal{H} \not\subset$ Z(GL(n, F)). Then $\mathcal{H} \supset$ SL(n, F).*

Theorem 6.5 implies Theorem 6.4. If $\mathcal{H}$ is a normal subgroup of SL(n, F) strictly containing Z(SL(n, F)), Theorem 6.5 gives that $\mathcal{H} =$ SL(n, F). Hence PSL$(n, F) =$ SL$(n, F)/$Z(SL(n, F)) is simple.

We will prove Theorem 6.5 in two separate cases. The proof for the case $n \geq 3$ is based on [**30**, §3.2]; the proof for the case $n = 2$ is based on [**24**, XIII.8]. To a large extent, the proof is about matrix computation. Elements of a particular type in SL(n, F), called transvections, play an important role in the proof. An element of SL(n, F) $(n \geq 2)$ of the form $I + aE_{ij}$, where $a \in F^*$ and $i \neq j$, is called a *transvection*. A transvection corresponds to a type III elementary operation of matrices. Note that $(I + aE_{ij})^{-1} = I - aE_{ij}$, which is also a transvection. We start with a few lemmas.

Lemma 6.6. SL(n, F) *is generated by transvections.*

Proof. Let $A = (a_{ij}) \in$ SL(n, F). We show that A is a product of transvections.

We may assume that $a_{21} \neq 0$. Otherwise, $a_{2k} \neq 0$ for some $k > 1$. Let $A' = A(I + E_{k1}) = (a'_{ij})$. Then $a'_{21} = a_{2k} \neq 0$.

Let $T = I + a_{21}^{-1}(1 - a_{11})E_{12}$. Then the $(1, 1)$-entry of TA is 1. Hence, there exist transvections $T_1, \dots, T_k$ and $S_1, \dots, S_l$ such that

$$T_k \cdots T_1 A S_1 \cdots S_l = \begin{bmatrix} 1 & 0 \\ 0 & A_1 \end{bmatrix},$$

where $A_1 \in$ SL$(n-1, F)$. By induction on n, A_1 is a product of transvections in SL$(n - 1, F)$. It follows easily that A is a product of transvections in SL(n, F). □

Lemma 6.7. *Let $n \geq 3$. Then all transvections in* SL(n, F) *are conjugate.*

Proof. We show that every transvection $I + aE_{ij}$, where $i \neq j$ and $a \in F^*$, is conjugate to $I + E_{12}$ in SL(n, F). There is an even permutation σ of $\{1, \dots, n\}$ such that $\{\sigma(i), \sigma(j)\} = \{1, 2\}$. Let P be the corresponding permutation matrix of σ. Then $P \in$ SL(n, F) and

$$P^{-1}(I + aE_{ij})P = I + aE_{12} \quad \text{or} \quad I + aE_{21}. \tag{6.6}$$

Since

$$\left(\begin{bmatrix} & 1 \\ -1 & \end{bmatrix} \oplus I_{n-2}\right)(I + aE_{21})\left(\begin{bmatrix} & 1 \\ -1 & \end{bmatrix} \oplus I_{n-2}\right)^{-1} = I - aE_{12},$$

we may assume that the right side of (6.6) is $I + aE_{12}$. We have

$$\begin{bmatrix} a^{-1} & & & \\ & 1 & & \\ & & a & \\ & & & I_{n-3} \end{bmatrix}\begin{bmatrix} 1 & a & & \\ & 1 & & \\ & & 1 & \\ & & & I_{n-3} \end{bmatrix}\begin{bmatrix} a & & & \\ & 1 & & \\ & & a^{-1} & \\ & & & I_{n-3} \end{bmatrix} = \begin{bmatrix} 1 & 1 & \\ & 1 & \\ & & I_{n-2} \end{bmatrix},$$

which completes the proof. □

Remark. When $n = 2$, Lemma 6.7 is true if and only if every element of F is a square (Exercise 6.2).

Before proceeding, we need to recall the rational canonical form of a square matrix from linear algebra. For $f = \mathtt{X}^n + a_{n-1}\mathtt{X}^{n-1} + \cdots + a_0 \in F[\mathtt{X}]$, the *companion matrix* of f is

(6.7)
$$M(f) = \begin{bmatrix} 0 & 1 & & & & \\ & 0 & 1 & & & \\ & & & \ddots & & \\ & & & & \ddots & \\ & & & & 0 & 1 \\ -a_0 & -a_1 & \cdot & \cdot & \cdot & -a_{n-1} \end{bmatrix}.$$

If A is an $n \times n$ matrix over F with elementary divisors (or invariant factors) $f_1, \dots, f_s$, then A is similar to

$$M(f_1) \oplus \cdots \oplus M(f_s),$$

which is called the *rational canonical form* of A. The companion matrix $M(f)$ is invertible if and only if $a_0 \neq 0$, and when $a_0 \neq 0$,

$$M(f)^{-1} = \begin{bmatrix} -\frac{a_1}{a_0} & \cdots & -\frac{a_{n-1}}{a_0} & -\frac{1}{a_0} \\ 1 & & & \\ & \ddots & & \\ & & 1 & 0 \end{bmatrix}.$$

Proof of Theorem 6.5 when $n \geq 3$. By Lemmas 6.6 and 6.7, it suffices to show that $\mathcal{H}$ contains a transvection. Choose $A \in \mathcal{H} \setminus \mathrm{Z}(\mathrm{GL}(n, F))$. Since $\mathrm{SL}(n, F) \lhd \mathrm{GL}(n, F)$, we may assume that A is a rational canonical form.

1° Assume that all the elementary divisors of A are of degree 1. Then A is diagonal but is not a scalar multiple of I, say

$$A = \begin{bmatrix} a_1 & & & \\ & a_2 & & \\ & & \ddots & \\ & & & a_n \end{bmatrix}, \qquad a_1 \neq a_2.$$

Then

$$\begin{aligned}
&(I + E_{12})A(I + E_{12})^{-1}A^{-1} \\
&= \begin{bmatrix} 1 & 1 & \\ & 1 & \\ & & I_{n-2} \end{bmatrix} \begin{bmatrix} a_1 & & & \\ & a_2 & & \\ & & \ddots & \\ & & & a_n \end{bmatrix} \begin{bmatrix} 1 & -1 & \\ & 1 & \\ & & I_{n-2} \end{bmatrix} \begin{bmatrix} a_1^{-1} & & & \\ & a_2^{-1} & & \\ & & \ddots & \\ & & & a_n^{-1} \end{bmatrix} \\
&= \begin{bmatrix} a_1 & a_2 - a_1 & & \\ & a_2 & & \\ & & \ddots & \\ & & & a_n \end{bmatrix} \begin{bmatrix} a_1^{-1} & & & \\ & a_2^{-1} & & \\ & & \ddots & \\ & & & a_n^{-1} \end{bmatrix} = I + \frac{a_2 - a_1}{a_2} E_{12},
\end{aligned}$$

which is a transvection contained in $\mathcal{H}$.

2° Assume that A has an elementary divisor (or an invariant factor) of degree $d \geq 3$. Then

$$A = \begin{bmatrix} 0 & 1 & & \\ & & \ddots & \\ & & & 1 \\ b_1 & b_2 & \cdots & b_d \end{bmatrix} \oplus A_1,$$

where $b_1 \neq 0$ and $A_1 \in \mathrm{GL}(n - d, F)$. Note that $\mathcal{H}$ contains

$$\begin{aligned}
&(I + E_{12})A(I + E_{12})^{-1}A^{-1} \\
&= \left(\begin{bmatrix} 1 & 1 & \\ & 1 & \\ & & I_{d-2} \end{bmatrix} \begin{bmatrix} 0 & 1 & & \\ & & \ddots & \\ & & & 1 \\ b_1 & b_2 & \cdots & b_d \end{bmatrix} \begin{bmatrix} 1 & -1 & \\ & 1 & \\ & & I_{d-2} \end{bmatrix} \begin{bmatrix} -\frac{b_2}{b_1} & \cdots & -\frac{b_d}{b_1} & \frac{1}{b_1} \\ 1 & & & \\ & \ddots & & \\ & & 1 & 0 \end{bmatrix} \right) \oplus I_{n-d}
\end{aligned}$$

$$= \begin{bmatrix} 1 & 1 & & \\ & 1 & & \\ & & \ddots & \\ -b_1 & & & 1 \end{bmatrix} \oplus I_{n-d}.$$

Denote the above matrix by B. Then $\mathcal{H}$ contains

$$B(I+E_{12})^{-1}B^{-1}(I+E_{12})$$
$$= \left(\begin{bmatrix} 1 & 1 & & \\ & 1 & & \\ & & \ddots & \\ -b_1 & & & 1 \end{bmatrix} \begin{bmatrix} 1 & -1 & \\ & 1 & \\ & & I_{d-2} \end{bmatrix} \begin{bmatrix} 1 & -1 & & \\ & 1 & & \\ & & \ddots & \\ b_1 & -b_1 & & 1 \end{bmatrix} \begin{bmatrix} 1 & 1 & \\ & 1 & \\ & & I_{d-2} \end{bmatrix} \right) \oplus I_{n-d}$$
$$= \begin{bmatrix} 1 & & & \\ & 1 & & \\ & & \ddots & \\ & b_1 & & 1 \end{bmatrix} \oplus I_{n-d},$$

which is a transvection.

3° Assume that A has an elementary divisor f of degree 2. Then $A = M(f) \oplus A_2$ for some $A_2 \in \mathrm{GL}(n-2, F)$. Since $M(f) \notin \mathrm{Z}(\mathrm{SL}(2, F))$, we can choose $R \in \mathrm{SL}(2, F)$ such that $M(f)\,R \neq R\,M(f)$. Let

$$C = A(R \oplus I_{n-2})A^{-1}(R \oplus I_{n-2})^{-1} = C_1 \oplus I_{n-2} \in \mathcal{H},$$

where $C_1 = M(f)\,R\,M(f)^{-1}R^{-1} \in \mathrm{SL}(2, F) \setminus \{I_2\}$. If the minimal polynomial of C_1 over F is of degree 1, we must have $C_1 = -I_2$ and $\operatorname{char} F \neq 2$; this case has been settled in 1°. If the minimal polynomial of C_1 is of degree 2, then C has an invariant factor of degree 3 and we are in 2°. □

To prove Theorem 6.5 with $n = 2$, we need a few more facts and lemmas. Let $\mathcal{B} = T(2, F) \cap \mathrm{SL}(2, F)$, where $T(2, F)$ is the group of 2×2 invertible lower triangular matrices over F, and let $\omega = \left[\begin{smallmatrix} 0 & 1 \\ -1 & 0 \end{smallmatrix}\right] \in \mathrm{SL}(2, F)$. We claim that

$$\mathcal{B}\omega\mathcal{B} = \left\{ \begin{bmatrix} a & b \\ c & d \end{bmatrix} \in \mathrm{SL}(2, F) : b \neq 0 \right\}.$$

If $B_i = \left[\begin{smallmatrix} a_i & 0 \\ b_i & a_i^{-1} \end{smallmatrix}\right] \in \mathcal{B}$, $i = 1, 2$, then

$$B_1\omega B_2 = \begin{bmatrix} a_1 & 0 \\ b_1 & a_1^{-1} \end{bmatrix} \begin{bmatrix} b_2 & a_2^{-1} \\ -a_2 & 0 \end{bmatrix} = \begin{bmatrix} * & a_1 a_2^{-1} \\ * & * \end{bmatrix},$$

where $a_1 a_2^{-1} \neq 0$. On the other hand, if $A = \left[\begin{smallmatrix} a & b \\ c & d \end{smallmatrix}\right] \in \mathrm{SL}(2, F)$, where $b \neq 0$, then $A(1 - (a/b)E_{21}) = \left[\begin{smallmatrix} 0 & * \\ * & * \end{smallmatrix}\right] \in \mathcal{B}\omega$, and hence $A \in \mathcal{B}\omega(1 + (a/b)E_{21}) \subset \mathcal{B}\omega\mathcal{B}$. Therefore the claim is proved. It follows that

$$\mathrm{SL}(2, F) = \mathcal{B} \,\dot{\cup}\, \mathcal{B}\omega\mathcal{B}, \tag{6.8}$$

where $\dot{\cup}$ means disjoint union. If $\mathcal{H}$ is a subgroup of $\mathrm{SL}(2, F)$ which strictly contains $\mathcal{B}$, then by (6.8), $\mathcal{H}$ contains ω, and hence $\mathcal{H} = \mathrm{SL}(2, F)$. We conclude that $\mathcal{B}$ is a maximal subgroup of $\mathrm{SL}(2, F)$.

Lemma 6.8. *Assume that* $|F| \geq 4$. *Then the commutator group of* $\mathrm{SL}(2, F)$ *is* $\mathrm{SL}(2, F)$.

Proof. By Lemma 6.6, it suffices to show that every transvection $T \in \mathrm{SL}(2, F)$ is a commutator of $\mathrm{SL}(2, F)$. Without loss of generality, assume that $T = I + aE_{12}$. Since $|F| \geq 4$, we can choose $b \in F^*$ such that $b^2 \neq 1$. Let $c = a/(b^2 - 1)$. Then

$$\begin{bmatrix} b & \\ & b^{-1} \end{bmatrix} \begin{bmatrix} 1 & c \\ 0 & 1 \end{bmatrix} \begin{bmatrix} b & \\ & b^{-1} \end{bmatrix}^{-1} \begin{bmatrix} 1 & c \\ 0 & 1 \end{bmatrix}^{-1} = \begin{bmatrix} 1 & c(b^2 - 1) \\ 0 & 1 \end{bmatrix} = T.$$

□

Lemma 6.9. *Assume that* $|F| \geq 4$. *Let* $\mathcal{H} \lhd \mathrm{SL}(2, F)$ *be such that* $\mathcal{H} \not\subset \mathrm{Z}\big(\mathrm{SL}(2, F)\big)$. *Then* $\mathcal{H} = \mathrm{SL}(2, F)$.

Proof. Since $\mathcal{H}\mathcal{B}$ is a subgroup of $\mathrm{SL}(2, F)$ containing $\mathcal{B}$, by the maximality of $\mathcal{B}$, we have $\mathcal{H}\mathcal{B} = \mathcal{B}$ or $\mathcal{H}\mathcal{B} = \mathrm{SL}(2, F)$.

If $\mathcal{H}\mathcal{B} = \mathcal{B}$, then $\mathcal{H} \subset \mathcal{B}$, so $\mathcal{H} \subset \bigcap_{\alpha \in \mathrm{SL}(2,F)} \alpha^{-1}\mathcal{B}\alpha$. Let $\gamma = \left[\begin{smallmatrix} 1 & 1 \\ 0 & 1 \end{smallmatrix}\right] \in \mathrm{SL}(2, F)$. It is easy to see that

$$\mathcal{B} \cap (\omega^{-1}\mathcal{B}\omega) = \left\{ \begin{bmatrix} a & \\ & a^{-1} \end{bmatrix} : a \in F^* \right\}$$

and

$$\mathcal{B} \cap (\omega^{-1}\mathcal{B}\omega) \cap (\gamma^{-1}\mathcal{B}\gamma) = \{\pm I\}.$$

Hence $\mathcal{H} \subset \{\pm I\}$, which is a contradiction.

Therefore, we must have $\mathcal{H}\mathcal{B} = \mathrm{SL}(2, F)$. In particular, $\omega = h\beta$ for some $h \in \mathcal{H}$ and $\beta \in \mathcal{B}$. Since $\beta\, T^*(2, F)\, \beta^{-1} = T^*(2, F)$, where $T^*(2, F)$ is the group of 2×2 lower triangular matrices over F with 1's on the diagonal, and since $\mathcal{H} \lhd \mathrm{SL}(2, F)$, we have

$$\omega\, T^*(2, F)\, \omega^{-1} = h\beta\, T^*(2, F)\, \beta^{-1}h^{-1} = h\, T^*(2, F)\, h^{-1} \subset \mathcal{H}\, T^*(2, F).$$

Therefore,

$$\mathcal{H}\, T^*(2, F) \supset T^*(2, F) \cup (\omega\, T^*(2, F)\, \omega^{-1}).$$

Since $T^*(2,F) \cup (\omega T^*(2,F)\omega^{-1})$ contains all transvections in $\mathrm{SL}(2,F)$, by Lemma 6.6, $\mathcal{H}\,T^*(2,F) = \mathrm{SL}(2,F)$. Thus

$$\mathrm{SL}(2,F)/\mathcal{H} \cong T^*(2,F)/\big(T^*(2,F) \cap \mathcal{H}\big),$$

which is abelian. Hence $\mathcal{H} \supset \mathrm{SL}(2,F)'$. However, by Lemma 6.8, $\mathrm{SL}(2,F)' = \mathrm{SL}(2,F)$. We conclude that $\mathcal{H} = \mathrm{SL}(2,F)$. □

Proof of Theorem 6.5 with $n = 2$. We claim that $\mathcal{H} \cap \mathrm{SL}(2,F) \not\subset \mathrm{Z}(\mathrm{SL}(2,F))$. Choose $A \in \mathcal{H} \setminus \mathrm{Z}\big(\mathrm{GL}(2,F)\big)$. We may assume that A is a rational canonical form. Then

$$A = \begin{bmatrix} a_1 & \\ & a_2 \end{bmatrix}, \quad a_1, a_2 \in F^*,\ a_1 \neq a_2, \quad \text{or} \quad A = \begin{bmatrix} 0 & 1 \\ a & b \end{bmatrix}, \quad a \neq 0.$$

In the first case,

$$\mathcal{H} \cap \mathrm{SL}(2,F) \ni \begin{bmatrix} 1 & 1 \\ 0 & 1 \end{bmatrix} A \begin{bmatrix} 1 & 1 \\ 0 & 1 \end{bmatrix}^{-1} A^{-1} = \begin{bmatrix} 1 & a_2^{-1}(a_2 - a_1) \\ 0 & 1 \end{bmatrix} \notin \mathrm{Z}\big(\mathrm{SL}(2,F)\big).$$

In the second case,

$$\mathcal{H} \cap \mathrm{SL}(2,F) \ni \begin{bmatrix} 1 & 1 \\ 0 & 1 \end{bmatrix} A \begin{bmatrix} 1 & 1 \\ 0 & 1 \end{bmatrix}^{-1} A^{-1} = \begin{bmatrix} 1-a & 1 \\ -a & 1 \end{bmatrix} \notin \mathrm{Z}\big(\mathrm{SL}(2,F)\big).$$

Thus the claim is proved. Since $\mathcal{H} \cap \mathrm{SL}(2,F) \lhd \mathrm{SL}(2,F)$, by Lemma 6.9, $\mathcal{H} \cap \mathrm{SL}(2,F) = \mathrm{SL}(2,F)$, i.e., $\mathcal{H} \supset \mathrm{SL}(2,F)$. □

The groups $\mathrm{PSL}(2,\mathbb{F}_2)$ and $\mathrm{PSL}(2,\mathbb{F}_3)$ are two exceptions to Theorem 6.4. Since $|\mathrm{PSL}(2,\mathbb{F}_2)| = 6$ and $|\mathrm{PSL}(2,\mathbb{F}_3)| = 12$, these two groups are not simple.

In general, $\mathrm{PGL}(n,F)$ and $\mathrm{PSL}(n,F)$ are embedded in $\mathfrak{S}_{\mathbb{P}^{n-1}(F)}$ since $\mathrm{PGL}(n,F)$ acts faithfully on $\mathbb{P}^{n-1}(F)$. Note that $|\mathbb{P}^1(\mathbb{F}_2)| = 3$, $|\mathbb{P}^1(\mathbb{F}_3)| = 4$ and $|\mathbb{P}^1(\mathbb{F}_4)| = 5$. It follows that $\mathrm{PGL}(2,\mathbb{F}_2) = \mathrm{PSL}(2,\mathbb{F}_2)$ is embedded in $\mathfrak{S}_3$, $\mathrm{PGL}(2,\mathbb{F}_3)$ and $\mathrm{PSL}(2,\mathbb{F}_3)$ are embedded in $\mathfrak{S}_4$, and $\mathrm{PGL}(2,\mathbb{F}_4) \cong \mathrm{PSL}(2,\mathbb{F}_4)$ are embedded in $\mathfrak{S}_5$. Since $|\mathrm{PGL}(2,\mathbb{F}_2)| = |\mathrm{PSL}(2,\mathbb{F}_2)| = 6 = 3!$, $|\mathrm{PGL}(2,\mathbb{F}_3)| = 24 = 4!$, $|\mathrm{PSL}(2,\mathbb{F}_3)| = 4!/2$, and $|\mathrm{PGL}(2,\mathbb{F}_4)| = |\mathrm{PSL}(2,\mathbb{F}_4)| = 60 = 5!/2$, we have

$$\begin{aligned} &\mathrm{PGL}(2,\mathbb{F}_2) = \mathrm{PSL}(2,\mathbb{F}_2) \cong \mathfrak{S}_3, \\ &\mathrm{PGL}(2,\mathbb{F}_3) \cong \mathfrak{S}_4, \qquad \mathrm{PSL}(2,\mathbb{F}_3) \cong A_4, \\ &\mathrm{PGL}(2,\mathbb{F}_4) \cong \mathrm{PSL}(2,\mathbb{F}_4) \cong A_5. \end{aligned}$$

Coincidentally, $|\mathrm{PSL}(2,\mathbb{F}_5)| = 60$, so we also have $\mathrm{PSL}(2,\mathbb{F}_5) \cong A_5$ since A_5 is the unique simple group of order 60. There are two more coincidental isomorphisms between a PSL group and an alternating group: $\mathrm{PSL}(2, \mathbb{F}_9) \cong A_6$ and $\mathrm{PSL}(4, \mathbb{F}_2) \cong A_8$; see [**1, 13**]. No other finite simple groups in the PSL family are isomorphic to an alternating group.

6.3. Conjugacy Classes of $\mathrm{GL}(n, \mathbb{F}_q)$

Let F be a field. It is well known from linear algebra that each similarity class of $\mathrm{M}_n(F)$ is determined by a list of elementary divisors (or invariant factors). The conjugacy classes of $\mathrm{GL}(n, F)$ are the similarity classes of $\mathrm{M}_n(F)$ whose elementary divisors (or invariant factors) are not multiples of $\mathtt{X}$. When $F = \mathbb{F}_q$, the conjugacy classes of $\mathrm{GL}(n, \mathbb{F}_q)$ are finite. The main purpose of this section is to develop an explicit formula for the cardinalities of the conjugacy classes of $\mathrm{GL}(n, \mathbb{F}_q)$. The formula is useful for computing the number of orbits when $\mathrm{GL}(n, \mathbb{F}_q)$ acts on a set.

For $A \in \mathrm{M}_n(F)$, let

$$[A]_{\mathrm{GL}(n,F)} = \{P^{-1}AP : P \in \mathrm{GL}(n, F)\}$$

be the similarity class of A and let

$$\begin{aligned} &\mathrm{C}_{\mathrm{M}_n(F)}(A) = \{X \in \mathrm{M}_n(F) : AX = XA\}, \\ &\mathrm{C}_{\mathrm{GL}(n,F)}(A) = \{P \in \mathrm{GL}(n, F) : AP = PA\} = \mathrm{C}_{\mathrm{M}_n(F)}(A) \cap \mathrm{GL}(n, F) \end{aligned}$$

be the centralizers of A in $\mathrm{M}_n(F)$ and $\mathrm{GL}(n, F)$, respectively. When $A \in \mathrm{GL}(n, F)$, $[A]_{\mathrm{GL}(n,F)}$ is the conjugacy class of A in $\mathrm{GL}(n, F)$. When $F = \mathbb{F}_q$,

$$\big|[A]_{\mathrm{GL}(n,\mathbb{F}_q)}\big| = \frac{|\mathrm{GL}(n, \mathbb{F}_q)|}{\big|\mathrm{C}_{\mathrm{GL}(n,\mathbb{F}_q)}(A)\big|}. \tag{6.9}$$

Therefore, to determine $\big|[A]_{\mathrm{GL}(n,\mathbb{F}_q)}\big|$, it suffices to determine $\big|\mathrm{C}_{\mathrm{GL}(n,\mathbb{F}_q)}(A)\big|$.

Lemma 6.10. *Let $A \in \mathrm{M}_m(F)$ and $B \in \mathrm{M}_n(F)$ be such that their characteristic polynomials are relatively prime. Then*

$$\mathrm{C}_{\mathrm{M}_{m+n}(F)}(A \oplus B) = \mathrm{C}_{\mathrm{M}_m(F)}(A) \oplus \mathrm{C}_{\mathrm{M}_n(F)}(B), \tag{6.10}$$

$$\mathrm{C}_{\mathrm{GL}(m+n,F)}(A \oplus B) = \mathrm{C}_{\mathrm{GL}(m,F)}(A) \oplus \mathrm{C}_{\mathrm{GL}(n,F)}(B). \tag{6.11}$$

Proof. Only (6.10) needs a proof since (6.11) immediately follows from (6.10). To prove (6.10), it suffices to show that the left side is contained in the right side. Assume that

$$\begin{bmatrix} P & C \\ D & Q \end{bmatrix} \in \mathrm{C}_{\mathrm{M}_{m+n}(F)}(A \oplus B),$$

where P is of size $m \times m$ and Q is of size $n \times n$. Then

$$\begin{bmatrix} PA & CB \\ DA & QB \end{bmatrix} = \begin{bmatrix} P & C \\ D & Q \end{bmatrix} \begin{bmatrix} A & \\ & B \end{bmatrix} = \begin{bmatrix} A & \\ & B \end{bmatrix} \begin{bmatrix} P & C \\ D & Q \end{bmatrix} = \begin{bmatrix} AP & AC \\ BD & BQ \end{bmatrix}.$$

It follows that $P \in \mathrm{C}_{\mathrm{M}_m(F)}(A)$, $Q \in \mathrm{C}_{\mathrm{M}_n(F)}(B)$, and

$$AC = CB, \qquad BD = DA.$$

Then, for any polynomial $h \in F[\mathtt{X}]$, $h(A)C = Ch(B)$ and $h(B)D = Dh(A)$. Let f and g be the characteristic polynomials of A and B, respectively, and choose $a, b \in F[\mathtt{X}]$ such that $af + bg = 1$. Then

$$C = \big[a(A)f(A) + b(A)g(A)\big]C = b(A)g(A)C = b(A)Cg(B) = 0.$$

In the same way, $D = 0$. Hence $\left[\begin{smallmatrix} P & C \\ D & Q \end{smallmatrix}\right] = P \oplus Q \in \mathrm{C}_{\mathrm{M}_m(F)}(A) \oplus \mathrm{C}_{\mathrm{M}_n(F)}(B)$. □

Lemma 6.11. *Let $A \in \mathrm{M}_n(F)$ be such that its characteristic polynomial f is also its minimal polynomial. Then for any $g \in F[\mathtt{X}]$,* $\operatorname{rank} g(A) = n - \deg \gcd(g, f)$.

Proof. Let $h = \gcd(g, f)$. Then $\operatorname{rank} g(A) \le \operatorname{rank} h(A)$. Write $h = ag + bf$ for some $a, b \in F[\mathtt{X}]$. Then $h(A) = a(A)g(A)$, so $\operatorname{rank} g(A) \ge \operatorname{rank} h(A)$. Hence $\operatorname{rank} g(A) = \operatorname{rank} h(A)$.

We may assume that A is a rational canonical form:

$$A = \begin{bmatrix} 0 & 1 & & \\ & & \ddots & \\ & & & 1 \\ * & * & \cdots & * \end{bmatrix}.$$

Note that

$$A^i = \begin{bmatrix} 0 & \cdots & 0 & 1 & & \\ \vdots & & \vdots & & \ddots & \\ 0 & \cdots & 0 & & & 1 \\ * & \cdots & * & * & \cdots & * \\ \vdots & & \vdots & \vdots & & \vdots \\ * & \cdots & * & * & \cdots & * \end{bmatrix} n-i, \qquad 0 \le i \le n.$$

Hence, the $(n - \deg h) \times (n - \deg h)$ submatrix at the upper right corner of $h(A)$ is invertible. Thus $\operatorname{rank} h(A) \ge n - \deg h$. On the other hand, since $h(A)(f/h)(A) = 0$, we have $\operatorname{rank} h(A) + \operatorname{rank}(f/h)(A) \le n$. Therefore, $\operatorname{rank} h(A) = n - \deg h$ and $\operatorname{rank}(f/h)(A) = \deg h$. □

For $f = \mathtt{X}^n + a_{n-1}\mathtt{X}^{n-1} + \cdots + a_0 \in F[\mathtt{X}]$, let

$$M'(f) = \begin{bmatrix} 0 & & & -a_0 \\ 1 & & & -a_1 \\ & \ddots & & \vdots \\ & & 1 & -a_{n-1} \end{bmatrix},$$

which is the transpose of the companion matrix $M(f)$ defined in (6.7) and is similar to $M(f)$. Here we choose to work with $M'(f)$ instead of $M(f)$ solely for the sake of notational convenience.

Lemma 6.12. *Let $f = \mathtt{X}^n + a_{n-1}\mathtt{X}^{n-1} + \cdots + a_0 \in F[\mathtt{X}]$, $A \in \mathrm{M}_m(F)$ and $X \in \mathrm{M}_{m\times n}(F)$. Then $AX = XM'(f)$ if and only if*

$$X = [x, Ax, \dots, A^{n-1}x]$$

for some $x \in \ker f(A)$.

Proof. Write $X = [x_1, \dots, x_n]$ where $x_1, \dots, x_n \in F^n$. Then the equation $AX = XM'(f)$ becomes

$$\begin{aligned} [Ax_1, \dots, Ax_n] &= [x_1, \dots, x_n] \begin{bmatrix} 0 & & & -a_0 \\ 1 & & & -a_1 \\ & \ddots & & \vdots \\ & & 1 & -a_{n-1} \end{bmatrix} \\ &= [x_2, \ \dots, \ x_n, \ -a_0x_1 - \cdots - a_{n-1}x_n], \end{aligned}$$

i.e.,

$$\begin{cases} Ax_1 = x_2, \\ \quad \vdots \\ Ax_{n-1} = x_n, \\ Ax_n = -a_0x_1 - \cdots - a_{n-1}x_n. \end{cases} \tag{6.12}$$

Clearly, (6.12) is equivalent to $x_i = A^{i-1}x_1$, $1 \le i \le n$, and $f(A)x_1 = 0$. □

Lemma 6.13. *Assume that $A \in \mathrm{M}_n(\mathbb{F}_q)$ has elementary divisors*

$$\underbrace{f^1, \dots, f^1}_{\mu_1}, \ \dots, \ \underbrace{f^s, \dots, f^s}_{\mu_s},$$

where $f \in \mathbb{F}_q[\mathtt{X}]$ is monic and irreducible with $\deg f = d$. Then

$$\left|\mathrm{C}_{\mathrm{GL}(n,\mathbb{F}_q)}(A)\right| = q^{d\sum_{1\le i,j\le s}\min(i,j)\mu_i\mu_j} \prod_{i=1}^{s}\prod_{u=1}^{\mu_i}(1 - q^{-du}).$$

Proof. We may assume that

$$A = \underbrace{M'(f) \oplus \cdots \oplus M'(f)}_{\mu_1} \oplus \cdots \oplus \underbrace{M'(f^s) \oplus \cdots \oplus M'(f^s)}_{\mu_s}.$$

By Lemma 6.11, for any integer $i \geq 0$,
(6.13)

$$\left|\ker f(A)^i\right| = \prod_{j=1}^{s} \left|\ker f(M'(f^j))^i\right|^{\mu_j} = \prod_{j=1}^{s} q^{\mu_j \deg \gcd(f^i, f^j)} = q^{d \sum_{j=1}^{s} \min(i,j)\mu_j}.$$

Let $X \in \mathrm{M}_n(\mathbb{F}_q)$ and write

$$X = \left[X_1^{(1)}, \dots, X_{\mu_1}^{(1)} ; \dots ; X_1^{(s)}, \dots, X_{\mu_s}^{(s)}\right],$$

where $X_u^{(i)} \in \mathrm{M}_{n \times di}(\mathbb{F}_q)$. Then $AX = XA$ if and only if

$$AX_u^{(i)} = X_u^{(i)} M'(f^i) \quad \text{for all } 1 \leq i \leq s \text{ and } 1 \leq u \leq \mu_i. \tag{6.14}$$

By Lemma 6.12, (6.14) holds if and only if

$$X_u^{(i)} = \left[x_u^{(i)}, Ax_u^{(i)}, \dots, A^{di-1} x_u^{(i)}\right] \tag{6.15}$$

for some $x_u^{(i)} \in \ker\left[f(A)^i\right]$. Therefore, $\left|C_{\mathrm{GL}(n,\mathbb{F}_q)}(A)\right|$ is the number of $x_u^{(i)} \in \ker\left[f(A)^i\right]$, $1 \leq i \leq s$, $1 \leq u \leq \mu_i$, such that the columns of

$$\left[X_1^{(1)}, \dots, X_{\mu_1}^{(1)} ; \dots ; X_1^{(s)}, \dots, X_{\mu_s}^{(s)}\right]$$

are linearly independent, where $X_u^{(i)}$ is given by (6.15).

The qualified vectors, in the order $x_1^{(s)}, \dots, x_{\mu_s}^{(s)}, \dots, x_1^{(1)}, \dots, x_{\mu_1}^{(1)}$, are constructed below. The construction provides counting formulas for these vectors. We always assume that $x_u^{(i)} \in \ker\left[f(A)^i\right]$.

First, we claim that the columns of $X_1^{(s)}$ are linearly *dependent* if and only if

$$x_1^{(s)} \equiv 0 \pmod{\ker\left[f(A)^{s-1}\right]}.$$

In fact, if $x_1^{(s)} \in \ker\left[f(A)^{s-1}\right]$, then $f(A)^{s-1} x_1^{(s)} = 0$. Since $\deg f^{s-1} < ds$, $f(A)^{s-1} x_1^{(s)}$ is a nontrivial linear combination of the columns of $X_1^{(s)}$, so the columns of $X_1^{(s)}$ are linearly dependent. On the other hand, if the columns of $X_1^{(s)}$ are linearly dependent, then $g(A)x_1^{(s)} = 0$ for some $0 \neq g \in \mathbb{F}_q[\mathtt{X}]$ with $\deg g < ds$. Let $h = \gcd(g, f^s)$. Then $h(A)x_1^{(s)} = 0$ and $h \mid f^{s-1}$. It follows that $f(A)^{s-1} x_1^{(s)} = 0$, i.e., $x_1^{(s)} \in \ker\left[f(A)^{s-1}\right]$. Hence the claim is proved. Consequently, the number of choices for $x_1^{(s)}$ is

$$\begin{aligned} &\left|\ker\left[f(A)^s\right]\right| - \left|\ker\left[f(A)^{s-1}\right]\right| \\ &= q^{d \sum_{j=1}^{s} j\mu_j} - q^{d \sum_{j=1}^{s} \min(s-1,j)\mu_j} \qquad \text{(by (6.13))} \end{aligned}$$

$$= q^{d\sum_{j=1}^{s} j\mu_j}(1 - q^{-d\mu_s}).$$

Assume that $x_1^{(s)}, \dots, x_{u-1}^{(s)}$, $u \le \mu_s$, have been chosen such that the columns of $\big[X_1^{(s)}, \dots, X_{u-1}^{(s)}\big]$ are linearly independent. We claim that the columns of $\big[X_1^{(s)}, \dots, X_{u-1}^{(s)}, X_u^{(s)}\big]$ are linearly dependent if and only if

$$x_u^{(s)} \equiv a_1(A)x_1^{(s)} + \cdots + a_{u-1}(A)x_{u-1}^{(s)} \pmod{\ker\big[f(A)^{s-1}\big]} \tag{6.16}$$

for some $a_1, \dots, a_{u-1} \in \mathbb{F}_q[\mathtt{X}]$ of degree $< d$. If (6.16) holds, then

$$f(A)^{s-1}x_u^{(s)} = b_1(A)x_1^{(s)} + \cdots + b_{u-1}(A)x_{u-1}^{(s)}, \tag{6.17}$$

where $b_l = f^{s-1}a_l$. The left side of (6.17) is a nontrivial linear combination of the columns of $X_u^{(s)}$, and the right side of (6.17) is a linear combination of the columns of $[X_1^{(s)}, \dots, X_{u-1}^{(s)}]$. Hence, the columns of $[X_1^{(s)}, \dots, X_u^{(s)}]$ are linearly dependent. On the other hand, assume that the columns of $[X_1^{(s)}, \dots, X_u^{(s)}]$ are linearly dependent. Then

$$g(A)x_u^{(s)} = c_1(A)x_1^{(s)} + \cdots + c_{u-1}(A)x_{u-1}^{(s)} \tag{6.18}$$

for some $c_1, \dots, c_{u-1}, g \in \mathbb{F}_q[\mathtt{X}]$ with $g \ne 0$ and $\deg g < ds$. Since $\gcd(g, f^s) \mid f^{s-1}$, there exists $a \in \mathbb{F}_q[\mathtt{X}]$ such that $ag \equiv f^{s-1} \pmod{f^s}$. Multiplying both sides of (6.18) by a gives

$$f(A)^{s-1}x_u^{(s)} = c_1'(A)x_1^{(s)} + \cdots + c_{u-1}'(A)x_{u-1}^{(s)}, \tag{6.19}$$

where $c_l' = ac_l$, $1 \le l \le u-1$. Then

$$(fc_1')(A)x_1^{(s)} + \cdots + (fc_{u-1}')(A)x_{u-1}^{(s)} = f(A)^s x_u^{(s)} = 0.$$

Since the columns of $[X_1^{(s)}, \dots, X_{u-1}^{(s)}]$ are linearly independent, we have $fc_l' \equiv 0 \pmod{f^s}$, i.e., $c_l' \equiv 0 \pmod{f^{s-1}}$, $1 \le l \le u-1$. Let $c_l' = f^{s-1}a_l$, where $a_l \in \mathbb{F}_q[\mathtt{X}]$, $1 \le l \le u-1$. Then (6.19) becomes

$$x_u^{(s)} \equiv a_1(A)x_1^{(s)} + \cdots + a_{u-1}(A)x_{u-1}^{(s)} \pmod{\ker\big[f(A)^{s-1}\big]}.$$

In the above, we may replace each a_l with its remainder modulo f since $f(A)x_l^{(s)} \in \ker[f(A)^{s-1}]$. Therefore, we may assume that $\deg a_l < d$ for all $1 \le l \le u-1$, so the claim is proved.

We further claim that the congruence class of the right side of (6.16) modulo $\ker[f(A)^{s-1}]$ is uniquely determined by $a_l \in \mathbb{F}_q[\mathtt{X}]$ with $\deg a_l < d$, $1 \le l \le u-1$. If $a_1(A)x_1^{(s)} + \cdots + a_{u-1}(A)x_{u-1}^{(s)} \equiv 0 \pmod{\ker\big[f(A)^{s-1}\big]}$, then

$$(f^{s-1}a_1)(A)x_1^{(s)} + \cdots + (f^{s-1}a_{u-1})(A)x_{u-1}^{(s)} = 0.$$

Since the columns of $[X_1^{(s)}, \dots, X_{u-1}^{(s)}]$ are linearly independent, this implies that $f^{s-1}a_l \equiv 0 \pmod{f^s}$, i.e., $a_l \equiv 0 \pmod{f}$, $1 \le l \le u-1$. Hence the additional claim is proved.

Now it is clear that the number of choices for $x_u^{(s)}$ is

$$\begin{aligned}&\left|\ker\left[f(A)^s\right]\right| - \left|\ker\left[f(A)^{s-1}\right]\right|q^{d(u-1)}\\ &= q^{d\sum_{j=1}^{s} j\mu_j} - q^{d\sum_{j=1}^{s}\min(s-1,j)\mu_j + d(u-1)}\\ &= q^{d\sum_{j=1}^{s} j\mu_j}(1 - q^{-d(\mu_s-u+1)}).\end{aligned}$$

To summarize, the number of choices for $x_1^{(s)}, \dots, x_{\mu_s}^{(s)}$ such that the columns of $[X_1^{(s)}, \dots, X_{\mu_s}^{(s)}]$ are linearly independent is

$$q^{d\mu_s\sum_{j=1}^{s} j\mu_j}\prod_{u=1}^{\mu_s}(1 - q^{-d(\mu_s-u+1)}) = q^{d\mu_s\sum_{j=1}^{s} j\mu_j}\prod_{u=1}^{\mu_s}(1 - q^{-du}).$$

More generally, assume that $x_1^{(s)}, \dots, x_{\mu_s}^{(s)}, \dots, x_1^{(i+1)}, \dots, x_{\mu_{i+1}}^{(i+1)}$ and $x_1^{(i)}, \dots, x_{u-1}^{(i)}$, $u < \mu_i$, have been chosen such that the columns of

$$[X_1^{(s)}, \dots, X_{\mu_s}^{(s)}\ ;\ \dots\ ;\ X_1^{(i+1)}, \dots, X_{\mu_{i+1}}^{(i+1)};\ X_1^{(i)}, \dots, X_{u-1}^{(i)}]$$

are linearly independent. By the same argument as above, we conclude that the columns of

$$[X_1^{(s)}, \dots, X_{\mu_s}^{(s)}\ ;\ \dots\ ;\ X_1^{(i+1)}, \dots, X_{\mu_{i+1}}^{(i+1)};\ X_1^{(i)}, \dots, X_{u-1}^{(i)}, X_u^{(i)}]$$

are linearly dependent if and only if

(6.20)

$$\begin{aligned}x_u^{(i)} &\equiv a_1^{(i)}(A)x_1^{(i)} + \cdots + a_{u-1}^{(i)}(A)x_{u-1}^{(i)}\\ &\quad + f(A)\left[a_1^{(i+1)}(A)x_1^{(i+1)} + \cdots + a_{\mu_{i+1}}^{(i+1)}(A)x_{\mu_{i+1}}^{(i+1)}\right]\\ &\quad + \cdots\\ &\quad + f(A)^{s-i}\left[a_1^{(s)}(A)x_1^{(s)} + \cdots + a_{\mu_s}^{(s)}(A)x_{\mu_s}^{(s)}\right] \pmod{\ker\left[f(A)^{i-1}\right]},\end{aligned}$$

where $a_l^{(k)} \in \mathbb{F}_q[\mathtt{X}]$ and $\deg a_l^{(k)} < d$. Moreover, the congruence class of the right side of (6.20) modulo $\ker\left[f(A)^{i-1}\right]$ is uniquely determined by the $a_l^{(k)}$'s with degree $< d$. Therefore, the number of choices for $x_u^{(i)}$ is

$$\begin{aligned}&\left|\ker\left[f(A)^i\right]\right| - \left|\ker\left[f(A)^{i-1}\right]\right|q^{d\sum_{j=i+1}^{s}\mu_j + d(u-1)}\\ &= q^{d\sum_{j=1}^{s}\min(i,j)\mu_j} - q^{d\sum_{j=1}^{s}\min(i-1,j)\mu_j + d\sum_{j=i+1}^{s}\mu_j + d(u-1)}\\ &= q^{d\sum_{j=1}^{s}\min(i,j)\mu_j}(1 - q^{-d(\mu_i-u+1)}).\end{aligned}$$

To summarize, when $x_1^{(s)}, \dots, x_{\mu_s}^{(s)}, \dots, x_1^{(i+1)}, \dots, x_{\mu_{i+1}}^{(i+1)}$ have been chosen, the number of choices for $x_1^{(i)}, \dots, x_{\mu_i}^{(i)}$ is

$$q^{d\mu_i\sum_{j=1}^{s}\min(i,j)\mu_j}\prod_{u=1}^{\mu_i}(1 - q^{-d(\mu_i-u+1)}) = q^{d\mu_i\sum_{j=1}^{s}\min(i,j)\mu_j}\prod_{u=1}^{\mu_i}(1 - q^{-du}).$$

Therefore,

$$\begin{aligned}
\left|\mathrm{C}_{\mathrm{GL}(n,\mathbb{F}_q)}(A)\right| &= \prod_{i=1}^{s} q^{d\mu_i \sum_{j=1}^{s} \min(i,j)\mu_j} \prod_{u=1}^{\mu_i} (1 - q^{-du}) \\
&= q^{d \sum_{1\le i,j\le s} \min(i,j)\mu_i\mu_j} \prod_{i=1}^{s} \prod_{u=1}^{\mu_i} (1 - q^{-du}).
\end{aligned}$$

□

Theorem 6.14. *Assume that the elementary divisors of* $A \in \mathrm{M}_n(\mathbb{F}_q)$ *are*

$$\underbrace{f_1^1, \dots, f_1^1}_{\mu_1^{(1)}}, \ \dots, \ \underbrace{f_1^{s_1}, \dots, f_1^{s_1}}_{\mu_{s_1}^{(1)}},$$

$$\cdots$$

$$\underbrace{f_t^1, \dots, f_t^1}_{\mu_1^{(t)}}, \ \dots, \ \underbrace{f_t^{s_t}, \dots, f_t^{s_t}}_{\mu_{s_t}^{(t)}},$$

where $f_1, \dots, f_t \in \mathbb{F}_q[\mathtt{X}]$ *are distinct monic irreducible polynomials with* $\deg f_k = d_k$. *Then*

$$\left|\mathrm{C}_{\mathrm{GL}(n,\mathbb{F}_q)}(A)\right| = q^{\sum_{k=1}^{t} d_k \sum_{1\le i,j\le s_k} \min(i,j)\mu_i^{(k)}\mu_j^{(k)}} \prod_{k=1}^{t} \prod_{i=1}^{s_k} \prod_{u=1}^{\mu_i^{(k)}} (1 - q^{-d_k u}).$$

Proof. This immediately follows from Lemmas 6.10 and 6.13. □

Remark. Theorem 6.14 can also be proved through a different approach, which requires the determination of the automorphism group of a finitely generated torsion module over a PID. Exercises 6.6–6.8 will guide interested readers through that proof.

Example 6.15. Let the elementary divisors of $A \in \mathrm{M}_{37}(\mathbb{F}_3)$ be

$$\begin{gathered}
\mathtt{X}^2 + 1,\ \mathtt{X}^2 + 1,\ (\mathtt{X}^2 + 1)^2,\ (\mathtt{X}^2 + 1)^2,\ (\mathtt{X}^2 + 1)^2, \\
(\mathtt{X}^3 - \mathtt{X} + 1)^2,\ (\mathtt{X}^3 - \mathtt{X} + 1)^2,\ (\mathtt{X}^3 - \mathtt{X} + 1)^3.
\end{gathered}$$

In the notation of Theorem 6.14,

$$\begin{aligned}
&d_1 = 2,\ \mu_1^{(1)} = 2,\ \mu_2^{(1)} = 3, \\
&d_2 = 3,\ \mu_1^{(2)} = 0,\ \mu_2^{(2)} = 2,\ \mu_3^{(2)} = 1.
\end{aligned}$$

Thus

$$\sum_k d_k \sum_{i,j} \min(i,j)\mu_i^{(k)}\mu_j^{(k)} = 2(1{\cdot}2^2{+}2{\cdot}3^3{+}2{\cdot}2{\cdot}3){+}3(2{\cdot}2^2{+}3{\cdot}1^2{+}2{\cdot}2{\cdot}2){=}125.$$

Hence,

$$\begin{aligned}|\mathrm{C}_{\mathrm{GL}(37,\mathbb{F}_3)}(A)| &= 3^{125}\prod_k\prod_i\prod_{u=1}^{\mu_i^{(k)}}(1-q^{-d_k u})\\ &= 3^{125}(1-3^{-2})(1-3^{-2\cdot 2})(1-3^{-2})(1-3^{-2\cdot 2})\\ &\quad\cdot(1-3^{-2\cdot 3})(1-3^{-3})(1-3^{-3\cdot 2})(1-3^{-3})\\ &= 2^{22}\cdot 3^{95}\cdot 5^2\cdot 7^2\cdot 13^4,\end{aligned}$$

and

$$|[A]_{\mathrm{GL}(37,\mathbb{F}_3)}| = \frac{|\mathrm{GL}(37,\mathbb{F}_3)|}{|\mathrm{C}_{\mathrm{GL}(37,\mathbb{F}_3)}(A)|}$$

$$= 2^6\cdot 3^{571}\cdot 5\cdot 11^4\cdot 23\cdot 41\cdot 61\cdot 73\cdot 757\cdot 1093\cdot 3851\cdot\prod_{i=13}^{37}(3^i-1).$$

6.4. Conjugacy Classes of AGL$(n,\mathbb{F}_q)$

The conjugacy classes of $\mathrm{AGL}(n,F)$ are less well known compared with those of $\mathrm{GL}(n,F)$. In this section, we first determine a system of representatives of the conjugacy classes of $\mathrm{AGL}(n,F)$. The representatives can be regarded as the *canonical forms* in $\mathrm{AGL}(n,F)$. When $F=\mathbb{F}_q$, we also determine the cardinalities of the conjugacy classes of $\mathrm{AGL}(n,\mathbb{F}_q)$. These results are useful for computing the number of orbits when $\mathrm{AGL}(n,\mathbb{F}_q)$ acts on a set.

6.4.1. Canonical forms in AGL(n,F).

For each $A\in\mathrm{GL}(n,F)$, let

$$M_A=\left\{\begin{bmatrix}A & \\ a & 1\end{bmatrix} : a\in F^n\right\}\subset\mathrm{AGL}(n,F)$$

and

$$G_A=\left\{\begin{bmatrix}P & \\ u & 1\end{bmatrix} : P\in\mathrm{C}_{\mathrm{GL}(n,F)}(A),\ u\in F^n\right\}<\mathrm{AGL}(n,F). \tag{6.21}$$

Note that the group G_A acts on M_A by conjugation. For $\left[\begin{smallmatrix}A & \\ a & 1\end{smallmatrix}\right]\in M_A$ and $\left[\begin{smallmatrix}P & \\ u & 1\end{smallmatrix}\right]\in G_A$, we have

$$\begin{bmatrix}P & \\ u & 1\end{bmatrix}^{-1}\begin{bmatrix}A & \\ a & 1\end{bmatrix}\begin{bmatrix}P & \\ u & 1\end{bmatrix}=\begin{bmatrix}A & \\ -u(A-I)+aP & 1\end{bmatrix}. \tag{6.22}$$

If $\sigma,\tau\in M_A$ are in the same G_A-orbit, we write $\sigma\overset{A}{\sim}\tau$.

Lemma 6.16. *Let $\mathcal{A}$ be a system of representatives of the conjugacy classes of $\mathrm{GL}(n, F)$. For each $A \in \mathcal{A}$, let $\mathcal{C}_A$ be a system of representatives of the G_A-orbits in M_A. Then*

$$\bigcup_{A \in \mathcal{A}} \mathcal{C}_A \tag{6.23}$$

is a system of representatives of the conjugacy classes of $\mathrm{AGL}(n, F)$.

Proof. 1° We show that every $\sigma \in \mathrm{AGL}(n, F)$ is conjugate to some element in (6.23). Let $\sigma = \left[\begin{smallmatrix} A & \\ a & 1 \end{smallmatrix}\right]$, where $A \in \mathrm{GL}(n, F)$ and $a \in F^n$. Choose $P \in \mathrm{GL}(n, F)$ such that $P^{-1}AP \in \mathcal{A}$. Then

$$\begin{bmatrix} P & \\ & 1 \end{bmatrix}^{-1} \begin{bmatrix} A & \\ a & 1 \end{bmatrix} \begin{bmatrix} P & \\ & 1 \end{bmatrix} = \begin{bmatrix} P^{-1}AP & \\ * & 1 \end{bmatrix}.$$

Hence, we may assume that $A \in \mathcal{A}$. Then σ is conjugate to some element in $\mathcal{C}_A$.

2° We show that any two distinct elements in (6.23) are not conjugate. Assume that $\sigma = \left[\begin{smallmatrix} A & \\ a & 1 \end{smallmatrix}\right]$ and $\tau = \left[\begin{smallmatrix} B & \\ b & 1 \end{smallmatrix}\right]$ are two elements of (6.23) which are conjugate. Then there exists $\left[\begin{smallmatrix} P & \\ u & 1 \end{smallmatrix}\right] \in \mathrm{AGL}(n, F)$ such that

$$\begin{bmatrix} B & \\ b & 1 \end{bmatrix} = \begin{bmatrix} P & \\ u & 1 \end{bmatrix}^{-1} \begin{bmatrix} A & \\ a & 1 \end{bmatrix} \begin{bmatrix} P & \\ u & 1 \end{bmatrix} = \begin{bmatrix} P^{-1}AP & \\ * & 1 \end{bmatrix}. \tag{6.24}$$

First, $B = P^{-1}AP$, i.e., A and B are conjugates in $\mathrm{GL}(n, F)$. Since $A, B \in \mathcal{A}$, we must have $A = B$. Then $\sigma, \tau \in M_A$ and $P \in \mathrm{C}_{\mathrm{GL}(n,F)}(A)$, i.e., $\left[\begin{smallmatrix} P & \\ u & 1 \end{smallmatrix}\right] \in G_A$. By (6.24), $\sigma \overset{A}{\sim} \tau$, and hence $\sigma = \tau$. □

For $\sigma = \left[\begin{smallmatrix} A & \\ a & 1 \end{smallmatrix}\right] \in \mathrm{AGL}(m, F)$ and $\tau = \left[\begin{smallmatrix} B & \\ b & 1 \end{smallmatrix}\right] \in \mathrm{AGL}(n, F)$, define

$$\sigma \boxplus \tau = \begin{bmatrix} A & & \\ & B & \\ a & b & 1 \end{bmatrix} \in \mathrm{AGL}(m+n, F).$$

Then $\mathrm{AGL}(m, F) \boxplus \mathrm{AGL}(n, F) < \mathrm{AGL}(m+n, F)$, and the map

$$\begin{array}{ccc} \mathrm{AGL}(m, F) \times \mathrm{AGL}(n, F) & \longrightarrow & \mathrm{AGL}(m, F) \boxplus \mathrm{AGL}(n, F) \\ (\sigma, \tau) & \longmapsto & \sigma \boxplus \tau \end{array} \tag{6.25}$$

is a group isomorphism. In fact, for $\sigma_i = \left[\begin{smallmatrix} A_i & \\ a_i & 1 \end{smallmatrix}\right] \in \mathrm{AGL}(m, F)$ and $\tau_i = \left[\begin{smallmatrix} B_i & \\ b_i & 1 \end{smallmatrix}\right] \in \mathrm{AGL}(n, F)$, $i = 1, 2$,

$$(\sigma_1 \boxplus \tau_1)(\sigma_2 \boxplus \tau_2)$$

$$= \begin{bmatrix} A_1 & & \\ & B_1 & \\ a_1 & b_1 & 1 \end{bmatrix} \begin{bmatrix} A_2 & & \\ & B_2 & \\ a_2 & b_2 & 1 \end{bmatrix} = \begin{bmatrix} A_1A_2 & & \\ & B_1B_2 & \\ a_1A_2 + a_2 & b_1B_2 + b_2 & 1 \end{bmatrix}$$

$$= \begin{bmatrix} A_1A_2 & \\ a_1A_2 + a_2 & 1 \end{bmatrix} \boxplus \begin{bmatrix} B_1B_2 & \\ b_1B_2 + b_2 & 1 \end{bmatrix} = (\sigma_1\sigma_2) \boxplus (\tau_1\tau_2).$$

Lemma 6.17. *Let $A \in \mathrm{GL}(m, F)$ and $B \in \mathrm{GL}(n, F)$ be such that their characteristic polynomials are relatively prime.*

(i) *Let $\mathcal{C}_A$ be a system of representatives of the G_A-orbits in M_A and $\mathcal{C}_B$ a system of representatives of the G_B-orbits in M_B. Then $\mathcal{C}_A \boxplus \mathcal{C}_B$ is a system of representatives of the $G_{A\oplus B}$-orbits in $M_{A\oplus B}$.*

(ii) *For any $\sigma \in M_A$ and $\tau \in M_B$,*

$$\mathrm{C}_{\mathrm{AGL}(m+n,F)}(\sigma \boxplus \tau) = \mathrm{C}_{\mathrm{AGL}(m,F)}(\sigma) \boxplus \mathrm{C}_{\mathrm{AGL}(n,F)}(\tau). \tag{6.26}$$

Proof. (i) By definition, $M_{A\oplus B} = M_A \boxplus M_B$. Thus, it suffices to prove that for $\sigma_i \in M_A$ and $\tau_i \in M_B$, $i = 1, 2$, $\sigma_1 \boxplus \tau_1 \overset{A\oplus B}{\sim} \sigma_2 \boxplus \tau_2$ if and only if $\sigma_1 \overset{A}{\sim} \sigma_2$ and $\tau_1 \overset{B}{\sim} \tau_2$. By Lemma 6.10, $\mathrm{C}_{\mathrm{GL}(m+n,F)}(A \oplus B) = \mathrm{C}_{\mathrm{GL}(m,F)}(A) \oplus \mathrm{C}_{\mathrm{GL}(n,F)}(B)$. Thus $G_{A\oplus B} = G_A \boxplus G_B$. By definition, $\sigma_1 \boxplus \tau_1 \overset{A\oplus B}{\sim} \sigma_2 \boxplus \tau_2$ if and only if there exists $\gamma \in G_{A\oplus B}$ such that $\gamma^{-1}(\sigma_1 \boxplus \tau_1)\gamma = \sigma_2 \boxplus \tau_2$. Write $\gamma = \alpha \boxplus \beta$, where $\alpha \in G_A$ and $\beta \in G_B$. Then by the isomorphism (6.25),

$$\gamma^{-1}(\sigma_1 \boxplus \tau_1)\gamma = (\alpha^{-1} \boxplus \beta^{-1})(\sigma_1 \boxplus \tau_1)(\alpha \boxplus \beta) = (\alpha^{-1}\sigma_1\alpha) \boxplus (\beta^{-1}\tau_1\beta).$$

Hence,

$$\begin{aligned} \sigma_1 \boxplus \tau_1 \overset{A\oplus B}{\sim} \sigma_2 \boxplus \tau_2 \quad &\Leftrightarrow \quad \alpha^{-1}\sigma_1\alpha = \sigma_2 \text{ and } \beta^{-1}\tau_1\beta = \tau_2 \\ &\qquad \text{for some } \alpha \in G_A \text{ and } \beta \in G_B \\ &\Leftrightarrow \quad \sigma_1 \overset{A}{\sim} \sigma_2 \text{ and } \tau_1 \overset{B}{\sim} \tau_2. \end{aligned}$$

(ii) To prove (6.26), it suffices to show that the left side is contained in the right side. By definition, $\mathrm{C}_{\mathrm{AGL}(m+n,F)}(\sigma \boxplus \tau) \subset G_{A\oplus B}$, where $G_{A\oplus B} = G_A \boxplus G_B$ (proved in (i)). Thus, every element in $\mathrm{C}_{\mathrm{AGL}(m+n,F)}(\sigma \boxplus \tau)$ is of the form $\alpha \boxplus \beta$ for some $\alpha \in G_A$ and $\beta \in G_B$. By the isomorphism (6.25), $\alpha \in \mathrm{C}_{\mathrm{AGL}(m,F)}(\sigma)$ and $\beta \in \mathrm{C}_{\mathrm{AGL}(n,F)}(\tau)$. □

Lemma 6.18. *Let $A \in \mathrm{GL}(n, F)$ be such that 1 is not its eigenvalue. Then there is only one G_A-orbit in M_A.*

Proof. Let $a \in F^n$ be arbitrary, and let $b = a(A - I)^{-1}$. Then

$$\begin{bmatrix} I & \\ b & 1 \end{bmatrix}^{-1} \begin{bmatrix} A & \\ a & 1 \end{bmatrix} \begin{bmatrix} I & \\ b & 1 \end{bmatrix} = \begin{bmatrix} A & \\ -b(A-I)+a & 1 \end{bmatrix} = \begin{bmatrix} A & \\ 0 & 1 \end{bmatrix}.$$

□

Next, we determine a system of representatives of the G_A-orbits in M_A where all the elementary divisors of A are powers of $\mathtt{X} - 1$. Let

$$N_n = \begin{bmatrix} 0 & 1 & & & \\ & \cdot & \cdot & & \\ & & \cdot & \cdot & \\ & & & \cdot & 1 \\ & & & & 0 \end{bmatrix} \in \mathrm{M}_n(F), \qquad J_n = I_n + N_n,$$

and

$$\epsilon_n = (1, 0, \dots, 0) \in F^n.$$

Lemma 6.19. *Let*

$$B = \begin{bmatrix} \underbrace{0}_{m-u} & b(N_u) \\ 0 & 0 \end{bmatrix}_{l \times m}, \qquad C = \begin{bmatrix} \underbrace{0}_{n-v} & c(N_v) \\ 0 & 0 \end{bmatrix}_{m \times n},$$

where $b, c \in F[\mathtt{X}]$. Then

$$BC = \begin{cases} 0 & \text{if } u + v \le m, \\ \begin{bmatrix} \underbrace{0}_{m+n-u-v} & (bc)(N_{u+v-m}) \\ 0 & 0 \end{bmatrix}_{l \times n} & \text{if } u + v > m. \end{cases}$$

Proof. Case 1. Assume that $m - u \ge v$. Then

$$BC = \begin{bmatrix} b(N_u) \\ 0 \end{bmatrix}_{l \times u} [0]_{u \times n} = 0.$$

Case 2. Assume that $m - u < v$. Then

$$\begin{aligned} BC &= \begin{bmatrix} b(N_u) \\ 0 \end{bmatrix}_{l \times u} \begin{bmatrix} 0 & c(N_{u+v-m}) \\ 0 & 0 \end{bmatrix}_{u \times n} \\ &= \begin{bmatrix} b(N_{u+v-m}) \\ 0 \end{bmatrix}_{l \times (u+v-m)} \begin{bmatrix} 0 & c(N_{u+v-m}) \end{bmatrix}_{(u+v-m) \times n} \end{aligned}$$

$$= \begin{bmatrix} 0 & (bc)(N_{u+v-m}) \\ 0 & 0 \end{bmatrix}_{l\times n}.$$

□

Lemma 6.20. *Let $n_1, \dots, n_s \in \mathbb{Z}^+$ and $A = N_{n_1} \oplus \cdots \oplus N_{n_s}$.*

(i) *The elements of $\mathrm{C}_{\mathrm{M}_{n_1+\cdots+n_s}(F)}(A)$ are precisely of the form*

$$P = \begin{bmatrix} P_{11} & \cdots & P_{1s} \\ \vdots & & \vdots \\ P_{s1} & \cdots & P_{ss} \end{bmatrix} \tag{6.27}$$

where $P_{ij} \in \mathrm{M}_{n_i \times n_j}(F)$ and

$$P_{ij} = \begin{cases} \begin{bmatrix} 0 & f_{ij}(N_{n_i}) \end{bmatrix} & \text{if } n_i \le n_j, \\ \begin{bmatrix} f_{ij}(N_{n_j}) \\ 0 \end{bmatrix} & \text{if } n_i > n_j \end{cases} \tag{6.28}$$

for some $f_{ij} \in F[\mathtt{X}]$ with $\deg f_{ij} < \min\{n_i, n_j\}$.

(ii) *Let P be the matrix in (6.27) and define*

$$g_{ij} = \mathtt{X}^{\max\{0, n_j - n_i\}} f_{ij}. \tag{6.29}$$

Then P is invertible if and only if

$$\begin{bmatrix} g_{11}(0) & \cdots & g_{1s}(0) \\ \vdots & & \vdots \\ g_{s1}(0) & \cdots & g_{ss}(0) \end{bmatrix} \tag{6.30}$$

is invertible.

Note. In (6.30), $g_{ij}(0)$ is the $(1,1)$-entry of P_{ij}. Thus, the matrix in (6.30) is a principal submatrix of P with row and column indices 1, n_1+1, $\dots$, $n_1 + \cdots + n_{s-1} + 1$.

Proof of Lemma 6.20. (i) Let $n = n_1 + \cdots + n_s$ and let $P \in \mathrm{M}_n(F)$ be partitioned as in (6.27). Then $P \in \mathrm{C}_{\mathrm{M}_n(F)}(A)$ if and only if

$$N_{n_i} P_{ij} = P_{ij} N_{n_j} \quad \text{for all } 1 \le i, j \le s. \tag{6.31}$$

Write $P_{ij} = (p_{kl})_{1\le k\le n_i, 1\le l\le n_j}$. Then (6.31) becomes

$$\begin{bmatrix} p_{21} & \cdots & p_{2,n_j} \\ \vdots & & \vdots \\ p_{n_i,1} & \cdots & p_{n_i,n_j} \\ 0 & \cdots & 0 \end{bmatrix} = \begin{bmatrix} 0 & p_{11} & \cdots & p_{1,n_j-1} \\ \vdots & \vdots & & \vdots \\ 0 & p_{n_i,1} & \cdots & p_{n_i,n_j-1} \end{bmatrix}. \tag{6.32}$$

If $n_i \le n_j$, a comparison of the entries of the above two matrices shows that $p_{kl} = 0$ when $l - k < n_j - n_i$ and that p_{kl} depends only on $l - k$ when $l - k \ge n_j - n_i$. Specifically,

$$P = \begin{bmatrix} 0 & \cdots & 0 & a_0 & a_1 & \cdot & \cdot & a_{n_i-1} \\ \cdot & & \cdot & & \cdot & \cdot & \cdot & \cdot \\ \cdot & & \cdot & & & \cdot & \cdot & \cdot \\ \cdot & & \cdot & & & & \cdot & a_1 \\ 0 & \cdots & 0 & & & & & a_0 \end{bmatrix} = \begin{bmatrix} 0 \ f(N_{n_i}) \end{bmatrix},$$

where $f = a_0 + a_1 \mathtt{X} + \cdots + a_{n_i-1}\mathtt{X}^{n_i-1}$. Similarly, if $n_i > n_j$, (6.32) holds if and only if

$$P = \begin{bmatrix} g(N_{n_j}) \\ 0 \end{bmatrix}$$

for some $g \in F[\mathtt{X}]$ with $\deg g < n_j$.

(ii) Without loss of generality, we may assume that $n_1 \le \cdots \le n_s$. Let

$$\mathcal{M} = \big\{(a_{ij}) \in \mathrm{M}_s(F) : a_{ij} = 0 \text{ if } n_i < n_j\big\}.$$

Elements of $\mathcal{M}$ are block lower triangular matrices with block sizes $\lambda_k = |\{i : n_i = k\}|$, $k = 1, 2, \dots$. Define

$$\begin{array}{rccc} \phi: & \mathrm{C}_{\mathrm{M}_n(F)}(A) & \longrightarrow & \mathcal{M} \\ & P & \longmapsto & (g_{ij}(0)), \end{array}$$

where P is given in (6.27) and g_{ij} is given in (6.29). We claim that ϕ is an onto ring homomorphism. For this claim, the only part that needs a proof is that ϕ preserves multiplication.

Let $P_l = (P_{ij}^{(l)}) \in \mathrm{C}_{\mathrm{M}_n(F)}(A)$, $l = 1, 2$, where $P_{ij}^{(l)} \in \mathrm{M}_{n_i \times n_j}(F)$ corresponds to $f_{ij}^{(l)} \in F[\mathtt{X}]$ as P_{ij} corresponds to f_{ij} in (6.28). Then $P_1 P_2 = (P_{ij})$, where

$$P_{ij} = \sum_k P_{ik}^{(1)} P_{kj}^{(2)}.$$

Let $i \ge j$. If $i \ge k \ge j$, by Lemma 6.19,

$$P_{ik}^{(1)} P_{kj}^{(2)} = \begin{bmatrix} f_{ik}^{(1)}(N_{n_k}) \\ 0 \end{bmatrix} \begin{bmatrix} f_{kj}^{(2)}(N_{n_j}) \\ 0 \end{bmatrix} = \begin{bmatrix} (f_{ik}^{(1)} f_{kj}^{(2)})(N_{n_j}) \\ 0 \end{bmatrix}.$$

If $k > i$ or $k < j$, the first column of either $P_{ik}^{(1)}$ or $P_{kj}^{(2)}$ is 0. By Lemma 6.19, the first column of $P_{ik}^{(1)} P_{kj}^{(2)}$ is 0, and we can write

$$P_{ik}^{(1)} P_{kj}^{(2)} = \begin{bmatrix} h_k(N_{n_j}) \\ 0 \end{bmatrix}$$

for some $h_k \in F[\mathtt{X}]$ with $h_k(0) = 0$. Now

$$P_{ij} = \begin{bmatrix} f_{ij}(N_{n_j}) \\ 0 \end{bmatrix},$$

where

$$f_{ij} = \sum_{i \geq k \geq j} f_{ik}^{(1)} f_{kj}^{(2)} + \sum_{k > i \text{ or } k < j} h_k.$$

Hence

$$f_{ij}(0) = \sum_{i \geq k \geq j} f_{ik}^{(1)}(0) f_{kj}^{(2)}(0),$$

which proves that $\phi(P_1P_2) = \phi(P_1)\phi(P_2)$. Therefore ϕ is an onto ring homomorphism.

Let $P \in \mathrm{C}_{\mathrm{M}_n(F)}(A)$. If P is invertible in $\mathrm{C}_{\mathrm{M}_n(F)}(A)$, then so is $\phi(P)$ in $\mathcal{M}$. On the other hand, assume that $\phi(P)$ is invertible in $\mathcal{M}$. Since ϕ is onto, there exists $Q \in \mathrm{C}_{\mathrm{M}_n(F)}(A)$ such that $\phi(P)^{-1} = \phi(Q)$, i.e., $\phi(PQ) = I_s = \phi(I_n)$. Let $S = PQ - I_n$. Then $\phi(S) = 0$. It follows that $S = (S_{ij})$, where

$$S_{ij} = \begin{cases} \begin{bmatrix} 0 & h_{ij}(N_{n_i}) \end{bmatrix} & \text{if } i \leq j, \\ \begin{bmatrix} h_{ij}(N_{n_j}) \\ 0 \end{bmatrix} & \text{if } i > j \end{cases}$$

for some $h_{ij} \in F[\mathtt{X}]$ with $(\mathtt{X}^{\max\{0, n_j - n_i\}} h_{ij})(0) = 0$. In particular, the first column of S_{ij} is 0. By Lemma 6.19, for any integer $k > 0$, $S^k = (S_{ij}^{(k)})$, where $S_{ij}^{(k)} \in \mathrm{M}_{n_i \times n_j}(F)$ and the first k columns of $S_{ij}^{(k)}$ are all 0. Thus S is nilpotent. Consequently, $PQ = I_n + S$ is invertible, and in particular, P is invertible. □

Lemma 6.21.

(i) *Let $a = (a_1, \dots, a_n) \in F^n$. Then*

$$\begin{bmatrix} J_n & \\ a & 1 \end{bmatrix} \overset{J_n}{\sim} \begin{cases} \begin{bmatrix} J_n & \\ \epsilon_n & 1 \end{bmatrix} & \textit{if } a_1 \neq 0, \\ \begin{bmatrix} J_n & \\ 0 & 1 \end{bmatrix} & \textit{if } a_1 = 0. \end{cases}$$

(ii) *Assume that $m \leq n$ and $a \in F^m$. Then*

$$\begin{bmatrix} J_m & \\ a & 1 \end{bmatrix} \boxplus \begin{bmatrix} J_n & \\ \epsilon_n & 1 \end{bmatrix} \overset{J_m \oplus J_n}{\sim} \begin{bmatrix} J_m & \\ 0 & 1 \end{bmatrix} \boxplus \begin{bmatrix} J_n & \\ \epsilon_n & 1 \end{bmatrix}.$$

(iii) *We have*

$$\begin{bmatrix} J_n & \\ \epsilon_n & 1 \end{bmatrix} \boxplus \begin{bmatrix} J_n & \\ 0 & 1 \end{bmatrix} \overset{J_n \oplus J_n}{\sim} \begin{bmatrix} J_n & \\ 0 & 1 \end{bmatrix} \boxplus \begin{bmatrix} J_n & \\ \epsilon_n & 1 \end{bmatrix}.$$

Proof. (i) First assume that $a_1 \neq 0$. By (6.22),

$$\begin{bmatrix} a_1^{-1}I_n & \\ u & 1 \end{bmatrix}^{-1} \begin{bmatrix} J_n & \\ a & 1 \end{bmatrix} \begin{bmatrix} a_1^{-1}I_n & \\ u & 1 \end{bmatrix} = \begin{bmatrix} J_n & \\ -uN_n + a_1^{-1}a & 1 \end{bmatrix} \quad \text{for all } u \in F^n.$$

We can choose $u \in F^n$ such that $-uN_n + a_1^{-1}a = (1,0,\dots,0) = \epsilon_n$.

Now assume that $a_1 = 0$. By (6.22) again,

$$\begin{bmatrix} I_n & \\ u & 1 \end{bmatrix}^{-1} \begin{bmatrix} J_n & \\ a & 1 \end{bmatrix} \begin{bmatrix} I_n & \\ u & 1 \end{bmatrix} = \begin{bmatrix} J_n & \\ -uN_n + a & 1 \end{bmatrix} \quad \text{for all } u \in F^n.$$

We can choose $u \in F^n$ such that $-uN_n + a = 0$.

(ii) Write $a = (a_0, \dots, a_{m-1})$ and let $g = -a_0 - \cdots - a_{m-1}\mathtt{X}^{m-1} \in F[\mathtt{X}]$. By Lemma 6.20 (i),

$$\begin{bmatrix} I_m & 0 \\ \begin{bmatrix} g(N_m) \\ 0 \end{bmatrix} & I_n \end{bmatrix} \in C_{\mathrm{GL}(m+n,F)}(J_m \oplus J_n),$$

and by (6.22),

$$\begin{bmatrix} I_m & & \\ \begin{bmatrix} g(N_m) \\ 0 \end{bmatrix} & I_n & \\ 0 & 0 & 1 \end{bmatrix}^{-1} \begin{bmatrix} J_m & & \\ & J_n & \\ a & \epsilon_n & 1 \end{bmatrix} \begin{bmatrix} I_m & & \\ \begin{bmatrix} g(N_m) \\ 0 \end{bmatrix} & I_n & \\ 0 & 0 & 1 \end{bmatrix}$$

$$= \begin{bmatrix} J_m & & \\ & J_n & \\ a + \epsilon_n \begin{bmatrix} g(N_m) \\ 0 \end{bmatrix} & \epsilon_n & 1 \end{bmatrix} = \begin{bmatrix} J_m & & \\ & J_n & \\ 0 & \epsilon_n & 1 \end{bmatrix}.$$

(iii) We have

$$\begin{bmatrix} 0 & I_n & \\ I_n & 0 & \\ 0 & 0 & 1 \end{bmatrix}^{-1} \begin{bmatrix} J_n & & \\ & J_n & \\ \epsilon_n & 0 & 1 \end{bmatrix} \begin{bmatrix} 0 & I_n & \\ I_n & 0 & \\ 0 & 0 & 1 \end{bmatrix} = \begin{bmatrix} J_n & & \\ & J_n & \\ 0 & \epsilon_n & 1 \end{bmatrix}.$$

□

A *partition* is a sequence $\lambda = (\lambda_1, \lambda_2, \dots)$ where $\lambda_i \in \mathbb{N}$, $i \geq 1$, and $\lambda_i \neq 0$ for only finitely many i. We define $|\lambda| = 1\lambda_1 + 2\lambda_2 + \cdots$. If $|\lambda| = n$, we say that λ is a partition of n and write $\lambda \vdash n$. Let $\mathcal{P}$ denote the set of all partitions. For $\lambda = (\lambda_1, \lambda_2, \dots) \in \mathcal{P}$, let $T(\lambda) = \{t \in \mathbb{Z}^+ : \lambda_t > 0\}$. Define

$$\sigma_\lambda = \underbrace{\begin{bmatrix} J_1 & \\ 0 & 1 \end{bmatrix} \boxplus \cdots \boxplus \begin{bmatrix} J_1 & \\ 0 & 1 \end{bmatrix}}_{\lambda_1} \boxplus \underbrace{\begin{bmatrix} J_2 & \\ 0 & 1 \end{bmatrix} \boxplus \cdots \boxplus \begin{bmatrix} J_2 & \\ 0 & 1 \end{bmatrix}}_{\lambda_2} \boxplus \cdots \in \mathrm{AGL}(|\lambda|, F)$$

and, for $t \in T(\lambda)$,

$$\begin{aligned} \sigma_{\lambda,t} = {} & \underbrace{\begin{bmatrix} J_1 & \\ 0 & 1 \end{bmatrix} \boxplus \cdots \boxplus \begin{bmatrix} J_1 & \\ 0 & 1 \end{bmatrix}}_{\lambda_1} \boxplus \cdots \boxplus \underbrace{\begin{bmatrix} J_{t-1} & \\ 0 & 1 \end{bmatrix} \boxplus \cdots \boxplus \begin{bmatrix} J_{t-1} & \\ 0 & 1 \end{bmatrix}}_{\lambda_{t-1}} \\ & \boxplus \underbrace{\begin{bmatrix} J_t & \\ \epsilon_t & 1 \end{bmatrix} \boxplus \begin{bmatrix} J_t & \\ 0 & 1 \end{bmatrix} \boxplus \cdots \boxplus \begin{bmatrix} J_t & \\ 0 & 1 \end{bmatrix}}_{\lambda_t} \\ & \boxplus \underbrace{\begin{bmatrix} J_{t+1} & \\ 0 & 1 \end{bmatrix} \boxplus \cdots \boxplus \begin{bmatrix} J_{t+1} & \\ 0 & 1 \end{bmatrix}}_{\lambda_{t+1}} \boxplus \cdots \in \mathrm{AGL}(|\lambda|, F). \end{aligned}$$

Lemma 6.22. *Let $\lambda = (\lambda_1, \lambda_2, \dots) \vdash n$ and*

$$A = \underbrace{J_1 \oplus \cdots \oplus J_1}_{\lambda_1} \oplus \underbrace{J_2 \oplus \cdots \oplus J_2}_{\lambda_2} \oplus \cdots \in \mathrm{GL}(n, F).$$

Then the following elements form a system of representatives of the G_A-orbits in M_A:

$$\sigma_\lambda \quad \textit{and} \quad \sigma_{\lambda,t}, \quad t \in T(\lambda). \tag{6.33}$$

Proof. By Lemma 6.21, every element in M_A is $\overset{A}{\sim}$ to an element in (6.33). It remains to show that the elements in (6.33) are in different G_A-orbits.

For $t \in T(\lambda)$, we have $\sigma_{\lambda,t} = \left[\begin{smallmatrix} A & \\ a_t & 1 \end{smallmatrix}\right]$, where $a_t \in F^n$ is the vector whose $(1\lambda_1 + \cdots + (t-1)\lambda_{t-1} + 1)$th component is 1 and whose other components are 0. By (6.22), elements in M_A which are $\overset{A}{\sim}$ to $\sigma_{\lambda,t}$ are precisely $\left[\begin{smallmatrix} A & \\ b & 1 \end{smallmatrix}\right]$ where

$$b = -u(A - I) + a_t P \tag{6.34}$$

for some $u \in F^n$ and $P \in C_{\mathrm{GL}(n,F)}(A)$. Let $s \in T(\lambda)$ be such that $s > t$. Note that the $(1\lambda_1 + \cdots + (s-1)\lambda_{s-1} + 1)$th column of $A - I$ is 0. From Lemma 6.20 (i), we can see that the $(1\lambda_1 + \cdots + (t-1)\lambda_{t-1} + 1,\ 1\lambda_1 + \cdots + (s-1)\lambda_{s-1} + 1)$-entry of P is 0. Then, by (6.34), the $(1\lambda_1 + \cdots + (s-1)\lambda_{s-1} + 1)$th component of b is 0. Thus $b \neq a_s$. Therefore, $\sigma_{\lambda,s}$ and $\sigma_{\lambda,t}$ are not in the same G_A-orbit.

Similarly, elements in M_A which are $\overset{A}{\sim}$ to σ_λ are precisely $\left[\begin{smallmatrix} A & \\ c & 1 \end{smallmatrix}\right]$ where

$$c = -v(A - I)$$

for some $v \in F^n$. The $(1\lambda_1 + \cdots + (t-1)\lambda_{t-1} + 1)$th component of c is 0. Thus $c \neq a_t$. Therefore, σ_λ and $\sigma_{\lambda,t}$ are not in the same G_A-orbit. □

Theorem 6.23. *For each $m \geq 0$, let $\mathcal{B}(m)$ be a system of representatives of the conjugacy classes of the elements in* $\mathrm{GL}(m,F)$ *which do not have eigenvalue* 1. *Then the following elements form a system of representatives of the conjugacy classes of* $\mathrm{AGL}(n,F)$:

(6.35)
$$\sigma_\lambda \boxplus \begin{bmatrix} B & \\ 0 & 1 \end{bmatrix} \text{ and } \sigma_{\lambda,t} \boxplus \begin{bmatrix} B & \\ 0 & 1 \end{bmatrix}, \quad \lambda \in \mathcal{P},\ |\lambda| \leq n,\ t \in T(\lambda),\ B \in \mathcal{B}(n-|\lambda|).$$

Proof. This follows immediately from Lemmas 6.16–6.18 and 6.22. □

We refer to the elements in (6.35) as the *canonical forms* in $\mathrm{AGL}(n,F)$.

6.4.2. Cardinalities of the conjugacy classes of $\mathrm{AGL}(n,\mathbb{F}_q)$.

Let $\left[\begin{smallmatrix} A & \\ a & 1 \end{smallmatrix}\right] \in \mathrm{AGL}(n,F)$. It follows from (6.22) that

(6.36)
$$\mathrm{C}_{\mathrm{AGL}(n,F)}\left(\begin{bmatrix} A & \\ a & 1 \end{bmatrix}\right) = \left\{\begin{bmatrix} P & \\ u & 1 \end{bmatrix} : P \in \mathrm{C}_{\mathrm{GL}(n,F)}(A),\ u(A-I) = a(P-I)\right\}.$$

Now let $\sigma \in \mathrm{AGL}(n,\mathbb{F}_q)$ be a canonical form in (6.35). The cardinality of the conjugacy class $[\sigma]_{\mathrm{AGL}(n,\mathbb{F}_q)}$ is given by

$$\left|[\sigma]_{\mathrm{AGL}(n,\mathbb{F}_q)}\right| = \frac{|\mathrm{AGL}(n,\mathbb{F}_q)|}{\left|\mathrm{C}_{\mathrm{AGL}(n,\mathbb{F}_q)}(\sigma)\right|}.$$

The next theorem determines the cardinality $\left|\mathrm{C}_{\mathrm{AGL}(n,\mathbb{F}_q)}(\sigma)\right|$.

Theorem 6.24. *Let $\lambda = (\lambda_1, \lambda_2, \dots) \in \mathcal{P}$ with $|\lambda| \leq n$, let $t \in T(\lambda)$, and let $B \in \mathrm{GL}(n-|\lambda|,\mathbb{F}_q)$ be such that* 1 *is not its eigenvalue. Then*

$$\begin{aligned}(6.37)\qquad &\left|\mathrm{C}_{\mathrm{AGL}(n,\mathbb{F}_q)}\left(\sigma_\lambda \boxplus \begin{bmatrix} B & \\ 0 & 1 \end{bmatrix}\right)\right| \\ &= \left|\mathrm{C}_{\mathrm{GL}(n-|\lambda|,\mathbb{F}_q)}(B)\right| q^{\sum_i \lambda_i + \sum_{i,j} \min(i,j)\lambda_i\lambda_j} \prod_i \prod_{l=1}^{\lambda_i} (1-q^{-l}),\end{aligned}$$

$$\begin{aligned}(6.38)\qquad &\left|\mathrm{C}_{\mathrm{AGL}(n,\mathbb{F}_q)}\left(\sigma_{\lambda,t} \boxplus \begin{bmatrix} B & \\ 0 & 1 \end{bmatrix}\right)\right| \\ &= \left|\mathrm{C}_{\mathrm{GL}(n-|\lambda|,\mathbb{F}_q)}(B)\right| \frac{1}{q^{\lambda_t}-1} q^{\sum_{i\geq t} \lambda_i + \sum_{i,j} \min(i,j)\lambda_i\lambda_j} \prod_i \prod_{l=1}^{\lambda_i} (1-q^{-l}).\end{aligned}$$

Proof. By Lemma 6.17 (ii),

(6.39)
$$\left|\mathrm{C}_{\mathrm{AGL}(n,\mathbb{F}_q)}\left(\sigma_\lambda \boxplus \begin{bmatrix} B & \\ 0 & 1 \end{bmatrix}\right)\right| = \left|\mathrm{C}_{\mathrm{AGL}(|\lambda|,\mathbb{F}_q)}(\sigma_\lambda)\right| \left|\mathrm{C}_{\mathrm{AGL}(n-|\lambda|,\mathbb{F}_q)}\left(\begin{bmatrix} B & \\ 0 & 1 \end{bmatrix}\right)\right|$$

and

(6.40)

$$\left|C_{\mathrm{AGL}(n,\mathbb{F}_q)}\left(\sigma_{\lambda,t}\boxplus\begin{bmatrix} B & \\ 0 & 1\end{bmatrix}\right)\right| = \left|C_{\mathrm{AGL}(|\lambda|,\mathbb{F}_q)}(\sigma_{\lambda,t})\right| \left|C_{\mathrm{AGL}(n-|\lambda|,\mathbb{F}_q)}\left(\begin{bmatrix} B & \\ 0 & 1\end{bmatrix}\right)\right|.$$

Since 1 is not an eigenvalue of B, it follows from (6.36) that

$$\left|C_{\mathrm{AGL}(n-|\lambda|,\mathbb{F}_q)}\left(\begin{bmatrix} B & \\ 0 & 1\end{bmatrix}\right)\right| = \left|C_{\mathrm{GL}(n-|\lambda|,\mathbb{F}_q)}(B)\right|. \tag{6.41}$$

Let $A = \underbrace{J_1\oplus\cdots\oplus J_1}_{\lambda_1}\oplus\underbrace{J_2\oplus\cdots\oplus J_2}_{\lambda_2}\oplus\cdots$. By (6.36),

$$C_{\mathrm{AGL}(|\lambda|,\mathbb{F}_q)}(\sigma_\lambda) = \left\{\begin{bmatrix} P & \\ u & 1\end{bmatrix} : P\in C_{\mathrm{GL}(|\lambda|,\mathbb{F}_q)}(A),\ u(A-I)=0\right\}.$$

Hence

$$\left|C_{\mathrm{AGL}(|\lambda|,\mathbb{F}_q)}(\sigma_\lambda)\right| = \left|C_{\mathrm{GL}(|\lambda|,\mathbb{F}_q)}(A)\right| q^{\mathrm{null}(A-I)},$$

where $\mathrm{null}(A-I) = |\lambda| - \mathrm{rank}(A-I) = \sum_i \lambda_i$ and $|C_{\mathrm{GL}(|\lambda|,\mathbb{F}_q)}(A)|$ is given by Lemma 6.13. Therefore

$$\left|C_{\mathrm{AGL}(|\lambda|,\mathbb{F}_q)}(\sigma_\lambda)\right| = q^{\sum_i\lambda_i+\sum_{i,j}\min(i,j)\lambda_i\lambda_j}\prod_i\prod_{l=1}^{\lambda_i}(1-q^{-l}). \tag{6.42}$$

Now (6.37) follows from (6.39), (6.41) and (6.42).

Next, we determine $\left|C_{\mathrm{AGL}(|\lambda|,\mathbb{F}_q)}(\sigma_{\lambda,t})\right|$. Recall that $\sigma_{\lambda,t} = \left[\begin{smallmatrix} A & \\ a & 1\end{smallmatrix}\right]$, where $a\in\mathbb{F}_q^{|\lambda|}$ is the vector whose $(1\lambda_1+\cdots+(t-1)\lambda_{t-1}+1)$th component is 1 and whose other components are 0. By (6.36),

$$\begin{aligned}&\left|C_{\mathrm{AGL}(|\lambda|,\mathbb{F}_q)}(\sigma_{\lambda,t})\right|\\ &= q^{\mathrm{null}(A-I)}\left|\left\{P\in C_{\mathrm{GL}(|\lambda|,\mathbb{F}_q)}(A) : u(A-I) = a(P-I)\text{ for some } u\in\mathbb{F}_q^{|\lambda|}\right\}\right|.\end{aligned}$$

By Lemma 6.20 (i), every $P\in C_{\mathrm{M}_{|\lambda|}(F)}(A)$ is of the form

$$P = \begin{bmatrix} P^{(1,1)} & P^{(1,2)} & \cdots\\ P^{(2,1)} & P^{(2,2)} & \cdots\\ \vdots & \vdots & \ddots\end{bmatrix},$$

where

$$P^{(i,j)} = \begin{bmatrix} P_{11}^{(i,j)} & \cdots & P_{1,\lambda_j}^{(i,j)}\\ \vdots & & \vdots\\ P_{\lambda_i,1}^{(i,j)} & \cdots & P_{\lambda_i,\lambda_j}^{(i,j)}\end{bmatrix},\qquad P_{kl}^{(i,j)}\in \mathrm{M}_{i\times j}(F),$$

and

$$P_{kl}^{(i,j)} = \begin{cases} \begin{bmatrix} 0 & f_{kl}^{(i,j)}(N_i)\end{bmatrix} & \text{if } i\le j,\\ \begin{bmatrix} f_{kl}^{(i,j)}(N_j)\\ 0\end{bmatrix} & \text{if } i>j\end{cases}$$

for some $f_{kl}^{(i,j)} \in \mathbb{F}_q[\mathtt{X}]$ with $\deg f_{kl}^{(i,j)} < \min\{i, j\}$. Note that aP is the first row of

$$\big[P_{11}^{(t,1)}, \dots, P_{1,\lambda_1}^{(t,1)};\ P_{11}^{(t,2)}, \dots, P_{1,\lambda_2}^{(t,2)};\ \dots\big].$$

Also note that vectors in the row space of $A - I$ are of the form

$$\big[\underbrace{0 \cdots 0}_{\lambda_1}\ \underbrace{\overbrace{0\ *}^{2} \cdots \overbrace{0\ *}^{2}}_{2\lambda_2}\ \underbrace{\overbrace{0\ *\ *}^{3} \cdots \overbrace{0\ *\ *}^{3}}_{3\lambda_3}\ \cdots\big].$$

Hence, $aP - a$ is in the row span of $A - I$ if and only if

(6.43)
$$\begin{aligned}&\big(f_{11}^{(t,1)}(0), \dots, f_{1,\lambda_1}^{(t,1)}(0);\ f_{11}^{(t,2)}(0), \dots, f_{1,\lambda_2}^{(t,2)}(0);\ \dots;\ f_{11}^{(t,t)}(0),\ \dots, f_{1,\lambda_t}^{(t,t)}(0)\big)\\ &= (\quad 0,\ \dots \qquad\qquad\qquad\qquad \dots,\ 0,\quad 1,\quad 0,\ \dots,\ 0\ ),\end{aligned}$$

that is, $f_{1l}^{(t,j)}(0) = 0$ for all $j \le t$ and $1 \le l \le \lambda_j$ except for $f_{11}^{(t,t)}(0)$ which is 1. By Lemma 6.20 (ii), P is invertible if and only if

$$\begin{bmatrix} f_{11}^{(i,i)}(0) & \cdots & f_{1,\lambda_i}^{(i,i)}(0) \\ \vdots & & \vdots \\ f_{\lambda_i,1}^{(i,i)}(0) & \cdots & f_{\lambda_i,\lambda_i}^{(i,i)}(0) \end{bmatrix} \in \mathrm{GL}(\lambda_i, \mathbb{F}_q) \tag{6.44}$$

for all $i \ge 1$.

For each fixed i, we first count the number of choices of $f_{kl}^{(i,i)}$, for all k, l, subject to (6.43) and (6.44). If $i \ne t$, the number of choices of $f_{kl}^{(i,i)}$ subject to (6.44) is

$$q^{(i-1)\lambda_i^2} |\mathrm{GL}(\lambda_i, \mathbb{F}_q)|.$$

If $i = t$, the number of choices of $f_{kl}^{(t,t)}$ subject to (6.43) and (6.44) is

$$q^{(t-1)\lambda_t^2} \frac{1}{q^{\lambda_t} - 1} |\mathrm{GL}(\lambda_t, \mathbb{F}_q)|.$$

Next, still with i fixed, we count the number of choices of $f_{kl}^{(i,j)}$, for all j, k, l with $j \ne i$, subject to (6.43). If $i \ne t$, the number of choices is

$$q^{\sum_{j \ne i} \min(i,j)\lambda_i\lambda_j}.$$

If $i = t$, the number of choices is

$$q^{\sum_{j \ne t} \min(t,j)\lambda_t\lambda_j - \sum_{j<t} \lambda_j}.$$

To summarize, we have

(6.45)
$$\begin{aligned}&\big|C_{\mathrm{AGL}(|\lambda|, \mathbb{F}_q)}(\sigma_{\lambda,t})\big|\\ &= \frac{1}{q^{\lambda_t} - 1} q^{\mathrm{null}(A-I) - \sum_{j<t} \lambda_j} \prod_{i \ge 1} q^{(i-1)\lambda_i^2} q^{\sum_{j \ne i} \min(i,j)\lambda_i\lambda_j} |\mathrm{GL}(\lambda_i, \mathbb{F}_q)|\end{aligned}$$

$$= \frac{1}{q^{\lambda_t}-1} q^{\sum_{i\ge t}\lambda_i} \prod_{i\ge 1} q^{(i-1)\lambda_i^2+\sum_{j\ne i}\min(i,j)\lambda_i\lambda_j+\lambda_i^2} \prod_{l=1}^{\lambda_i}(1-q^{-l})$$

$$= \frac{1}{q^{\lambda_t}-1} q^{\sum_{i\ge t}\lambda_i+\sum_{i,j}\min(i,j)\lambda_i\lambda_j} \prod_{i\ge 1}\prod_{l=1}^{\lambda_i}(1-q^{-l}).$$

Now (6.38) follows from (6.40), (6.41) and (6.45). □

Example 6.25. We apply Theorems 6.23 and 6.24 to $\mathrm{AGL}(4,\mathbb{F}_2)$. The canonical forms and the cardinalities of the conjugacy classes are given in Table 6.1. In Table 6.1, $\sigma_{\lambda,t,B} = \sigma_{\lambda,t} \boxplus \left[\begin{smallmatrix} B & \\ 0 & 1\end{smallmatrix}\right]$; if the parameter t is absent, $\sigma_{\lambda,t}$ is treated as σ_λ.

6.5. Bilinear Forms, Hermitian Forms, and Quadratic Forms

6.5.1. Symmetric, alternating, and hermitian forms.

Let V be a vector space over a field F and let $\sigma \in \mathrm{Aut}(F)$. A function $l : V \times V \to F$ is called a *σ-sesquilinear form* on V if for all $x, y, z \in V$ and $a, b \in F$,

$$\begin{cases} l(ax+by,z) = al(x,z) + bl(y,z), \\ l(x,ay+bz) = \sigma(a)l(x,y) + \sigma(b)l(x,z). \end{cases} \tag{6.46}$$

When $\sigma = \mathrm{id}$, f is called a *bilinear form*. The pair (V, l) is called a *σ-sesquilinear space*. Two σ-sesquilinear spaces (V_1, l_1) and (V_2, l_2) are said to be *isomorphic*, denoted by $(V_1, l_1) \cong (V_2, l_2)$, and the sesquilinear forms l_1 and l_2 are said to be *equivalent*, denoted by $l_1 \cong l_2$, if there exists an F-isomorphism $\alpha : V_1 \to V_2$ such that $l_2(\alpha(x), \alpha(y)) = l_1(x, y)$ for all $x, y \in V_1$.

We are interested in the sesquilinear forms l with the property that $l(x, y) = 0$ implies $l(y, x) = 0$. Such sesquilinear forms are determined in the following proposition.

Proposition 6.26. *Let $l : V \times V \to F$ be a nonzero σ-sesquilinear form such that $l(x, y) = 0$ if and only if $l(y, x) = 0$. Then $\sigma^2 = \mathrm{id}$, and there exists $\epsilon \in F^*$ such that $\epsilon\sigma(\epsilon) = 1$ and*

$$l(y,x) = \epsilon\sigma\big(l(x,y)\big) \quad \textit{for all } x, y \in V.$$

Proof. 1° For all $x, y, z \in V$, we have $l\big(l(y,z)x - l(x,z)y, z\big) = 0$. Thus $l\big(z, l(y,z)x - l(x,z)y\big) = 0$, i.e.,

$$l(z,x)\sigma\big(l(y,z)\big) - l(z,y)\sigma\big(l(x,z)\big) = 0.$$

Table 6.1. $\mathrm{AGL}(4, \mathbb{F}_2)$: canonical forms and cardinalities of the conjugacy classes

canonical form $\sigma_{\lambda,t,B}$			$\lvert \mathrm{C}_{\mathrm{AGL}(4,\mathbb{F}_2)}(\sigma_{\lambda,t,B}) \rvert$	$\lvert [\sigma_{\lambda,t,B}]_{\mathrm{AGL}(4,\mathbb{F}_2)} \rvert$
λ	B	t		
(0)	$\begin{bmatrix} 0 & 1 & & \\ 1 & 1 & & \\ & & 0 & 1 \\ & & 1 & 1 \end{bmatrix}$		$2^2 \cdot 3^2 \cdot 5$	$2^8 \cdot 7$
	$\begin{bmatrix} & 1 & & \\ & & 1 & \\ & & & 1 \\ 1 & 0 & 1 & 0 \end{bmatrix}$		$2^2 \cdot 3$	$2^8 \cdot 3 \cdot 5 \cdot 7$
	$\begin{bmatrix} & 1 & & \\ & & 1 & \\ & & & 1 \\ 1 & 1 & 0 & 0 \end{bmatrix}$		$3 \cdot 5$	$2^{10} \cdot 3 \cdot 7$
	$\begin{bmatrix} & 1 & & \\ & & 1 & \\ & & & 1 \\ 1 & 0 & 0 & 1 \end{bmatrix}$		$3 \cdot 5$	$2^{10} \cdot 3 \cdot 7$
	$\begin{bmatrix} & 1 & & \\ & & 1 & \\ & & & 1 \\ 1 & 1 & 1 & 1 \end{bmatrix}$		$3 \cdot 5$	$2^{10} \cdot 3 \cdot 7$
(1)	$\begin{bmatrix} & 1 & \\ & & 1 \\ 1 & 1 & 0 \end{bmatrix}$		$2 \cdot 7$	$2^9 \cdot 3^2 \cdot 5$
		1	$2 \cdot 7$	$2^9 \cdot 3^2 \cdot 5$
	$\begin{bmatrix} & 1 & \\ & & 1 \\ 1 & 0 & 1 \end{bmatrix}$		$2 \cdot 7$	$2^9 \cdot 3^2 \cdot 5$
		1	$2 \cdot 7$	$2^9 \cdot 3^2 \cdot 5$
(2)	$\begin{bmatrix} 0 & 1 \\ 1 & 1 \end{bmatrix}$		$2^3 \cdot 3^2$	$2^7 \cdot 5 \cdot 7$
		1	$2^3 \cdot 3$	$2^7 \cdot 3 \cdot 5 \cdot 7$
$(0, 1)$	$\begin{bmatrix} 0 & 1 \\ 1 & 1 \end{bmatrix}$		$2^2 \cdot 3$	$2^8 \cdot 3 \cdot 5 \cdot 7$
		2	$2^2 \cdot 3$	$2^8 \cdot 3 \cdot 5 \cdot 7$
(4)			$2^{10} \cdot 3^2 \cdot 5 \cdot 7$	1
		1	$2^{10} \cdot 3 \cdot 7$	$3 \cdot 5$
$(2, 1)$			$2^9 \cdot 3$	$2 \cdot 3 \cdot 5 \cdot 7$
		1	2^9	$2 \cdot 3^2 \cdot 5 \cdot 7$
		2	$2^7 \cdot 3$	$2^3 \cdot 3 \cdot 5 \cdot 7$
$(1, 0, 1)$			2^6	$2^4 \cdot 3^2 \cdot 5 \cdot 7$
		1	2^6	$2^4 \cdot 3^2 \cdot 5 \cdot 7$
		3	2^5	$2^5 \cdot 3^2 \cdot 5 \cdot 7$
$(0, 2)$			$2^7 \cdot 3$	$2^3 \cdot 3 \cdot 5 \cdot 7$
		2	2^7	$2^3 \cdot 3^2 \cdot 5 \cdot 7$
$(0, 0, 0, 1)$			2^4	$2^6 \cdot 3^2 \cdot 5 \cdot 7$
		4	2^4	$2^6 \cdot 3^2 \cdot 5 \cdot 7$

Choose $x, z \in V$ such that $l(x, z) \neq 0$. Then

$$\sigma\Big(\frac{l(y,z)}{l(x,z)}\Big) = \frac{l(z,y)}{l(z,x)}. \tag{6.47}$$

Hence

$$\sigma^2\Big(\frac{l(y,z)}{l(x,z)}\Big) = \frac{l(y,z)}{l(x,z)} \quad \text{for all } y \in V,$$

which gives $\sigma^2 = \mathrm{id}$.

2° We claim that if $x, y \in V$ are such $l(x, y) \neq 0$, then

$$\frac{l(y,x)}{\sigma\big(l(x,y)\big)} = \epsilon,$$

where $\epsilon \in F^*$ is a constant not depending on x and y. Assume that $x', y' \in V$ are such that $l(x', y') \neq 0$. If $l(x', y) \neq 0$, by (6.47) and the fact that $\sigma^2 = \mathrm{id}$,

$$\sigma\Big(\frac{l(x,y)}{l(x',y')}\Big) = \sigma\Big(\frac{l(x,y)}{l(x',y)}\Big)\sigma\Big(\frac{l(x',y)}{l(x',y')}\Big) = \frac{l(y,x)}{l(y,x')} \cdot \frac{l(y,x')}{l(y',x')} = \frac{l(y,x)}{l(y',x')},$$

i.e.,

$$\frac{l(y,x)}{\sigma(l(x,y))} = \frac{l(y',x')}{\sigma(l(x',y'))}.$$

If $l(x, y') \neq 0$, by symmetry, the above equation also holds. If $l(x', y) = l(x, y') = 0$, then $l(x, y + y') \neq 0$ and $l(x', y + y') \neq 0$, and we have

$$\frac{l(y,x)}{\sigma(l(x,y))} = \frac{l(y+y',x)}{\sigma(l(x,y+y'))} = \frac{l(y',x')}{\sigma(l(x',y'))}.$$

Therefore,

$$l(y,x) = \epsilon\sigma\big(l(x,y)\big) \quad \text{for all } x, y \in V.$$

Using the above equation twice, we have

$$l(y,x) = \epsilon\sigma\big[\epsilon\sigma(l(y,x))\big] = \epsilon\sigma(\epsilon)l(y,x),$$

and hence $\epsilon\sigma(\epsilon) = 1$. □

In Proposition 6.26, if $\sigma = \mathrm{id}$, then $\epsilon = \pm 1$. If $o(\sigma) = 2$, F is a Galois extension over $E = \{a \in F : \sigma(a) = a\}$ with $\mathrm{Aut}(F/E) = \langle\sigma\rangle$. Since $\mathrm{N}_{F/E}(\epsilon) = \epsilon\sigma(\epsilon) = 1$, by Hilbert's Theorem 90, $\epsilon = \sigma(\delta)/\delta$ for some $\delta \in F^*$. Let $h(x, y) = \delta l(x, y)$, $x, y \in V$. Then h is a σ-sesquilinear form such that

$$h(y,x) = \sigma\big(h(x,y)\big) \quad \text{for all } x, y \in V.$$

Definition 6.27. Let V be a vector space over a field F.

(i) A bilinear form $l : V \times V \to F$ is said to be

$$\begin{cases} \textit{symmetric} & \text{if } l(y,x) = l(x,y) \text{ for all } x, y \in V, \\ \textit{skew symmetric} & \text{if } l(y,x) = -l(x,y) \text{ for all } x, y \in V, \\ \textit{alternating} & \text{if } l(x,x) = 0 \text{ for all } x \in V. \end{cases}$$

(ii) Let $\sigma \in \mathrm{Aut}(F)$ be such that $o(\sigma) = 2$. A σ-sesquilinear form $l : V \times V \to F$ is said to be *σ-hermitian* if $l(y, x) = \sigma(l(x, y))$ for all $x, y \in V$.

Lemma 6.28. *Let l be a bilinear form on V.*

(i) *If l is alternating, then l is skew symmetric.*

(ii) *If $\mathrm{char}\, F = 2$, then l is skew symmetric if and only if l is symmetric.*

(iii) *If $\mathrm{char}\, F \neq 2$, then l is skew symmetric if and only if l is alternating.*

Proof. (i) For all $x, y \in V$,

$$0 = l(x + y, x + y) = l(x, x) + l(x, y) + l(y, x) + l(y, y) = l(x, y) + l(y, x),$$

that is, $l(y, x) = -l(x, y)$.

(ii) Obvious.

(iii) Assume that l is skew symmetric. Then for all $x \in V$, $l(x, x) = -l(x, x)$. Since $\mathrm{char}\, F \neq 2$, it follows that $l(x, x) = 0$. Hence l is alternating. □

Because of Lemma 6.28 (ii) and (iii), to study the sesquilinear forms with the property that $l(x, y) = 0$ implies $l(y, x) = 0$, it suffices to consider symmetric, alternating, and hermitian forms. The set of all symmetric (alternating, σ-hermitian, resp.) forms on V is denoted by $\mathcal{S}(V)$ ($\mathcal{A}(V)$, $\mathcal{S}^\sigma(V)$, resp.). A sesquilinear space (V, l) is called a

$$\begin{cases} \textit{symmetric bilinear space} & \text{if } l \in \mathcal{S}(V), \\ \textit{symplectic space} & \text{if } l \in \mathcal{A}(V), \\ \sigma\textit{-unitary space} & \text{if } l \in \mathcal{S}^\sigma(V). \end{cases}$$

For $l \in \mathcal{S}(V)$ (or $\mathcal{A}(V)$, $\mathcal{S}^\sigma(V)$), define $\ker l = \{x \in V : l(x, y) = 0 \text{ for all } y \in V\}$. If $\ker l = \{0\}$, l is said to be *nondegenerate*.

Matrices of symmetric, alternating, and hermitian forms. Let V be an n-dimensional vector space over F, $l : V \times V \to F$ a σ-sesquilinear form, and $\mathcal{E} = (\epsilon_1, \dots, \epsilon_n)$ a basis of V. Let $A(l, \mathcal{E}) = (l(\epsilon_i, \epsilon_j)) \in \mathrm{M}_n(F)$, which is called the *matrix of l* with respect to $\mathcal{E}$. For

$$x = (x_1, \dots, x_n) \begin{bmatrix} \epsilon_1 \\ \vdots \\ \epsilon_n \end{bmatrix}, \quad y = (y_1, \dots, y_n) \begin{bmatrix} \epsilon_1 \\ \vdots \\ \epsilon_n \end{bmatrix} \in V, \tag{6.48}$$

we have

$$l(x, y) = (x_1, \dots, x_n) A(l, \mathcal{E}) (y_1, \dots, y_n)^{\sigma T},$$

where $(y_1, \ldots, y_n)^\sigma = (\sigma(y_1), \ldots, \sigma(y_n))$. Let $\mathcal{E}' = (\epsilon_1', \ldots, \epsilon_n')$ be another basis of V with

$$\begin{bmatrix} \epsilon_1' \\ \vdots \\ \epsilon_n' \end{bmatrix} = P \begin{bmatrix} \epsilon_1 \\ \vdots \\ \epsilon_n \end{bmatrix},$$

where $P \in \mathrm{GL}(n, F)$, and let $\alpha \in \mathrm{Aut}_F(V)$ be such that $\alpha(\epsilon_i) = \epsilon_i'$, $1 \le i \le n$. Then for $x, y \in V$ given in (6.48),

$$\alpha(x) = (x_1, \ldots, x_n) \begin{bmatrix} \epsilon_1' \\ \vdots \\ \epsilon_n' \end{bmatrix} = (x_1, \ldots, x_n) P \begin{bmatrix} \epsilon_1 \\ \vdots \\ \epsilon_n \end{bmatrix},$$

$$\alpha(y) = (y_1, \ldots, y_n) \begin{bmatrix} \epsilon_1' \\ \vdots \\ \epsilon_n' \end{bmatrix} = (y_1, \ldots, y_n) P \begin{bmatrix} \epsilon_1 \\ \vdots \\ \epsilon_n \end{bmatrix}.$$

Thus

$$\begin{aligned} l(\alpha(x), \alpha(y)) &= (x_1, \ldots, x_n) A(l, \mathcal{E}') (y_1, \ldots, y_n)^{\sigma T} \\ &= (x_1, \ldots, x_n) P A(l, \mathcal{E}) P^{\sigma T} (y_1, \ldots, y_n)^{\sigma T}, \end{aligned}$$

where P^σ is the matrix obtained by applying σ to P entrywise. Let l^α denote the σ-sesquilinear form $l(\alpha(\ ), \alpha(\ ))$. Then the above equation can be stated as

$$A(l^\alpha, \mathcal{E}) = A(l, \mathcal{E}') = P A(l, \mathcal{E}) P^{\sigma T}. \tag{6.49}$$

Define

$$S_n(F) = \{A \in \mathrm{M}_n(F) : A^T = A\},$$

$$\Lambda_n(F) = \{A \in \mathrm{M}_n(F) : A^T = -A \text{ and all diagonal entries of } A \text{ are } 0\}.$$

For $\sigma \in \mathrm{Aut}(F)$ with $o(\sigma) = 2$, define

$$S_n^\sigma(F) = \{A \in \mathrm{M}_n(F) : A^T = A^\sigma\}.$$

Matrices in $S_n(F)$, $\Lambda_n(F)$, and $S_n^\sigma(F)$ are called *symmetric*, *alternating*, and *σ-hermitian* matrices, respectively.

Lemma 6.29. *Let V be an n-dimensional vector space over F with a basis $\mathcal{E}$. The mapping $l \mapsto A(l, \mathcal{E})$ is a bijection between the set of all symmetric (alternating, σ-hermitian, resp.) forms on V and $S_n(F)$ ($\Lambda_n(F)$, $S_n^\sigma(F)$, resp.).*

Proof. We only prove the case of symmetric forms; the other two cases are similar. Let $\mathcal{E} = (\epsilon_1, \ldots, \epsilon_n)$. For each $l \in \mathcal{S}(V)$, clearly $A(l, \mathcal{E}) \in S_n(F)$. On the other hand, given $A \in S_n(F)$, define

$$\begin{aligned} l_A : \quad V \times V &\longrightarrow F \\ (x, y) &\longmapsto (x_1, \ldots, x_n) A (y_1, \ldots, y_n)^T, \end{aligned}$$

where x and y are given in (6.48). Then $l_A \in \mathcal{S}(V)$. Moreover, the mappings $\mathcal{S}(V) \to S_n(F)$, $l \mapsto A(l, \mathcal{E})$, and $S_n(F) \to \mathcal{S}(V)$, $A \mapsto l_A$, are inverses of each other. □

Two matrices $A, B \in \mathrm{M}_n(F)$ are said to be *congruent*, denoted by $A \cong B$, if $B = PAP^{\sigma T}$ for some $P \in \mathrm{GL}(n, F)$, where $\sigma \in \mathrm{Aut}(F)$ is defined in the context. Two matrices $A, B \in S_n(F)$ or $\Lambda_n(F)$ are congruent if $B = PAP^T$ for some $P \in \mathrm{GL}(n, F)$; two matrices $A, B \in S_n^\sigma(F)$ are congruent if $B = PAP^{\sigma T}$ for some $P \in \mathrm{GL}(n, F)$. By (6.49), two forms (symmetric, alternating, or hermitian) on V are equivalent if and only if their matrices (with respect to any basis) are congruent. Therefore, to classify symmetric, alternating, and hermitian forms on V under equivalence, it suffices to determine the canonical forms of the matrices in $S_n(F)$, $\Lambda_n(F)$, and $S_n^\sigma(F)$ under congruence.

6.5.2. Some matrix theory.

Let $\mathrm{diag}\,(a_1, \ldots, a_n)$ denote the $n \times n$ diagonal matrix with diagonal entries $a_1, \ldots, a_n$. A field is said to be *perfect* if all its algebraic extensions are separable. A field F is perfect if and only if either $\mathrm{char}\, F = 0$, or $\mathrm{char}\, F = p$ and every element of F is a pth power.

Theorem 6.30 (Canonical forms of symmetric and alternating matrices).

(i) *If* $\mathrm{char}\, F \neq 2$, *every* $A \in S_n(F)$ *is congruent to a diagonal matrix. If* $F = \mathbb{F}_q$, *where* q *is odd and* d *is any nonsquare in* $\mathbb{F}_q$, *then every* $A \in S_n(\mathbb{F}_q)$ *is congruent to exactly one of the following:*

$$\mathrm{diag}\,(\underbrace{1, \ldots, 1}_{r}, 0, \ldots, 0), \quad 0 \le r \le n, \tag{6.50}$$

$$\mathrm{diag}\,(\underbrace{1, \ldots, 1, d}_{r}, 0, \ldots, 0), \quad 1 \le r \le n. \tag{6.51}$$

(ii) *If* $\mathrm{char}\, F = 2$, *every* $A \in S_n(F) \setminus \Lambda_n(F)$ *is congruent to a diagonal matrix. If* F *is a perfect field of characteristic* 2, *then every* $A \in S_n(F) \setminus \Lambda_n(F)$ *of rank* r *is congruent to* $I_r \oplus 0$.

(iii) *Every* $A \in \Lambda_n(F)$ *is congruent to*

$$\underbrace{\begin{bmatrix} 0 & 1 \\ -1 & 0 \end{bmatrix} \oplus \cdots \oplus \begin{bmatrix} 0 & 1 \\ -1 & 0 \end{bmatrix}}_{r} \oplus 0,$$

where $2r = \operatorname{rank} A$.

Proof. (i) Let $A = (a_{ij}) \in S_n(F)$ and assume that $A \neq 0$. We show that A is congruent to a diagonal matrix. First assume that $a_{11}, a_{22}, \dots, a_{nn}$ are not all 0. By a suitable transposition of rows and columns, we may assume $a_{11} \neq 0$. Then

(6.52)

$$\begin{bmatrix} 1 & & & \\ -a_{12}a_{11}^{-1} & 0 & & \\ \vdots & & \ddots & \\ -a_{1n}a_{11}^{-1} & & & 0 \end{bmatrix} \begin{bmatrix} a_{11} & \cdots & a_{1n} \\ \vdots & & \vdots \\ a_{n1} & \cdots & a_{nn} \end{bmatrix} \begin{bmatrix} 1 & -a_{12}a_{11}^{-1} & \cdots & -a_{1n}a_{11}^{-1} \\ & 0 & & \\ & & \ddots & \\ & & & 0 \end{bmatrix}$$
$$= [a_{11}] \oplus A_1,$$

where $A_1 \in S_{n-1}(F)$. The conclusion follows by induction.

If $a_{11}, a_{22}, \dots, a_{nn}$ are all 0, then one of the nondiagonal entries of A, say a_{12}, is nonzero. We have

$$\begin{bmatrix} 1 & 1 & & \\ & 1 & & \\ & & \ddots & \\ & & & 1 \end{bmatrix} \begin{bmatrix} a_{11} & \cdots & a_{1n} \\ \vdots & & \vdots \\ a_{n1} & \cdots & a_{nn} \end{bmatrix} \begin{bmatrix} 1 & & & \\ 1 & 1 & & \\ & & \ddots & \\ & & & 1 \end{bmatrix} = \begin{bmatrix} 2a_{12} & * & \cdots & * \\ * & * & \cdots & * \\ \vdots & \vdots & & \vdots \\ * & * & \cdots & * \end{bmatrix},$$

where $2a_{12} \neq 0$. This brings us to the above case. Therefore, we have proved that A is congruent to a diagonal matrix.

Now assume that $F = \mathbb{F}_q$, where q is odd, and that d is a nonsquare of $\mathbb{F}_q$. Every nonzero element of $\mathbb{F}_q$ is either c^2 or $c^2 d$ for some $c \in \mathbb{F}_q$. Hence A is congruent to a diagonal matrix whose diagonal entries are 0, 1, or d. By Exercise 1.5, $d = a^2 + b^2$ for some $a, b \in \mathbb{F}_q$. Then

$$\begin{bmatrix} d & \\ & d \end{bmatrix} = \begin{bmatrix} a & -b \\ b & a \end{bmatrix} \begin{bmatrix} a & b \\ -b & a \end{bmatrix} \cong \begin{bmatrix} 1 & \\ & 1 \end{bmatrix}.$$

Using the above congruence repeatedly, we conclude that A is congruent to one of the matrices in (6.50) or (6.51).

Next, we show that the matrices in (6.50) and (6.51) are pairwise nonequivalent. Since the rank is invariant under congruence, it suffices to show that

$$\operatorname{diag}(\underbrace{1,\dots,1}_{r},0,\dots,0) \not\cong \operatorname{diag}(\underbrace{1,\dots,1,d}_{r},0,\dots,0).$$

Assume the contrary. Then by the next lemma, $I_r \cong I_{r-1} \oplus [d]$, i.e., $I_{r-1} \oplus [d] = PP^T$ for some $P \in \mathrm{GL}(r, \mathbb{F}_q)$. It follows that $d = \det(PP^T) = (\det P)^2$ is a square, which is a contradiction.

(iii) Let $A = (a_{ij}) \in \Lambda_n(F)$ and assume that $A \neq 0$. By a suitable permutation of rows and columns, we may assume that $a_{12} \neq 0$. Since

$$\begin{bmatrix} 1 &&&& \\ & a_{12}^{-1} &&& \\ && 1 && \\ &&& \ddots & \\ &&&& 1 \end{bmatrix} \begin{bmatrix} 0 & a_{12} & * & \cdots & * \\ -a_{12} & 0 & * & \cdots & * \\ * & * & 0 & \cdots & * \\ \vdots & \vdots & \vdots & \ddots & \vdots \\ * & * & * & \cdots & 0 \end{bmatrix} \begin{bmatrix} 1 &&&& \\ & a_{12}^{-1} &&& \\ && 1 && \\ &&& \ddots & \\ &&&& 1 \end{bmatrix} = \begin{bmatrix} 0 & 1 & * & \cdots & * \\ -1 & 0 & * & \cdots & * \\ * & * & 0 & \cdots & * \\ \vdots & \vdots & \vdots & \ddots & \vdots \\ * & * & * & \cdots & 0 \end{bmatrix},$$

we may assume that $a_{12} = 1$. We then have

$$\begin{bmatrix} 1 &&&& \\ & 1 &&& \\ a_{23} & -a_{13} & 1 && \\ \vdots & \vdots && \ddots & \\ a_{2n} & -a_{1n} &&& 1 \end{bmatrix} \begin{bmatrix} 0 & 1 & a_{13} & \cdots & a_{1n} \\ -1 & 0 & a_{23} & \cdots & a_{2n} \\ -a_{13} & -a_{23} & 0 & \cdots & * \\ \vdots & \vdots & \vdots & \ddots & \vdots \\ -a_{1n} & -a_{2n} & * & \cdots & 0 \end{bmatrix} \begin{bmatrix} 1 && a_{23} & \cdots & a_{2n} \\ & 1 & -a_{13} & \cdots & -a_{1n} \\ && 1 && \\ &&& \ddots & \\ &&&& 1 \end{bmatrix}$$
$$= \begin{bmatrix} 0 & 1 & \\ -1 & 0 & \\ && A_1 \end{bmatrix},$$

where $A_1 \in \Lambda_{n-2}(F)$. The conclusion follows by induction.

(ii) Let $A = (a_{ij}) \in S_n(F) \setminus \Lambda_n(F)$. Since the diagonal entries of A are not all 0, by a suitable transposition of rows and columns, we may assume that $a_{11} \neq 0$. By (6.52),

$$A \cong [a_{11}] \oplus A_1,$$

where $A_1 \in S_{n-1}(F)$. If $A_1 = 0$ or $A_1 \in S_{n-1}(F) \setminus \Lambda_{n-1}(F)$, the conclusion follows by induction. If $0 \neq A_1 \in \Lambda_{n-1}(F)$, by (iii), we may assume that

$$A = \begin{bmatrix} a_{11} && \\ & 0 & 1 \\ & 1 & 0 \end{bmatrix} \oplus A_2, \quad A_2 \in \Lambda_{n-3}(F).$$

Since

$$\begin{bmatrix} 1 & 0 & a_{11} \\ 1 & 1 & a_{11} \\ 0 & 0 & 1 \end{bmatrix} \begin{bmatrix} a_{11} && \\ & 0 & 1 \\ & 1 & 0 \end{bmatrix} \begin{bmatrix} 1 & 1 & 0 \\ 0 & 1 & 0 \\ a_{11} & a_{11} & 1 \end{bmatrix} = \begin{bmatrix} a_{11} && \\ & a_{11} & 1 \\ & 1 & 0 \end{bmatrix},$$

we have

$$A \cong [a_{11}] \oplus \begin{bmatrix} a_{11} & 1 \\ 1 & 0 \end{bmatrix} \oplus A_2,$$

where $\left[\begin{smallmatrix} a_{11} & 1 \\ 1 & 0 \end{smallmatrix}\right] \oplus A_2 \in S_{n-1}(F) \setminus \Lambda_{n-1}(F)$. Again, the conclusion follows by induction.

If F is a perfect field of characteristic 2, every element of F is a square. Hence, every diagonal matrix of rank r over F is congruent to $I_r \oplus 0$. □

Lemma 6.31. *Let $A, B \in \mathrm{M}_m(F)$ and let 0_n be the $n\times n$ zero matrix. With respect to a fixed $\sigma \in \mathrm{Aut}(F)$, $A \oplus 0_n \cong B \oplus 0_n$ if and only if $A \cong B$.*

Proof. We only have to prove the "only if" part. Since $A \oplus 0_n \cong B \oplus 0_n$, there exists $P \in \mathrm{GL}(m+n, F)$ such that $P(A \oplus 0_n)P^* = B \oplus 0_n$, where $(\ )^* = (\ )^{\sigma T}$. Write

$$P = \begin{bmatrix} P_{11} & P_{12} \\ P_{21} & P_{22} \end{bmatrix},$$

where P_{11} is of size $m \times m$. Then we have

$$\begin{bmatrix} P_{11}AP_{11}^* & P_{11}AP_{21}^* \\ P_{21}AP_{11}^* & P_{21}AP_{21}^* \end{bmatrix} = \begin{bmatrix} B & 0 \\ 0 & 0 \end{bmatrix},$$

i.e.,

$$P_{11}AP_{11}^* = B, \quad P_{11}AP_{21}^* = 0, \quad P_{21}AP_{11}^* = 0, \quad P_{21}AP_{21}^* = 0.$$

Since rank $\left[\begin{smallmatrix} P_{11} \\ P_{21} \end{smallmatrix}\right] = m$, the last two equations in the above imply that $P_{21}A = 0$. Moreover, there exists $C \in \mathrm{M}_{m\times n}(F)$ such that $Q = P_{11} + CP_{21}$ is invertible. Using the above equations, we have $QAQ^* = B$ and hence $A \cong B$. □

Theorem 6.32 (Canonical forms of hermitian matrices)**.** *Let $\sigma \in \mathrm{Aut}(F)$ be such that $o(\sigma) = 2$, and let $E = \{a \in F : \sigma(a) = a\}$.*

(i) *Every $A \in S_n^\sigma(F)$ is congruent to a diagonal matrix in $\mathrm{M}_n(E)$.*

(ii) *If the norm map $\mathrm{N}_{F/E} : F \to E$ is onto, then every $A \in S_n^\sigma(F)$ is congruent to I_r, where $r = \mathrm{rank}\, A$. In particular, if $\sigma \in \mathrm{Aut}(\mathbb{F}_{q^2}/\mathbb{F}_q)$, $\sigma \neq \mathrm{id}$, then every $A \in S_n^\sigma(\mathbb{F}_{q^2})$ is congruent to I_r, where $r = \mathrm{rank}\, A$.*

Proof. (i) Let $A = (a_{ij}) \in S_n^\sigma(F)$ and assume that $A \neq 0$. First assume that $a_{11}, a_{22}, \dots, a_{nn}$ are not all 0. By a suitable transposition of rows and columns, we may assume that $a_{11} \neq 0$. Write $A = \left[\begin{smallmatrix} a_{11} & \alpha \\ \alpha^* & * \end{smallmatrix}\right]$, where $(\ )^* = (\ )^{\sigma T}$. Then

$$\begin{bmatrix} 1 & 0 \\ -a_{11}^{-1}\alpha^* & I \end{bmatrix} \begin{bmatrix} a_{11} & \alpha \\ \alpha^* & * \end{bmatrix} \begin{bmatrix} 1 & -a_{11}^{-1}\alpha \\ 0 & I \end{bmatrix} = \begin{bmatrix} a_{11} & 0 \\ 0 & A_1 \end{bmatrix},$$

where $a_{11} \in E$ (since $a_{11}^* = a_{11}$) and $A_1 \in S_{n-1}^\sigma(F)$. The conclusion follows by induction.

If $a_{11}, a_{22}, \dots, a_{nn}$ are all 0, then one of the nondiagonal entries of A, say a_{12}, is nonzero. We claim that $\sigma + \mathrm{id} \neq 0$. (Otherwise, $2 = (\sigma + \mathrm{id})(1) = 0$. Then $\operatorname{char} F = 2$ and $\sigma = \mathrm{id}$, which is a contradiction.) Let $x \in F$ be such that $\sigma(x) + x \neq 0$. Then

$$\begin{bmatrix} 1 & (xa_{12}^{-1})^* & & \\ & 1 & & \\ & & \ddots & \\ & & & 1 \end{bmatrix} \begin{bmatrix} 0 & a_{12} & * \\ a_{12}^* & 0 & \\ * & & * \end{bmatrix} \begin{bmatrix} 1 & & & \\ xa_{12}^{-1} & 1 & & \\ & & \ddots & \\ & & & 1 \end{bmatrix} = \begin{bmatrix} x + x^* & * \\ * & * \end{bmatrix},$$

which brings us to the above case.

(ii) By (i), $A \cong \operatorname{diag}(a_1, \dots, a_r, 0, \dots, 0)$, where $a_1, \dots, a_r \in E^*$. Write $a_i = \mathrm{N}_{F/E}(b_i) = b\sigma(b_i)$, $b_i \in F^*$, $1 \le i \le r$. Then

$$\operatorname{diag}(b_1^{-1}, \dots, b_r^{-1} 1, \dots, 1)\operatorname{diag}(a_1, \dots, a_r, 0, \dots, 0)\operatorname{diag}(b_1^{-1}, \dots, b_r^{-1} 1, \dots, 1)^*$$
$$= I_r \oplus 0.$$

□

6.5.3. Quadratic forms.

Definition 6.33. Let F be a field and V a finite-dimensional vector space over F. A *quadratic form* on V is a function $f : V \to F$ satisfying the following conditions:

(i) $f(ax) = a^2 f(x)$ for all $x \in V$ and $a \in F$;

(ii) the map $l_f : V \times V \to F$, $l_f(x, y) = f(x + y) - f(x) - f(y)$, is a symmetric bilinear form on V.

Note. When $\operatorname{char} F \neq 2$, conditions (i) and (ii) in Definition 6.33 imply that $f(x) = 2^{-1} l_f(x, x)$ for all $x \in V$, so a quadratic form f is determined by its symmetric bilinear form l_f. However, the statement is false when $\operatorname{char} F = 2$.

Let $\mathcal{Q}(V)$ denote the set of all quadratic forms on V. A pair (V, f), where $f \in \mathcal{Q}(V)$, is called a *quadratic space*. Two quadratic spaces (V_1, f_1) and (V_2, f_2) are said to be *isomorphic*, denoted by $(V_1, f_1) \cong (V_2, f_2)$, and the quadratic forms f_1 and f_2 are said to be *equivalent*, denoted by $f_1 \cong f_2$, if there exists an F-isomorphism $\alpha : V_1 \to V_2$ such that $f_1 = f_2 \circ \alpha$. The map α is called an *isometry* from (V_1, f_1) to (V_2, f_2).

For $f \in \mathcal{Q}(V)$, recall that

$$\ker l_f = \{x \in V : l_f(x, y) = 0 \text{ for all } y \in V\}$$

$$= \{x \in V : f(x+y) = f(x) + f(y) \text{ for all } y \in V\}.$$

We define

$$\ker f = \{x \in V : f(x+y) = f(y) \text{ for all } y \in V\},$$

which is a subspace of V. (To see that $\ker f$ is closed under scalar multiplication, let $x \in \ker f$, $y \in V$ and $a \in F^*$, and compute $f(ax+y) = a^2 f(x + a^{-1}y) = a^2 f(a^{-1}y) = f(y)$.) Further define

$$\operatorname{rank} f = \dim_F V - \dim_F \ker f. \tag{6.53}$$

A quadratic form $f \in \mathcal{Q}(V)$ is said to be *nondegenerate* on V if $\ker f = 0$.

Let $\epsilon_1, \dots, \epsilon_n$ be a basis of V and let $x = \sum_{i=1}^n x_i \epsilon_i \in V$, $x_i \in F$. Using (i) and (ii) in Definition 6.33 repeatedly, we have

$$f(x) = \sum_{i=1}^n f(\epsilon_i) x_i^2 + \sum_{1 \le i < j \le n} l_f(\epsilon_i, \epsilon_j) x_i x_j = \sum_{1 \le i \le j \le n} a_{ij} x_i x_j, \tag{6.54}$$

where

$$a_{ij} = \begin{cases} f(\epsilon_i) & \text{if } i = j, \\ l_f(\epsilon_i, \epsilon_j) & \text{if } i < j. \end{cases}$$

By choosing the standard basis of F^n, it is clear from (6.54) that $\mathcal{Q}(F^n)$ is in one-to-one correspondence with the set of all homogeneous polynomials of degree 2 in $F[\mathtt{X}_1, \dots, \mathtt{X}_n]$. When it causes no confusion, we will treat elements of $\mathcal{Q}(F^n)$ as homogeneous polynomials of degree 2.

There is a left action of $\mathrm{GL}(n, F)$ on $\mathcal{Q}(F^n)$: For $P \in \mathrm{GL}(n, F)$ and $f \in \mathcal{Q}(F^n)$, $P(f) \in \mathcal{Q}(F^n)$ is defined as $P(f) = f((\mathtt{X}_1, \dots, \mathtt{X}_n)P)$. The $\mathrm{GL}(n, F)$-orbits in $\mathcal{Q}(F^n)$ are precisely the equivalence classes of quadratic forms on F^n.

Quadratic forms on F^n behave quite differently depending on whether $\operatorname{char} F$ is 2. When $\operatorname{char} F \ne 2$, a quadratic form $f \in \mathcal{Q}(F^n)$ is determined by its symmetric bilinear form l_f. In this case, the classification of quadratic forms on F^n under equivalence is the same as the classification of symmetric forms on F^n under equivalence, which is also the same as the classification of matrices in $S_n(F)$ under congruence. When $\operatorname{char} F = 2$, additional information is needed to classify quadratic forms on F^n.

Let $f = \mathtt{X} A \mathtt{X}^T \in \mathcal{Q}(F^n)$, where $\mathtt{X} = (\mathtt{X}_1, \dots, \mathtt{X}_n)$ and $A \in \mathrm{M}_n(F)$. Let $\mathtt{Y} = (\mathtt{Y}_1, \dots, \mathtt{Y}_n)$ and treat the symmetric form l_f as a polynomial in $(\mathtt{X}, \mathtt{Y})$. Then

$$l_f = f(\mathtt{X} + \mathtt{Y}) - f(\mathtt{X}) - f(\mathtt{Y}) = \mathtt{X}(A + A^T)\mathtt{Y}^T.$$

Thus

$$\ker l_f = \{x \in F^n : x(A + A^T) = 0\}. \tag{6.55}$$

For $x \in F^n$, we have

$$\begin{aligned} x \in \ker f &\Leftrightarrow f(x+\mathtt{Y}) = f(\mathtt{Y}) \\ &\Leftrightarrow (x+\mathtt{Y})A(x+\mathtt{Y})^T = \mathtt{Y}A\mathtt{Y}^T \\ &\Leftrightarrow xAx^T + x(A+A^T)\mathtt{Y}^T = 0 \\ &\Leftrightarrow xAx^T = 0 \text{ and } x(A+A^T) = 0. \end{aligned}$$

Hence

$$\ker f = \{x \in F^n : f(x) = 0 \text{ and } x(A+A^T) = 0\}. \tag{6.56}$$

By (6.55) and (6.56), $\ker f \subset \ker l_f$. Define the *type* of f as

$$\operatorname{type} f = \dim_F(\ker l_f / \ker f). \tag{6.57}$$

Since $\dim_F \ker l_f = n - \operatorname{rank}(A+A^T)$, it follows from (6.53) and (6.57) that

$$\operatorname{rank} f = \operatorname{rank}(A+A^T) + \operatorname{type} f. \tag{6.58}$$

Lemma 6.34. *Let $f \in \mathcal{Q}(F^n)$.*

(i) *If* $\operatorname{char} F \neq 2$, *then* $\ker f = \ker l_f$, *i.e.*, $\operatorname{type} f = 0$.

(ii) *If* $\operatorname{char} F = 2$ *and F is perfect, then*

$$\operatorname{type} f = \begin{cases} 0 & \textit{if } f(\ker l_f) = \{0\}, \\ 1 & \textit{if } f(\ker l_f) \neq \{0\}. \end{cases}$$

Proof. (i) Let $f = \mathtt{X}A\mathtt{X}^T$, where $\mathtt{X} = (\mathtt{X}_1, \dots, \mathtt{X}_n)$ and $A \in \mathrm{M}_n(F)$. For each $x \in \ker l_f$, $x(A+A^T) = 0$. Thus $f(x) = 2^{-1}x(A+A^T)x^T = 0$, so $x \in \ker f$.

(ii) Since F is perfect with $\operatorname{char} F = 2$, the map $\sigma(\ ) = (\ )^2$ is an automorphism of F. Moreover, $\sigma^{-1} \circ f : \ker l_f \to F$ is an F-map with kernel $\ker f$. Thus

$$\dim_F(\ker l_f / \ker f) = \begin{cases} 0 & \text{if } f(\ker l_f) = \{0\}, \\ 1 & \text{if } f(\ker l_f) \neq \{0\}. \end{cases}$$

□

Example 6.35. Assume that $\operatorname{char} F = 2$ and let $f = \mathtt{X}_1^2 + \cdots + \mathtt{X}_n^2 \in \mathcal{Q}(F^n)$. Then $l_f = 0$, and hence $\ker l_f = F^n$. Clearly, $\ker f = \{(x_1, \dots, x_n) \in F^n : x_1 + \cdots + x_n = 0\}$, so $\operatorname{type} f = 1$.

Example 6.36. Let $F = K(t_1, \dots, t_n)$, where $\operatorname{char} K = 2$ and $t_1, \dots, t_n$ are algebraically independent over K. Let $f = t_1\mathtt{X}_1^2 + \cdots + t_n\mathtt{X}_n^2 \in \mathcal{Q}(F^n)$. Then $\ker l_f = F^n$. Since $t_1, \dots, t_n$ are linearly independent over $K(t_1^2, \dots, t_n^2)$, we have $\ker f = \{0\}$, and hence $\operatorname{type} f = n$.

6.5.4. Quadratic forms with char $F \neq 2$.

Throughout this subsection, it is assumed that char $F \neq 2$.

Lemma 6.37. *The following maps are all F-isomorphisms.*

$$\begin{array}{ccccccc} S_n(F) & \xrightarrow{\cong} & \mathcal{Q}(F^n) & \xrightarrow{\cong} & \mathcal{S}(F^n) & \xrightarrow{\cong} & S_n(F) \\ A & \longmapsto & \mathtt{X}A\mathtt{X}^T & \longmapsto & 2\mathtt{X}A\mathtt{Y}^T & \longmapsto & A \\ & & \| & & \| & & \| \\ & & f & & l_f & & 2^{-1}A(l_f, \mathcal{E}) \end{array}$$

In the above, $\mathtt{X} = (\mathtt{X}_1, \dots, \mathtt{X}_n)$, $\mathtt{Y} = (\mathtt{Y}_1, \dots, \mathtt{Y}_n)$, $\mathtt{X}A\mathtt{Y}^T$ is treated as a function from $F^n \times F^n$ to F, $\mathcal{E}$ is the standard basis of F^n, and $A(l_f, \mathcal{E})$ is the matrix of l_f with respect to $\mathcal{E}$.

Proof. To see that the first map is onto, consider any $f \in \mathcal{Q}(F^n)$. Write $f = \mathtt{X}B\mathtt{X}^T$ for some $B \in \mathrm{M}_n(F)$. Then $f = \mathtt{X}A\mathtt{X}^T$, where $A = 2^{-1}(B+B^T) \in S_n(F)$. The rest of the lemma is obvious. $\square$

Lemma 6.38. *Two quadratic forms $f, g \in \mathcal{Q}(F^n)$ are equivalent if and only if their corresponding matrices in $S_n(F)$ are congruent.*

Proof. Write $f = \mathtt{X}A\mathtt{X}^T$ and $g = \mathtt{X}B\mathtt{X}^T$, where $A, B \in S_n(F)$. Then $f \cong g$ if and only if $f = P(g)$ for some $P \in \mathrm{GL}(n, F)$. Moreover, $f = P(g)$ if and only if $\mathtt{X}A\mathtt{X}^T = (\mathtt{X}P)B(\mathtt{X}P)^T = \mathtt{X}(PBP^T)\mathtt{X}^T$, i.e., $A = PBP^T$. $\square$

Discriminant. Let $f \in \mathcal{Q}(F^n)$ and let $l = l_f$ be the symmetric bilinear form of f. Clearly, l induces a symmetric form

$$\begin{array}{rccc} \bar{l}: & (F^n/\ker l) \times (F^n/\ker l) & \longrightarrow & F \\ & (x + \ker l,\ y + \ker l) & \longmapsto & l(x, y) \end{array} \tag{6.59}$$

with $\ker \bar{l} = \{0\}$. Let $A \in S_r(F)$ be the matrix of $\bar{l}$ with respect to any basis of $F^n/\ker l$, where $r = n - \dim_F \ker l = \operatorname{rank} f$. Then $\det A \in F^*$. Let $F^{*2} = \{a^2; a \in F^*\} < F^*$ and define $d(f) = d(l) = (\det A)F^{*2} \in F^*/F^{*2}$ as the *discriminant* of f and l. Let $f_1 \in \mathcal{Q}(F^n)$ be such that $f_1 \cong f$, and let $l_1 = l_{f_1}$. Then $l_1 \cong l$, which implies that $\bar{l} \cong \bar{l}_1$. If A_1 is the matrix of $\bar{l}_1$ with respect to a basis of $F^n/\ker l_1$, then $A_1 = PAP^T$ for some $P \in \mathrm{GL}(r, F)$. Thus $(\det A_1)F^{*2} = (\det A)(\det P)^2 F^{*2} = (\det A)F^{*2}$. This means that the discriminant $d(f)$ is independent of the basis of $F^n/\ker l$ and depends only on the equivalence class of f.

Classification of quadratic forms on $\mathbb{F}_q$, q odd. The canonical forms of the matrices in $S_n(\mathbb{F}_q)$, where q is odd, have been determined in Theorem 6.30 (i). The following is a restatement of that result.

Theorem 6.39. *Let q be odd and let $d \in \mathbb{F}_q^*$ be a nonsquare. Then every $f \in \mathcal{Q}(\mathbb{F}_q^n)$ is equivalent to exactly one of the following:*

$$\mathtt{X}_1^2 + \cdots + \mathtt{X}_r^2, \quad 0 \le r \le n, \tag{6.60}$$

$$\mathtt{X}_1^2 + \cdots + \mathtt{X}_{r-1}^2 + d\,\mathtt{X}_r^2, \quad 1 \le r \le n. \tag{6.61}$$

Remark. In Theorem 6.39, the quadratic form (6.60) has rank r and discriminant $1\mathbb{F}_q^{*2}$; the quadratic form (6.61) has rank r and discriminant $d\,\mathbb{F}_q^{*2}$. Hence the equivalence classes in $\mathcal{Q}(\mathbb{F}_q^n)$ are determined by the rank and the discriminant.

6.5.5. Quadratic forms with char $F = 2$.

In this subsection, it is assumed that char $F = 2$.

Lemma 6.40. *The following are well-defined F-maps, where $\mathtt{X} = (\mathtt{X}_1, \dots, \mathtt{X}_n)$, $\mathtt{Y} = (\mathtt{Y}_1, \dots, \mathtt{Y}_n)$, $\mathcal{E}$ is the standard basis of F^n, and $A(l_f, \mathcal{E})$ is the matrix of l_f with respect to $\mathcal{E}$.*

$$\begin{array}{ccccccc}
\mathrm{M}_n(F)/\Lambda_n(F) & \underset{\cong}{\xrightarrow{\alpha}} & \mathcal{Q}(F^n) & \xrightarrow{\beta} & \mathcal{A}(F^n) & \underset{\cong}{\xrightarrow{\gamma}} & \Lambda_n(F) \\
A + \Lambda_n(F) & \longmapsto & \mathtt{X}A\mathtt{X}^T & \longmapsto & \mathtt{X}(A + A^T)\mathtt{Y}^T & \longmapsto & A + A^T \\
 & & \| & & \| & & \| \\
 & & f & & l_f & & A(l_f, \mathcal{E})
\end{array}$$

Moreover, α and γ are isomorphisms and β is onto with $\ker\beta = \{a_1\mathtt{X}_1^2 + \cdots + a_n\mathtt{X}_n^2 : a_i \in F\}$, *the set of all diagonal quadratic forms.*

Proof. Consider $\phi : \mathrm{M}_n(F) \to \mathcal{Q}(F^n)$, $\phi(A) = \mathtt{X}A\mathtt{X}^T$. Clearly, ϕ is an onto F-map. For $A \in \mathrm{M}_n(F)$, $\mathtt{X}A\mathtt{X}^T = 0$ if and only if $A \in \Lambda_n(F)$, i.e., $\ker\phi = \Lambda_n(F)$. Hence ϕ induces the isomorphism $\alpha : \mathrm{M}_n(F)/\Lambda_n(F) \to \mathcal{Q}(F^n)$. (Note that the characteristic of F is irrelevant to α.) The claims about β and γ are left for the reader to verify. □

Remark 6.41. Two elements $A + \Lambda_n(F)$, $B + \Lambda_n(F) \in \mathrm{M}_n(F)/\Lambda_n(F)$ are said to be *congruent*, denoted by $A + \Lambda_n(F) \cong B + \Lambda_n(F)$, if there exists $P \in \mathrm{GL}(n, F)$ such that $A \equiv PBP^T \pmod{\Lambda_n(F)}$. Since $P\Lambda_n(F)P^T = \Lambda_n(F)$, the relation on $\mathrm{M}_n(F)/\Lambda_n(F)$ is well-defined. It is clear that two quadratic forms $\mathtt{X}A\mathtt{X}^T, \mathtt{X}B\mathtt{X}^T \in \mathcal{Q}(F^n)$ are equivalent if and only if their corresponding matrix classes $A + \Lambda_n(F)$, $B + \Lambda_n(F) \in \mathrm{M}_n(F)/\Lambda_n(F)$ are congruent.

The Arf invariant. Let $F_0 = \{c^2 + c : c \in F\}$, which is a subgroup of $(F, +)$. Let $l \in \mathcal{A}(F^{2r})$ be nondegenerate. A basis $\epsilon_1, \dots, \epsilon_{2r}$ of F^{2r} is called a *symplectic basis* of F^{2r} with respect to l if

$$(l_f(\epsilon_i, \epsilon_j)) = \underbrace{\begin{bmatrix} 0 & 1 \\ -1 & 0 \end{bmatrix} \oplus \cdots \oplus \begin{bmatrix} 0 & 1 \\ -1 & 0 \end{bmatrix}}_{r}.$$

The existence of a symplectic basis is guaranteed by Theorem 6.30 (iii).

Lemma 6.42. *Let $f \in \mathcal{Q}(F^{2r})$ be such that $l_f \in \mathcal{A}(F^{2r})$ is nondegenerate. Let $\epsilon_1, \dots, \epsilon_{2r}$ and $\delta_1, \dots, \delta_{2r}$ be two symplectic bases of F^{2r} with respect to l_f. Then*

$$\sum_{i=1}^{r} f(\epsilon_{2i-1}) f(\epsilon_{2i}) \equiv \sum_{i=1}^{r} f(\delta_{2i-1}) f(\delta_{2i}) \pmod{F_0}.$$

Proof. There exists $\tau \in \mathrm{Aut}_F(F^{2r})$ such that $\tau(\epsilon_i) = \delta_i$, $1 \le i \le 2r$. By Theorem 6.51, τ is a product of *symplectic transvections*, which are elements of $\mathrm{Aut}_F(F^{2r})$ of the form

$$\begin{aligned} \sigma : \ F^{2r} &\longrightarrow F^{2r} \\ x &\longmapsto x + \lambda l_f(x, a) a \end{aligned}$$

for some $\lambda \in F$ and $a \in F^{2r}$. Therefore, to prove the lemma, we may assume that $\delta_i = \sigma(\epsilon_i)$, $1 \le i \le 2r$, where σ is given above.

For each $x \in F^{2r}$,

(6.62)
$$\begin{aligned} f(\sigma(x)) &= f(x + \lambda l_f(x, a) a) = f(x) + f(\lambda l_f(x, a) a) + l_f(x,\ \lambda l_f(x, a) a) \\ &= f(x) + \lambda^2 l_f(x, a)^2 f(a) + \lambda l_f(x, a)^2 = f(x) + \mu l_f(x, a)^2, \end{aligned}$$

where

(6.63)
$$\mu = \lambda^2 f(a) + \lambda.$$

Write $a = \sum_{i=1}^{2r} a_i \epsilon_i$. Then

(6.64)
$$\begin{aligned} f(a) &= f\Big(\sum_{i=1}^{2r} a_i \epsilon_i\Big) = \sum_{i=1}^{2r} a_i^2 f(\epsilon_i) + \sum_{i<j} a_i a_j l_f(\epsilon_i, \epsilon_j) \\ &= \sum_{i=1}^{2r} a_i^2 f(\epsilon_i) + \sum_{i=1}^{r} a_{2i-1} a_{2i}. \end{aligned}$$

Thus,

$$\sum_{i=1}^{r} f(\delta_{2i-1}) f(\delta_{2i}) = \sum_{i=1}^{r} f(\sigma(\epsilon_{2i-1})) f(\sigma(\epsilon_{2i}))$$

$$
\begin{aligned}
&= \sum_{i=1}^{r} \big[f(\epsilon_{2i-1}) + \mu l_f(\epsilon_{2i-1}, a)^2\big]\big[f(\epsilon_{2i}) + \mu l_f(\epsilon_{2i}, a)^2\big] && \text{(by (6.62))} \\
&= \sum_{i=1}^{r} f(\epsilon_{2i-1})f(\epsilon_{2i}) + \mu \sum_{i=1}^{r}\big[a_{2i}^2 f(\epsilon_{2i}) + a_{2i-1}^2 f(\epsilon_{2i-1})\big] + \mu^2 \sum_{i=1}^{r} a_{2i-1}^2 a_{2i}^2 \\
&= \sum_{i=1}^{r} f(\epsilon_{2i-1})f(\epsilon_{2i}) + \mu\Big[f(a) + \sum_{i=1}^{r} a_{2i-1}a_{2i}\Big] + \Big[\mu \sum_{i=1}^{r} a_{2i-1}a_{2i}\Big]^2 && \text{(by (6.64))} \\
&\equiv \sum_{i=1}^{r} f(\epsilon_{2i-1})f(\epsilon_{2i}) + \mu f(a) \pmod{F_0} \\
&= \sum_{i=1}^{r} f(\epsilon_{2i-1})f(\epsilon_{2i}) + \lambda^2 f(a)^2 + \lambda f(a) && \text{(by (6.63))} \\
&\equiv \sum_{i=1}^{r} f(\epsilon_{2i-1})f(\epsilon_{2i}) \pmod{F_0}.
\end{aligned}
$$

□

Definition 6.43. Let $f \in \mathcal{Q}(F^{2r})$ be such that $l_f \in \mathcal{A}(F^{2r})$ is nondegenerate, and let $\epsilon_1, \dots, \epsilon_{2r}$ be a symplectic basis of F^{2r} with respect to l_f. The *Arf invariant* of f is defined to be

$$\mathrm{Arf}(f) = \sum_{i=1}^{r} f(\epsilon_{2i-1})f(\epsilon_{2i}) + F_0 \in F/F_0.$$

By Lemma 6.42, $\mathrm{Arf}(f)$ is independent of the choice of the symplectic basis.

Remark 6.44. When $F = \mathbb{F}_q$, with q even, F_0 is the kernel of the trace map $\mathrm{Tr}_{q/2}$. In this case, it is more convenient to define

$$\mathrm{Arf}(f) = \mathrm{Tr}_{q/2}\Big(\sum_{i=1}^{r} f(\epsilon_{2i-1})f(\epsilon_{2i})\Big) \in \mathbb{F}_2.$$

In Definition 6.43, the Arf invariant is defined for $f \in \mathcal{Q}(F^{2r})$ with $\ker l_f = \{0\}$. The definition can be extended to all $f \in \mathcal{Q}(F^n)$ with $\operatorname{type} f = 0$. Let $f \in \mathcal{Q}(F^n)$ be such that $\operatorname{type} f = 0$. Then for all $x \in F^n$ and $y \in \ker l_f$,

$$f(x+y) = f(x) + f(y) + l_f(x,y) = f(x).$$

Thus

$$
\begin{array}{rccc}
\bar{f}: & F^n/\ker l_f & \longrightarrow & F \\
& x + \ker l_f & \longmapsto & f(x)
\end{array}
$$

is a well-defined function. Clearly, $\bar{f}$ is a quadratic form on $F^n/\ker l_f$ with a nondegenerate alternating form $l_{\bar{f}} = \bar{l}_f$. (See (6.59) for the definition of $\bar{l}_f$.) We define the Arf invariant of f to be $\mathrm{Arf}(f) = \mathrm{Arf}(\bar{f})$.

Theorem 6.45. *The Arf invariant of a quadratic form of type* 0 *in* $\mathcal{Q}(F^n)$ *is invariant under equivalence.*

Proof. Let $f, g \in \mathcal{Q}(F^n)$ be two equivalent quadratic forms of type 0. Clearly, their induced quadratic forms $\bar{f}$ and $\bar{g}$ are also equivalent. Let $V_f = F^n/\ker l_f$ and $V_g = F^n/\ker l_g$, and let $\sigma : V_f \to V_g$ be an F-isomorphism such that $\bar{f} = \bar{g} \circ \sigma$. Let $\epsilon_1, \dots, \epsilon_{2r}$ be a symplectic basis of V_f with respect to $l_{\bar{f}}$. Then $\sigma(\epsilon_1), \dots, \sigma(\epsilon_{2r})$ is a symplectic basis of V_g with respect to $l_{\bar{g}}$. Thus

$$\begin{aligned}\mathrm{Arf}(f) = \mathrm{Arf}(\bar{f}) &= \sum_{i=1}^{r} \bar{f}(\epsilon_{2i-1})\bar{f}(\epsilon_{2i}) + F_0 \\ &= \sum_{i=1}^{r} \bar{g}\big((\sigma(\epsilon_{2i-1})\big)\bar{g}\big(\sigma(\epsilon_{2i})\big) + F_0 = \mathrm{Arf}(\bar{g}) = \mathrm{Arf}(g).\end{aligned}$$

□

Classification of quadratic forms over a perfect field of characteristic 2. Recall that F is a field with $\operatorname{char} F = 2$ and $F_0 = \{c^2 + c : c \in F\}$. Let Γ be a system of coset representatives of F_0 in F. By Remark 6.41, the classification of quadratic forms in $\mathcal{Q}(F^n)$ under equivalence is the same as the classification of elements in $\mathrm{M}_n(F)/\Lambda_n(F)$ under congruence.

Theorem 6.46. *Assume that F is perfect with* $\operatorname{char} F = 2$. *Every nonzero element of* $\mathrm{M}_n(F)/\Lambda_n(F)$ *is congruent to exactly one of the following:*

$$\underbrace{\begin{bmatrix} 0 & 1 \\ 0 & 0 \end{bmatrix} \oplus \cdots \oplus \begin{bmatrix} 0 & 1 \\ 0 & 0 \end{bmatrix}}_{r} \oplus [1] \oplus 0 + \Lambda_n(F), \quad 0 \le r \le (n-1)/2, \tag{6.65}$$

$$\underbrace{\begin{bmatrix} 0 & 1 \\ 0 & 0 \end{bmatrix} \oplus \cdots \oplus \begin{bmatrix} 0 & 1 \\ 0 & 0 \end{bmatrix}}_{r-1} \oplus \begin{bmatrix} 1 & 1 \\ 0 & a \end{bmatrix} \oplus 0 + \Lambda_n(F), \quad 1 \le r \le n/2,\ a \in \Gamma. \tag{6.66}$$

Proof. For $A, B \in \mathrm{M}_n(F)$, we write $A \approx B$ to mean that $A \equiv PBP^T \pmod{\Lambda_n(F)}$ for some $P \in \mathrm{GL}(n, F)$.

Let $A + \Lambda_n(F) \in \mathrm{M}_n(F)/\Lambda_n(F)$, where

$$A = \begin{bmatrix} a_{11} & a_{12} & \cdots & a_{1n} \\ 0 & a_{22} & \cdots & a_{2n} \\ \vdots & \vdots & \ddots & \vdots \\ 0 & 0 & \cdots & a_{nn} \end{bmatrix} \in \mathrm{M}_n(F) \setminus \Lambda_n(F).$$

We first show that A is $\approx$ to one of the "canonical" matrices in (6.65) or (6.66). In fact, we give an algorithm for finding the "canonical" form of A.

Case 1. Assume that one of $a_{11}, a_{22}, \dots, a_{nn}$ is 0, say $a_{11} = 0$.

Case 1.1. Assume that $(a_{12}, \dots, a_{1n}) = (0, \dots, 0)$. Then

$$A = [0] \oplus A_1, \quad A_1 \in \mathrm{M}_{n-1}(F).$$

Continue to determine the canonical form of A_1.

Case 1.2. Assume $(a_{12}, \dots, a_{1n}) \neq (0, \dots, 0)$, say $a_{12} \neq 0$. We may assume that $a_{12} = 1$. Subtract a_{1j} times column 2 from column j, $3 \le j \le n$, and perform the same row operations; then

$$A \approx \begin{bmatrix} 0 & 1 & 0 & \cdots & 0 \\ 0 & a_{22} & * & \cdots & * \\ 0 & * & * & \cdots & * \\ \vdots & \vdots & \vdots & & \vdots \\ 0 & * & * & \cdots & * \end{bmatrix} \equiv \begin{bmatrix} 0 & 1 & 0 & \cdots & 0 \\ 0 & a'_{22} & 0 & \cdots & 0 \\ 0 & a'_{32} & * & \cdots & * \\ \vdots & \vdots & \vdots & & \vdots \\ 0 & a'_{n2} & * & \cdots & * \end{bmatrix} \pmod{\Lambda_n(F)}.$$

Subtract a'_{i2} times row 1 from row i, $2 \le i \le n$, and perform the same column operations; this gives

$$A \approx \begin{bmatrix} 0 & 1 \\ 0 & 0 \end{bmatrix} \oplus A_2, \quad A_2 \in \mathrm{M}_{n-2}(F).$$

Continue to determine the canonical form of A_2.

Case 2. Assume that $a_{11}, a_{22}, \dots, a_{nn}$ are all nonzero. Since F is perfect, each a_{ii} is a square in F^*; hence we may assume that $a_{11} = a_{22} = \cdots = a_{nn} = 1$.

Case 2.1. Assume that $a_{ij} = 0$ for some $1 \le i < j \le n$, say $a_{12} = 0$. Then

$$A = \begin{bmatrix} 1 & 0 & \\ 0 & 1 & * \\ & 0 & * \end{bmatrix}$$

$$\approx \begin{bmatrix} 1 & 1 & \\ 0 & 1 & \\ & & I_{n-2} \end{bmatrix} \begin{bmatrix} 1 & 0 & \\ 0 & 1 & * \\ & 0 & * \end{bmatrix} \begin{bmatrix} 1 & 0 & \\ 1 & 1 & \\ & & I_{n-2} \end{bmatrix} \equiv \begin{bmatrix} 0 & 0 & \\ 0 & 1 & * \\ & 0 & * \end{bmatrix} \pmod{\Lambda_n(F)},$$

which takes us back to Case 1.

Case 2.2. Assume that $n \ge 3$ and $a_{ij} \ne 0$ for all $1 \le i < j \le n$. Subtract $a_{12}a_{13}^{-1}$ times column 3 from column 2 and perform the same row operation; then

$$A \approx \begin{bmatrix} 1 & 0 & * & \cdots & * \\ 0 & * & * & \cdots & * \\ 0 & * & * & \cdots & * \\ \vdots & \vdots & \vdots & \ddots & \vdots \\ 0 & 0 & 0 & \cdots & * \end{bmatrix} \equiv \begin{bmatrix} 1 & 0 & * & \cdots & * \\ 0 & * & * & \cdots & * \\ 0 & 0 & * & \cdots & * \\ \vdots & \vdots & \vdots & \ddots & \vdots \\ 0 & 0 & 0 & \cdots & * \end{bmatrix} \pmod{\Lambda_n(F)}.$$

Thus, we are in either Case 1 or Case 2.1.

Case 2.3. Assume that $n = 2$ and $a_{12} \ne 0$. Let $b = a_{12}^{-1}$. Then

$$A = \begin{bmatrix} 1 & a_{12} \\ 0 & 1 \end{bmatrix} \approx \begin{bmatrix} 1 & \\ & b \end{bmatrix} \begin{bmatrix} 1 & a_{12} \\ 0 & 1 \end{bmatrix} \begin{bmatrix} 1 & \\ & b \end{bmatrix} = \begin{bmatrix} 1 & 1 \\ 0 & b^2 \end{bmatrix}.$$

Write $b^2 = a + c^2 + c$ for some $a \in \Gamma$ and $c \in F$. Then

$$A \approx \begin{bmatrix} 1 & 0 \\ c & 1 \end{bmatrix} \begin{bmatrix} 1 & 1 \\ 0 & b^2 \end{bmatrix} \begin{bmatrix} 1 & c \\ 0 & 1 \end{bmatrix} \equiv \begin{bmatrix} 1 & 1 \\ 0 & a \end{bmatrix} \pmod{\Lambda_2(F)}.$$

Case 2.4. Assume that $n = 1$. Then $A = [1]$ and nothing needs to be done.

Therefore, we have proved that every nonzero element in $\mathrm{M}_n(F)/\Lambda_n(F)$ is congruent to one of the elements in (6.65) and (6.66).

It remains to show that the elements in (6.65) and (6.66) are pairwise noncongruent. The corresponding quadratic form of (6.65) has rank $2r+1$ and type 1; the corresponding quadratic form of (6.66) has rank $2r$, type 0, and Arf invariant $a + F_0$. Since the rank, the type, and the Arf invariant are invariants of quadratic forms, the corresponding quadratic forms of (6.65) and (6.66) are pairwise nonequivalent. This completes the proof of the theorem. □

Theorem 6.47. *Assume that F is perfect with* $\operatorname{char} F = 2$. *Let* $0 \neq f \in \mathcal{Q}(F^n)$. *If f has rank $2r+1$ and type 1, where $0 \le r \le (n-1)/2$, then f is equivalent to*

$$\mathtt{X}_1\mathtt{X}_2 + \mathtt{X}_3\mathtt{X}_4 + \cdots + \mathtt{X}_{2r-1}\mathtt{X}_{2r} + \mathtt{X}_{2r+1}^2. \tag{6.67}$$

If f has rank $2r$, type 0, and Arf invariant $a + F_0$, where $1 \le r \le n/2$ and $a \in \Gamma$, then f is equivalent to

$$\mathtt{X}_1\mathtt{X}_2 + \mathtt{X}_3\mathtt{X}_4 + \cdots + \mathtt{X}_{2r-1}\mathtt{X}_{2r} + \mathtt{X}_{2r-1}^2 + a\mathtt{X}_{2r}^2. \tag{6.68}$$

Consequently, two quadratic forms in $\mathcal{Q}(F^n)$ are equivalent if and only if they have the same rank, the same type, and, in the case of type 0, the same Arf invariant.

Proof. By Theorem 6.46, every $0 \neq f \in \mathcal{Q}(F^n)$ is equivalent to one of the quadratic forms in (6.67) with $0 \le r \le (n-1)/2$ or (6.68) with $1 \le r \le n/2$ and $a \in \Gamma$. The quadratic form in (6.67) has rank $2r+1$ and type 1; the quadratic form in (6.68) has rank $2r$, type 0, and Arf invariant $a + F_0$. Thus, f must be equivalent to the quadratic form in (6.67) or (6.68) with the same rank and type and, in the case of type 0, the same Arf invariant. □

Recall that when $F = \mathbb{F}_q$ with q even, the Arf invariant is an element of $\mathbb{F}_2$; see Remark 6.44.

Corollary 6.48. *Let* $0 \neq f \in \mathcal{Q}(\mathbb{F}_q^n)$, *where q is even. If f has rank $2r+1$ and type 1, where $0 \le r \le (n-1)/2$, then f is equivalent to*

$$\mathtt{X}_1\mathtt{X}_2 + \mathtt{X}_3\mathtt{X}_4 + \cdots + \mathtt{X}_{2r-1}\mathtt{X}_{2r} + \mathtt{X}_{2r+1}^2.$$

If f has rank $2r$ and type 0, where $1 \le r \le n/2$, then f is equivalent to

$$\begin{cases} \mathtt{X}_1\mathtt{X}_2 + \mathtt{X}_3\mathtt{X}_4 + \cdots + \mathtt{X}_{2r-1}\mathtt{X}_{2r} & \text{if } \operatorname{Arf}(f) = 0, \\ \mathtt{X}_1\mathtt{X}_2 + \mathtt{X}_3\mathtt{X}_4 + \cdots + \mathtt{X}_{2r-1}\mathtt{X}_{2r} + \mathtt{X}_{2r-1}^2 + a\mathtt{X}_{2r}^2 & \text{if } \operatorname{Arf}(f) = 1, \end{cases}$$

where $a \in \mathbb{F}_q$ is such that $\operatorname{Tr}_{q/2}(a) = 1$.

Proof. This follows from Theorem 6.47 and Remark 6.44. Note that if f has rank $2r$, type 0, and Arf invariant 0, then $f \cong \mathtt{X}_1\mathtt{X}_2 + \mathtt{X}_3\mathtt{X}_4 + \cdots + \mathtt{X}_{2r-1}\mathtt{X}_{2r} + \mathtt{X}_{2r-1}^2 \cong \mathtt{X}_1\mathtt{X}_2 + \mathtt{X}_3\mathtt{X}_4 + \cdots + \mathtt{X}_{2r-1}\mathtt{X}_{2r}$. □

Remark. With a fixed r, the quadratic forms in Corollary 6.48 have different exponential sums (Exercise 6.10), which allow us to distinguish these quadratic forms without using the type and the Arf invariant.

Table 6.2. Classification of forms and matrices

forms/matrices	field F	invariants	reference
$\Lambda_n(F)$, $\mathcal{A}(F^n)$	arbitrary	rank	Thm 6.30 (iii)
$S_n^\sigma(F)$, $\mathcal{S}^\sigma(F^n)$	$\mathrm{N}_{F/E}: F \to E$ onto, $E = \{x \in F : \sigma(x) = x\}$	rank	Thm 6.32 (ii)
$S_n(\mathbb{F}_q)$, $\mathcal{S}(\mathbb{F}_q^n)$, $\mathcal{Q}(\mathbb{F}_q^n)$	$\mathbb{F}_q$, q odd	rank and discriminant	Thm 6.30 (i) Thm 6.39
$S_n(F) \setminus \Lambda_n(F)$, $\mathcal{S}(F^n) \setminus \mathcal{A}(F^n)$	F perfect and $\operatorname{char} F = 2$	rank	Thm 6.30 (ii)
$\mathrm{M}_n(F)/\Lambda_n(F)$, $\mathcal{Q}(F^n)$	F perfect and $\operatorname{char} F = 2$	rank, type, and Arf invariant	Thm 6.46

6.5.6. Summary and comments.

In this section, we have classified several types of forms under equivalence and their corresponding matrices (in one case, matrix classes) under congruence. Table 6.2 provides a summary of these results for easy reference.

The classification of $\mathcal{A}(F^n)$ requires no assumption on the field. However, classifications of other forms usually depend on the field F. For example, when F is an ordered field, quadratic forms on F^n are classified by Sylvester's inertia theorem; see [**36**, §139]. When $F = \mathbb{Q}$, two quadratic forms on $\mathbb{Q}^n$ are equivalent if and only if they are equivalent in $\mathcal{Q}(\mathbb{R}^n)$ and in $\mathcal{Q}(\mathbb{Q}_p^n)$ for all primes p, where $\mathbb{Q}_p$ is the field of p-adic numbers; see [**31**, Ch. IV, §3].

When $\operatorname{char} F \neq 2$, a quadratic form $f \in \mathcal{Q}(F^n)$ and its bilinear form $l_f \in \mathcal{S}(F^n)$ determine each other. When $\operatorname{char} F = 2$, f contains more information than l_f does; in this case, the bilinear form l_f is alternating and does not capture the diagonal terms of f.

6.6. Groups of Spaces Equipped with Forms

The main objective of this section is to determine the orders of the classical groups associated to various forms (symmetric, alternating, unitary, and quadratic) over finite fields. However, for a considerable portion of this section, we allow the field to be more general since no additional effort is required.

Throughout this section, V is always a finite-dimensional vector space over F. Let l be a σ-sesquilinear form on V. We define the *automorphism group* of the sesquilinear space (V,l), also called the *group* of (V,l), as

$$\mathcal{G}(V,l)=\{\alpha\in \operatorname{Aut}_F(V): l^\alpha=l\},$$

where l^α is the sesquilinear form on V defined by $l^\alpha(x,y)=l(\alpha(x),\alpha(y))$ for all $x,y\in V$. If $\tau:(V_1,l_1)\to(V_2,l_2)$ is an isomorphism of two sesquilinear spaces, we have

$$\mathcal{G}(V_1,l_1)=\tau^{-1}\mathcal{G}(V_2,l_2)\tau.$$

For $f\in\mathcal{Q}(V)$, we define the *automorphism group* of the quadratic space (V,f), also called the *group* of (V,f), as

$$\mathcal{G}(V,f)=\{\alpha\in \operatorname{Aut}_F(V): f\circ\alpha=f\}.$$

If $\tau:(V_1,f_1)\to(V_2,f_2)$ is an isomorphism of two quadratic spaces, we also have

$$\mathcal{G}(V_1,f_1)=\tau^{-1}\mathcal{G}(V_2,f_2)\tau.$$

Let $A\in \mathrm{M}_n(F)$ and $\sigma\in \operatorname{Aut}(F)$, and define the *group* of A (with respect to σ) to be

$$\mathcal{G}(A,\sigma)=\{P\in \mathrm{GL}(n,F): PAP^{\sigma T}=A\}.$$

If $\sigma=\mathrm{id}$, we write $\mathcal{G}(A,\sigma)=\mathcal{G}(A)$. If $A_1,A_2\in \mathrm{M}_n(F)$ are such that $A_1\cong A_2$, that is, $A_1=QA_2Q^{\sigma T}$ for some $Q\in \mathrm{GL}(n,F)$, then

$$\mathcal{G}(A_2,\sigma)=Q^{-1}\mathcal{G}(A_1,\sigma)Q.$$

Proposition 6.49.

(i) *Let l be a σ-sesquilinear form on F^n, and let $\mathcal{E}=(\epsilon_1,\dots,\epsilon_n)$ be the standard basis of F^n. The map*

$$\begin{array}{rccc}\phi: & \mathcal{G}(F^n,l) & \longrightarrow & \mathcal{G}(A(l,\mathcal{E}),\sigma)\\ & \alpha & \longmapsto & P_\alpha^{-1}\end{array}$$

is a group isomorphism, where $P_\alpha\in \mathrm{GL}(n,F)$ is defined by

$$\begin{bmatrix}\alpha(\epsilon_1)\\ \vdots\\ \alpha(\epsilon_n)\end{bmatrix}=P_\alpha\begin{bmatrix}\epsilon_1\\ \vdots\\ \epsilon_n\end{bmatrix}. \tag{6.69}$$

(ii) *For each $f\in\mathcal{Q}(F^n)$, $\mathcal{G}(F^n,f)<\mathcal{G}(F^n,l_f)$. If $\operatorname{char}F\neq 2$, then $\mathcal{G}(F^n,f)=\mathcal{G}(F^n,l_f)$.*

(iii) *Assume that $\operatorname{char}F=2$ and $f=\mathtt{X}A\mathtt{X}^T\in\mathcal{Q}(F^n)$, where $\mathtt{X}=(\mathtt{X}_1,\dots,\mathtt{X}_n)$ and $A\in \mathrm{M}_n(F)$. Define*

$$\mathcal{G}(A+\Lambda_n(F))=\{P\in \mathrm{GL}(n,F): PAP^T\equiv A \pmod{\Lambda_n(F)}\}.$$

Then

$$\begin{array}{rccc}\psi: & \mathcal{G}(F^n,f) & \longrightarrow & \mathcal{G}(A+\Lambda_n(F))\\ & \alpha & \longmapsto & P_\alpha^{-1}\end{array}$$

is a group isomorphism, where $P_\alpha \in \mathrm{GL}(n,F)$ is defined in (6.69).

Proof. (i) Let $\alpha \in \mathcal{G}(F^n, l)$. By (6.49), $A(l^\alpha, \mathcal{E}) = P_\alpha A(l, \mathcal{E}) P_\alpha^{\sigma T}$, and hence $P_\alpha \in \mathcal{G}(A(l,\mathcal{E}), \sigma)$. For $\alpha, \beta \in \mathcal{G}(F^n, l)$,

$$\alpha\beta\left(\begin{bmatrix}\epsilon_1\\ \vdots\\ \epsilon_n\end{bmatrix}\right) = \alpha\left(P_\beta\begin{bmatrix}\epsilon_1\\ \vdots\\ \epsilon_n\end{bmatrix}\right) = P_\beta P_\alpha \begin{bmatrix}\epsilon_1\\ \vdots\\ \epsilon_n\end{bmatrix}.$$

Hence $P_{\alpha\beta} = P_\beta P_\alpha$, i.e., ϕ is a group homomorphism. Clearly, $\ker\phi = \{\mathrm{id}\}$. Given $P \in \mathcal{G}(A(l,\mathcal{E}), \sigma)$, let $\alpha \in \mathrm{Aut}_F(F^n)$ be defined by

$$\alpha\left(\begin{bmatrix}\epsilon_1\\ \vdots\\ \epsilon_n\end{bmatrix}\right) = P^{-1}\begin{bmatrix}\epsilon_1\\ \vdots\\ \epsilon_n\end{bmatrix}.$$

It is straightforward to verify that $\alpha \in \mathcal{G}(F^n, l)$ and $\phi(\alpha) = P$. Therefore ϕ is onto and hence is an isomorphism.

(ii) Obvious.

(iii) Let $\alpha \in \mathrm{Aut}_F(F^n)$ and let P_α be as in (6.69). Then

$$\begin{aligned}\alpha \in \mathcal{G}(F^n, f) \;&\Leftrightarrow\; (\mathtt{X}P_\alpha)A(\mathtt{X}P_\alpha)^T = \mathtt{X}A\mathtt{X}^T\\ &\Leftrightarrow\; \mathtt{X}(P_\alpha A P_\alpha^T)\mathtt{X}^T = \mathtt{X}A\mathtt{X}^T\\ &\Leftrightarrow\; P_\alpha A P_\alpha^T \equiv A \pmod{\Lambda_n(F)}\\ &\Leftrightarrow\; P_\alpha \in \mathcal{G}(A + \Lambda_n(F)).\end{aligned}$$

Hence ψ is a bijection. By the proof of (i), ψ is a group homomorphism. □

Let (V, l) be a nondegenerate sesquilinear space which is either symmetric bilinear, or symplectic, or unitary. Two elements $x, y \in V$ are said to be *orthogonal*, denoted by $x \perp y$, if $l(x,y) = 0$. The *orthogonal complement* of a subspace $U \subset V$ is defined as $U^\perp = \{x \in V : l(x,u) = 0 \text{ for all } u \in U\}$. Two subspaces U and W of V are said to be orthogonal, denoted by $U \perp W$, if $u \perp w$ for all $u \in U$ and $w \in W$. If $U \perp W$ and $U \cap W = \{0\}$, we write $U + W = U \oplus W$.

Lemma 6.50. *Let $A \in \mathrm{GL}(m,F)$ and $\sigma \in \mathrm{Aut}(F)$. Then*

$$\mathcal{G}(A \oplus 0_n, \sigma) = \left\{\begin{bmatrix}P_{11} & P_{12}\\ 0 & P_{22}\end{bmatrix} : P_{11} \in \mathcal{G}(A,\sigma),\; P_{22} \in \mathrm{GL}(n,F)\right\}.$$

Proof. It suffices to prove that the left side of the above equation is contained in the right side. Let

$$\begin{bmatrix} P_{11} & P_{12} \\ P_{21} & P_{22} \end{bmatrix} \in \mathcal{G}(A \oplus 0_n, \sigma),$$

where P_{11} is of size $m \times m$. Then

$$\begin{bmatrix} A & \\ & 0_n \end{bmatrix} = \begin{bmatrix} P_{11} & P_{12} \\ P_{21} & P_{22} \end{bmatrix} \begin{bmatrix} A & \\ & 0_n \end{bmatrix} \begin{bmatrix} P_{11}^* & P_{21}^* \\ P_{12}^* & P_{22}^* \end{bmatrix} = \begin{bmatrix} P_{11}AP_{11}^* & P_{11}AP_{21}^* \\ P_{21}AP_{11}^* & P_{21}AP_{21}^* \end{bmatrix},$$

where $(\ )^* = (\ )^{\sigma T}$. Hence $P_{21}A\,[P_{11}^*\ P_{21}^*] = 0$. Since $\operatorname{rank}\bigl(A\,[P_{11}^*\ P_{21}^*]\bigr) = m$, we have $P_{21} = 0$. Then obviously, $P_{11} \in \mathcal{G}(A,\sigma)$ and $P_{22} \in \mathrm{GL}(n,F)$. □

6.6.1. Groups of symplectic spaces.

Let (V,l) be a nondegenerate symplectic space over F, i.e., $l \in \mathcal{A}(V)$ and $\ker l = \{0\}$. By Theorem 6.30 (iii), $\dim V$ is even. A *symplectic basis* of (V,l) is a basis $\epsilon_1,\dots,\epsilon_{2r}$ of V such that

$$(l(\epsilon_i,\epsilon_j)) = \underbrace{\begin{bmatrix} 0 & 1 \\ -1 & 0 \end{bmatrix} \oplus \cdots \oplus \begin{bmatrix} 0 & 1 \\ -1 & 0 \end{bmatrix}}_{r}.$$

The existence of a symplectic basis of a nondegenerate symplectic space is guaranteed by Theorem 6.30 (iii). For $a \in V$ and $\lambda \in F$, the F-map

$$\begin{array}{rccl} \sigma_{a,\lambda}: & V & \longrightarrow & V \\ & x & \longmapsto & x + \lambda l(x,a)a \end{array}$$

is called a *symplectic transvection* in the direction a. It is easy to see that $\sigma_{a,\lambda}\sigma_{a,-\lambda} = \mathrm{id}$, and hence $\sigma_{a,\lambda} \in \mathrm{Aut}_F(V)$. Moreover, for $x, y \in V$,

$$\begin{aligned} l\bigl(\sigma_{a,\lambda}(x),\, \sigma_{a,\lambda}(y)\bigr) &= l\bigl(x + \lambda l(x,a)a,\ y + \lambda l(y,a)a\bigr) \\ &= l(x,y) + \lambda l(y,a)l(x,a) + \lambda l(x,a)l(a,y) \\ &= l(x,y). \end{aligned}$$

Hence $\sigma_{a,\lambda} \in \mathcal{G}(V,l)$.

Theorem 6.51. *Let (V,l) be a nondegenerate symplectic space. Then $\mathcal{G}(V,l)$ is generated by symplectic transvections.*

This theorem has been used in the proof of Lemma 6.42, but its own proof has been deferred until now.

Proof of Theorem 6.51. 1° Let $a, b \in V \setminus \{0\}$. We claim that there is a product of symplectic transvections which maps a to b. If $l(a,b) \neq 0$, then the symplectic transvection

$$\sigma(x) = x + \frac{1}{l(a,b)} l(x, b-a)(b-a), \quad x \in V,$$

maps a to b. In general, since a vector space is never a union of two proper subspaces, we have $\langle a\rangle^{\perp}\cup\langle b\rangle^{\perp}\neq V$. Choose $c\in V$ such that $l(a,c)\neq 0$ and $l(b,c)\neq 0$. From the above, there exist symplectic transvections σ and τ such that $\sigma(a)=c$ and $\tau(c)=b$.

2° Two elements $a_1,a_2\in V$ are called a *hyperbolic pair* if $l(a_1,a_2)=1$. Let (a_1,a_2) and (b_1,b_2) be two hyperbolic pairs of V. We claim that there is a product of symplectic transvections which maps a_1 to b_1 and a_2 to b_2.

By 1°, we may assume $a_1=b_1$. If $l(a_2,b_2)\neq 0$, by 1°, there exists a symplectic transvection σ in the direction b_2-a_2 such that $\sigma(a_2)=b_2$. Since $l(a_1,b_2-a_2)=1-1=0$, we have $\sigma(a_1)=a_1$. If $l(a_2,b_2)=0$, then (a_1,a_2), (a_1,a_1+a_2), and (a_1,b_2) are three hyperbolic pairs with $l(a_2,a_1+a_2)\neq 0$ and $l(a_1+a_2,b_2)\neq 0$. Thus, there exist two symplectic transvections σ and τ such that $(\sigma(a_1),\sigma(a_2))=(a_1,a_1+a_2)$ and $(\tau(a_1),\tau(a_1+a_2))=(a_1,b_2)$.

3° Let $\sigma\in\mathcal{G}(V,l)$ be arbitrary. We show that σ is a product of symplectic transvections.

Let $\epsilon_1,\dots,\epsilon_{2r}$ be a symplectic basis of V. By 2°, we may assume that $\sigma(\epsilon_1)=\epsilon_1$ and $\sigma(\epsilon_2)=\epsilon_2$. We have $V=\langle\epsilon_1,\epsilon_2\rangle\oplus V_1$, where $V_1=\langle\epsilon_3,\dots,\epsilon_{2r}\rangle$. Clearly, $(V_1,\, l|_{V_1\times V_1})$ is a nondegenerate symplectic space, and $\sigma|_{V_1}\in\mathcal{G}(V_1,l|_{V_1\times V_1})$. Using induction on $\dim V$, we may assume that

$$\sigma|_{V_1}=\tau_1\cdots\tau_k,$$

where each τ_i is a symplectic transvection of V_1. A symplectic transvection of V_1 is the restriction of a symplectic transvection of V in the same direction. Thus, $\tau_i=\sigma_i|_{V_1}$, where σ_i is a symplectic transvection of V whose direction is in V_1. It follows that $\sigma_i|_{\langle\epsilon_1,\epsilon_2\rangle}=\mathrm{id}$. Then σ and $\sigma_1\cdots\sigma_k$ are identical on both $\langle\epsilon_1,\epsilon_2\rangle$ and V_1; hence $\sigma=\sigma_1\cdots\sigma_k$. □

By Theorem 6.30 (iii), the canonical forms in $\Lambda_n(F)$ are

$$K_{n,r}=\underbrace{\begin{bmatrix}0&1\\-1&0\end{bmatrix}\oplus\cdots\oplus\begin{bmatrix}0&1\\-1&0\end{bmatrix}}_{r}\oplus 0_{n-2r},\quad 0\le r\le n/2. \tag{6.70}$$

Theorem 6.52. *Let $F=\mathbb{F}_q$. For $0\le r\le n/2$,*

$$|\mathcal{G}(K_{n,r})|=q^{2r(n-2r)}|\mathrm{GL}(n-2r,\mathbb{F}_q)|\,q^{r^2}\prod_{i=1}^{r}(q^{2i}-1).$$

Proof. By Lemma 6.50, we have

$$|\mathcal{G}(K_{n,r})|=q^{2r(n-2r)}|\mathrm{GL}(n-2r,\mathbb{F}_q)|\,|\mathcal{G}(K_{2r,r})|,$$

so we only have to determine $|\mathcal{G}(K_{2r,r})|$. Let $l=\mathtt{X}K_{2r,r}\mathtt{Y}^T\in\mathcal{A}(\mathbb{F}_q^{2r})$. An element $\alpha\in\mathrm{Aut}_{\mathbb{F}_q}(\mathbb{F}_q^{2r})$ belongs to $\mathcal{G}(\mathbb{F}_q^{2r},l)$ if and only if it maps a symplectic basis of $(\mathbb{F}_q^{2r},l)$ to a symplectic basis of $(\mathbb{F}_q^{2r},l)$. Therefore, $|\mathcal{G}(K_{2r,r})|$

$(= |\mathcal{G}(\mathbb{F}_q^{2r}, l)|\,)$ is the number of symplectic bases of $(\mathbb{F}_q^{2r}, l)$. A symplectic basis $\epsilon_1, \dots, \epsilon_{2r}$ of $(\mathbb{F}_q^{2r}, l)$ can be obtained as follows. First choose $\epsilon_1, \epsilon_2 \in \mathbb{F}_q^{2r}$ such that $l(\epsilon_1, \epsilon_2) = 1$, and let $V_1 = \langle \epsilon_1, \epsilon_2 \rangle^{\perp}$. Then $(V_1, l|_{V_1 \times V_1})$ is a $2(r-1)$-dimensional nondegenerate symplectic space. Choose $\epsilon_3, \dots, \epsilon_{2r}$ as a symplectic basis of $(V_1, l|_{V_1 \times V_1})$. The number of choices for ϵ_1 is $q^{2r} - 1$; the number of choices for ϵ_2 is q^{2r-1}; the number of choices for $\epsilon_3, \dots, \epsilon_{2r}$ is $|\mathcal{G}(K_{2(r-1),r-1})|$. Therefore,

$$|\mathcal{G}(K_{2r,r})| = (q^{2r} - 1)q^{2r-1}|\mathcal{G}(K_{2(r-1),r-1})|.$$

By induction,

$$|\mathcal{G}(K_{2r,r})| = \prod_{i=1}^{r}(q^{2i} - 1)q^{2i-1} = q^{r^2}\prod_{i=1}^{r}(q^{2i} - 1).$$

□

The group $\mathcal{G}(K_{2r,r})$ is called the *symplectic group* of degree $2r$ over F and is denoted by $\mathrm{Sp}(2r, F)$.

6.6.2. Groups of unitary spaces.

In this subsection, assume that $F = \mathbb{F}_{q^2}$ and let $\sigma \in \mathrm{Aut}(\mathbb{F}_{q^2}/\mathbb{F}_q) \setminus \{\mathrm{id}\}$. Let (V, l) be a nondegenerate unitary space over $\mathbb{F}_{q^2}$ with $\dim V = r$. An *orthonormal basis* of (V, l) is a basis $\epsilon_1, \dots, \epsilon_r$ of V such that $(l(\epsilon_i, \epsilon_j)) = I_r$. By Theorem 6.32 (ii), (V, l) has an orthonormal basis.

Theorem 6.53. *Let $F = \mathbb{F}_{q^2}$ and $\sigma \in \mathrm{Aut}(\mathbb{F}_{q^2}/\mathbb{F}_q) \setminus \{\mathrm{id}\}$. For $0 \le r \le n$,*

$$|\mathcal{G}(I_r \oplus 0_{n-r}, \sigma)| = q^{rn - r(r+1)/2}|\mathrm{GL}(n-r, \mathbb{F}_{q^2})|\prod_{i=1}^{r}(q^i - (-1)^i).$$

Proof. By Lemma 6.50,

$$|\mathcal{G}(I_r \oplus 0_{n-r}, \sigma)| = q^{r(n-r)}|\mathrm{GL}(n-r, \mathbb{F}_{q^2})|\,|\mathcal{G}(I_r, \sigma)|,$$

so it remains to determine $|\mathcal{G}(I_r, \sigma)|$. Let $l = \mathtt{X}_1\mathtt{Y}_1^q + \cdots + \mathtt{X}_r\mathtt{Y}_r^q \in \mathcal{S}^\sigma(\mathbb{F}_{q^2}^r)$. (The matrix of l with respect to the standard basis of $\mathbb{F}_{q^2}^r$ is I_r.) An element $\alpha \in \mathrm{Aut}_{\mathbb{F}_{q^2}}(\mathbb{F}_{q^2}^r)$ belongs to $\mathcal{G}(I_r, \sigma)$ if and only if it maps an orthonormal basis of $(\mathbb{F}_{q^2}^r, l)$ to an orthonormal basis of $(\mathbb{F}_{q^2}^r, l)$. Thus, $|\mathcal{G}(I_r, \sigma)|$ $(= |\mathcal{G}(\mathbb{F}_{q^2}^r, l)|\,)$ is the number of orthonormal bases of $(\mathbb{F}_{q^2}^r, l)$. Each orthonormal basis of $(\mathbb{F}_{q^2}^r, l)$ can be obtained as follows. First choose $\epsilon_1 \in \mathbb{F}_{q^2}$ such that $l(\epsilon_1, \epsilon_1) = 1$, and let $V_1 = \langle \epsilon_1 \rangle^{\perp}$. Then $(V_1, l|_{V_1 \times V_1})$ is an $(r-1)$-dimensional nondegenerate unitary space. Choose $\epsilon_2, \dots, \epsilon_r$ as an orthonormal basis of $(V_1, l|_{V_1 \times V_1})$. By Lemma 6.54 below, the number of choices for ϵ_1 is $q^{r-1}(q^r - (-1)^r)$; the number of choices for $\epsilon_2, \dots, \epsilon_r$ is $|\mathcal{G}(I_{r-1}, \sigma)|$. Hence

$$|\mathcal{G}(I_r, \sigma)| = q^{r-1}(q^r - (-1)^r)|\mathcal{G}(I_{r-1}, \sigma)|.$$

By induction,

$$|\mathcal{G}(I_r,\sigma)| = q^{r(r-1)/2}\prod_{i=1}^{r}(q^i-(-1)^i).$$

□

Lemma 6.54. *Let* $f = \mathtt{X}_1^{1+q}+\cdots+\mathtt{X}_r^{1+q}\in\mathbb{F}_{q^2}[\mathtt{X}_1,\dots,\mathtt{X}_r]$, *and let* $Z(f-1)$ *denote the set of zeros of* $f-1$ *in* $\mathbb{F}_{q^2}^r$. *Then*

$$|Z(f-1)| = q^{r-1}(q^r-(-1)^r).$$

Proof. Let $p=\operatorname{char}\mathbb{F}_q$, $\zeta_p = e^{2\pi i/p}$, and $e(x)=\zeta_p^{\operatorname{Tr}_{q^2/p}(x)}$ for $x\in\mathbb{F}_{q^2}$. Then

$$\begin{aligned}|Z(f-1)| &= q^{-2}\sum_{y,x_1,\dots,x_r\in\mathbb{F}_{q^2}} e\big(y(f(x_1,\dots,x_r)-1)\big)\\ &= q^{-2}\sum_{y,x_1,\dots,x_r\in\mathbb{F}_{q^2}} e(yx_1^{1+q}+\cdots+yx_r^{1+q}-y)\\ &= q^{-2}\sum_{y\in\mathbb{F}_{q^2}} e(-y)\Big(\sum_{x\in\mathbb{F}_{q^2}} e(yx^{1+q})\Big)^r.\end{aligned}$$

In the above,

$$\begin{aligned}\sum_{x\in\mathbb{F}_{q^2}} e(yx^{1+q}) &= \sum_{x\in\mathbb{F}_{q^2}} \zeta_p^{\operatorname{Tr}_{q^2/p}(yx^{1+q})} = \sum_{x\in\mathbb{F}_{q^2}} \zeta_p^{\operatorname{Tr}_{q/p}(x^{1+q}\operatorname{Tr}_{q^2/q}(y))}\\ &= -q+(1+q)\sum_{x\in\mathbb{F}_q}\zeta_p^{\operatorname{Tr}_{q/p}(x\operatorname{Tr}_{q^2/q}(y))}\\ &= \begin{cases} q^2 & \text{if } \operatorname{Tr}_{q^2/q}(y)=0,\\ -q & \text{if } \operatorname{Tr}_{q^2/q}(y)\neq 0.\end{cases}\end{aligned}$$

Hence

$$\begin{aligned}|Z(f-1)| &= q^{-2}\Big[q^{2r}\sum_{\operatorname{Tr}_{q^2/q}(y)=0} e(-y) + (-q)^r\sum_{\operatorname{Tr}_{q^2/q}(y)\neq 0} e(-y)\Big]\\ &= q^{-2}\Big[q^{2r}\cdot q + (-q)^r\Big(\sum_y - \sum_{\operatorname{Tr}_{q^2/q}(y)=0}\Big)e(-y)\Big]\\ &= q^{-2}\big(q^{2r+1}-(-q)^r q\big) = q^{r-1}(q^r-(-1)^r).\end{aligned}$$

□

When $r=n$ in Theorem 6.53, the group $\mathcal{G}(I_n,\sigma)$ is called the *unitary group* of degree n over $\mathbb{F}_{q^2}$ and is denoted by $\mathrm{U}(n,\mathbb{F}_{q^2})$.

6.6.3. Groups of symmetric bilinear forms over $\mathbb{F}_q$, q odd.

In this subsection, assume that $F = \mathbb{F}_q$, where q is odd, and fix a nonsquare d of $\mathbb{F}_q$. Let (V, l) be a nondegenerate symmetric bilinear form over $\mathbb{F}_q$ with discriminant $d(l) = \delta\mathbb{F}_q^{*2}$, where $\delta \in \{1, d\}$. A basis $\epsilon_1, \dots, \epsilon_r$ of V is called an *orthonormal basis* of (V, l) if $(l(\epsilon_i, \epsilon_j)) = I_{r-1} \oplus [\delta]$. By Theorem 6.30 (i), (V, l) has an orthonormal basis.

Lemma 6.55. *Let η be the quadratic character of $\mathbb{F}_q$. Let $f \in \mathcal{Q}(\mathbb{F}_q^n)$ be a nondegenerate quadratic form with discriminant $\delta\mathbb{F}_q^{*2}$, where $\delta \in \{1, d\}$. For each $b \in \mathbb{F}_q$, the number of zeros of $f - b$ is given by*

$$|Z(f-b)| = \begin{cases} \left.\begin{array}{ll} q^{n-1} + (q-1)q^{n/2-1}\,\eta\big((-1)^{n/2}\delta\big) & \text{if } b = 0 \\ q^{n-1} - q^{n/2-1}\,\eta\big((-1)^{n/2}\delta\big) & \text{if } b \neq 0 \end{array}\right\} & \text{if } n \text{ is even,} \\ q^{n-1} + q^{(n-1)/2}\,\eta\big((-1)^{(n-1)/2}b\delta\big) & \text{if } n \text{ is odd.} \end{cases}$$

Proof. We may assume that $f = \mathtt{X}_1^2 + \cdots + \mathtt{X}_{n-1}^2 + \delta\mathtt{X}_n^2$. Let $\operatorname{char}\mathbb{F}_q = p$, $\zeta_p = e^{2\pi i/p}$, and $e(x) = \zeta_p^{\operatorname{Tr}_{q/p}(x)}$, $x \in \mathbb{F}_q$. Then

$$\begin{aligned} |Z(f-b)| &= q^{-1} \sum_{y, x_1, \dots, x_n \in \mathbb{F}_q} e\big(y(f(x_1, \dots, x_n) - b)\big) \\ &= q^{n-1} + q^{-1} \sum_{y \in \mathbb{F}_q^*,\, x_1, \dots, x_n \in \mathbb{F}_q} e\big(y(x_1^2 + \cdots + x_{n-1}^2 + \delta x_n^2) - b\big) \\ &= q^{n-1} + q^{-1} \sum_{y \in \mathbb{F}_q^*} \Big(\sum_{x \in \mathbb{F}_q} e(yx^2)\Big)^{n-1} \Big(\sum_{x \in \mathbb{F}_q} e(y\delta x^2)\Big) e(-yb). \end{aligned}$$

In the above, we claim that

$$\sum_{x \in \mathbb{F}_q} e(yx^2) = \eta(y) g(\eta),$$

where $g(\eta)$ is the Gauss quadratic sum over $\mathbb{F}_q$. If y is a square in $\mathbb{F}_q^*$, then $\sum_{x \in \mathbb{F}_q} e(yx^2) = \sum_{x \in \mathbb{F}_q} e(x^2) = g(\eta)$. If y is a nonsquare in $\mathbb{F}_q^*$, then $\sum_{x \in \mathbb{F}_q} e(yx^2) = 2\sum_{x \in \mathbb{F}_q} e(x) - \sum_{x \in \mathbb{F}_q} e(x^2) = -g(\eta)$. Thus the claim is proved. Now we have

$$|Z(f-b)| = q^{n-1} + q^{-1}\eta(\delta) g(\eta)^n \sum_{y \in \mathbb{F}_q^*} \eta(y)^n e(-yb).$$

First assume that n is even. We have $g(\eta)^n = \eta(-1)^{n/2} q^{n/2}$ (by Theorem 3.11) and

$$\sum_{y \in \mathbb{F}_q^*} \eta(y)^n e(-yb) = \sum_{y \in \mathbb{F}_q^*} e(-yb) = \begin{cases} q-1 & \text{if } b = 0, \\ -1 & \text{if } b \neq 0. \end{cases}$$

Therefore,

$$|Z(f-b)| = \begin{cases} q^{n-1} + (q-1)q^{n/2-1}\,\eta\big((-1)^{n/2}\,\delta\big) & \text{if } b=0, \\ q^{n-1} - q^{n/2-1}\,\eta\big((-1)^{n/2}\,\delta\big) & \text{if } b \neq 0. \end{cases}$$

Now assume that n is odd. We have

$$\sum_{y\in\mathbb{F}_q^*} \eta(y)^n e(-yb) = \sum_{y\in\mathbb{F}_q^*} \eta(y)e(-by) = \eta(-b)g(\eta)$$

and $g(\eta)^{n+1} = \eta(-1)^{(n+1)/2} q^{(n+1)/2}$. Thus,

$$|Z(f-b)| = q^{n-1} + q^{(n-1)/2}\,\eta\big((-1)^{(n-1)/2}\,b\delta\big).$$

□

By Theorem 6.30 (i), the canonical forms in $S_n(\mathbb{F}_q)$ are

$$I_r \oplus 0_{n-r}, \quad 0 \le r \le n,$$

and

$$I_{r-1} \oplus [d] \oplus 0_{n-r}, \quad 1 \le r \le n.$$

Lemma 6.56. *Assume that r is odd and $1 \le r \le n$. Then*

$$\mathcal{G}(I_r \oplus 0_{n-r}) \cong \mathcal{G}(I_{r-1} \oplus [d] \oplus 0_{n-r}).$$

Proof. Since r is odd, both $\det(I_{r-1}\oplus[d]) = d$ and $\det(dI_r) = d^r$ are nonsquares in $\mathbb{F}_q^*$. By Theorem 6.30 (i), $I_{r-1}\oplus[d] \cong dI_r$, and hence $I_{r-1}\oplus [d]\oplus 0_{n-r} \cong dI_r \oplus 0_{n-r}$. Therefore,

$$\mathcal{G}(I_{r-1}\oplus[d]\oplus 0_{n-r}) = \mathcal{G}(dI_r\oplus 0_{n-r}) \cong \mathcal{G}(I_r\oplus 0_{n-r}).$$

□

Theorem 6.57. *For $0 \le r \le n$,*

$$|\mathcal{G}(I_r\oplus 0_{n-r})| = q^{r(n-r)}|\mathrm{GL}(n-r,\mathbb{F}_q)| \cdot \begin{cases} 2q^{r(r-2)/4}\big(q^{r/2}+\eta(-1)^{r/2}\big)^{-1}\displaystyle\prod_{i=1}^{r/2}(q^{2i}-1) & \textit{if } r \textit{ is even}, \\ 2q^{(r-1)^2/4}\displaystyle\prod_{i=1}^{(r-1)/2}(q^{2i}-1) & \textit{if } r \textit{ is odd}. \end{cases}$$

For $1 \le r \le n$,

$$|\mathcal{G}(I_{r-1} \oplus [d] \oplus 0_{n-r})| = q^{r(n-r)} |\mathrm{GL}(n-r, \mathbb{F}_q)| \cdot \begin{cases} 2q^{r(r-2)/4}\left(q^{r/2} + \eta(-1)^{r/2}\right) \prod_{i=1}^{r/2-1} (q^{2i}-1) & \text{if } r \text{ is even,} \\ 2q^{(r-1)^2/4} \prod_{i=1}^{(r-1)/2} (q^{2i}-1) & \text{if } r \text{ is odd.} \end{cases}$$

Proof. By Lemma 6.50, we only have to determine $|\mathcal{G}(I_r)|$ and $|\mathcal{G}(I_{r-1} \oplus [d])|$. Let $l = \mathtt{X} I_r \mathtt{Y} \in \mathcal{S}(\mathbb{F}_q^r)$. Then $|\mathcal{G}(I_r)|$ is the number of orthonormal bases $\epsilon_1, \dots, \epsilon_r$ of $(\mathbb{F}_q^r, l)$. The number of choices for ϵ_1 is $|Z(\mathtt{X}_1^2 + \cdots + \mathtt{X}_r^2 - 1)|$. When ϵ_1 is chosen, let $V_1 = \langle \epsilon_1 \rangle^{\perp}$. Then $(V_1, l|_{V_1 \times V_1})$ is an $(r-1)$-dimensional nondegenerate symmetric bilinear space with discriminant $1\mathbb{F}_q^{*2}$. Hence, the number of choices for $\epsilon_2, \dots, \epsilon_r$ is $|\mathcal{G}(I_{r-1})|$. It follows from Lemma 6.55 that

$$\begin{aligned} |\mathcal{G}(I_r)| &= |Z(\mathtt{X}_1^2 + \cdots + \mathtt{X}_r^2 - 1)|\, |\mathcal{G}(I_{r-1})| \\ &= |\mathcal{G}(I_{r-1})| \cdot \begin{cases} q^{r-1} - q^{r/2-1}\,\eta(-1)^{r/2} & \text{if } r \text{ is even,} \\ q^{r-1} + q^{(r-1)/2}\,\eta(-1)^{(r-1)/2} & \text{if } r \text{ is odd.} \end{cases} \end{aligned}$$

By induction,

$$\begin{aligned} |\mathcal{G}(I_r)| &= \begin{cases} \prod_{i=1}^{r/2} \left(q^{2i-1} - q^{i-1}\eta(-1)^i\right)\left(q^{2i-2} + q^{i-1}\eta(-1)^{i-1}\right) & \text{if } r \text{ is even,} \\ 2 \prod_{i=1}^{(r-1)/2} \left(q^{2i} + q^i\eta(-1)^i\right)\left(q^{2i-1} - q^{i-1}\eta(-1)^i\right) & \text{if } r \text{ is odd} \end{cases} \\ &= \begin{cases} 2q^{r(r-2)/4}\left(q^{r/2} + \eta(-1)^{r/2}\right)^{-1} \prod_{i=1}^{r/2} (q^{2i}-1) & \text{if } r \text{ is even,} \\ 2q^{(r-1)^2/4} \prod_{i=1}^{(r-1)/2} (q^{2i}-1) & \text{if } r \text{ is odd.} \end{cases} \end{aligned}$$

The order $|\mathcal{G}(I_{r-1} \oplus [d])|$ is determined in a similar manner. Let $g = \mathtt{X}(I_{r-1} \oplus [d])\mathtt{Y}^T \in \mathcal{S}(\mathbb{F}_q^r)$. Then $|\mathcal{G}(I_{r-1} \oplus [d])|$ is the number of orthonormal bases $\epsilon_1, \dots, \epsilon_r$ of $(\mathbb{F}_q^r, g)$. By the above argument, but choosing ϵ_r first, we have

$$\begin{aligned} |\mathcal{G}(I_{r-1} \oplus [d])| &= |Z(\mathtt{X}_1^2 + \cdots + \mathtt{X}_{r-1}^2 + d\mathtt{X}_r^2 - d)|\, |\mathcal{G}(I_{r-1})| \\ &= |\mathcal{G}(I_{r-1})| \cdot \begin{cases} q^{r-1} + q^{r/2-1}\,\eta(-1)^{r/2} & \text{if } r \text{ is even,} \\ q^{r-1} + q^{(r-1)/2}\,\eta(-1)^{(r-1)/2} & \text{if } r \text{ is odd} \end{cases} \end{aligned}$$

$$= \begin{cases} 2q^{r(r-2)/4}\left(q^{r/2} + \eta(-1)^{r/2}\right) \prod_{i=1}^{r/2-1} (q^{2i} - 1) & \text{if } r \text{ is even,} \\ 2q^{(r-1)^2/4} \prod_{i=1}^{(r-1)/2} (q^{2i} - 1) & \text{if } r \text{ is odd.} \end{cases}$$

□

When $n = 2k + 1$, the group $\mathcal{G}(I_{2k+1})$ ($\cong \mathcal{G}(I_{2k} \oplus [d]\,)$ is denoted by $\mathrm{O}(2k + 1, \mathbb{F}_q)$. When $n = 2k$ and $\delta \in \{1, d\}$, the group $\mathcal{G}(I_{2k-1} \oplus [\delta])$ is denoted by

$$\begin{cases} \mathrm{O}_+(2k, \mathbb{F}_q) & \text{if } \eta((-1)^k\delta) = 1, \\ \mathrm{O}_-(2k, \mathbb{F}_q) & \text{if } \eta((-1)^k\delta) = -1. \end{cases}$$

The groups $\mathrm{O}(2k + 1, \mathbb{F}_q)$ and $\mathrm{O}_\pm(2k, \mathbb{F}_q)$ are all called *orthogonal groups* over $\mathbb{F}_q$.

6.6.4. Groups of symmetric bilinear forms over $\mathbb{F}_q$, q even.

In this subsection, assume that $F = \mathbb{F}_q$, where q is even. In this case, we only have to consider symmetric bilinear forms that are not alternating. Let (V, l) be a nondegenerate symmetric bilinear space such that $l \in \mathcal{S}(V)\backslash\mathcal{A}(V)$. An *orthonormal basis* of (V, l) is a basis $\epsilon_1, \dots, \epsilon_r$ of V such that $(l(\epsilon_i, \epsilon_j)) = I_r$. By Theorem 6.30 (ii), (V, l) has an orthonormal basis. By the same theorem, the canonical forms of $S_n(\mathbb{F}_q) \setminus \Lambda_n(\mathbb{F}_q)$ are $I_r \oplus 0_{n-r}$, $0 \le r \le n$.

Theorem 6.58. *For $0 \le r \le n$,*

$$|\mathcal{G}(I_r \oplus 0_{n-r})| = q^{r(n-r)}|\mathrm{GL}(n-r, \mathbb{F}_q)| \cdot \begin{cases} q^{r^2/4} \prod_{i=1}^{r/2-1} (q^{2i} - 1) & \textit{if } r \textit{ is even,} \\ q^{(r-1)^2/4} \prod_{i=1}^{(r-1)/2} (q^{2i} - 1) & \textit{if } r \textit{ is odd.} \end{cases}$$

Proof. By Lemma 6.50, we only have to determine $|\mathcal{G}(I_r)|$. Let $l = \mathtt{XY}^T \in \mathcal{S}(\mathbb{F}_q^r)$. The order $|\mathcal{G}(I_r)|$ is the number of orthonormal bases of $(\mathbb{F}_q^r, l)$. Assume that $r > 1$. An orthonormal basis $\epsilon_1, \dots, \epsilon_r$ of $(\mathbb{F}_q^r, l)$ can be constructed as follows. First choose $\epsilon_1 \in \mathbb{F}_q^r$ such that $l(\epsilon_1, \epsilon_1) = 1$. We must require $\epsilon_1 \neq (1, \dots, 1)$ because otherwise there does not exist $\epsilon_2 \in \mathbb{F}_q^r$ such that $l(\epsilon_1, \epsilon_2) = 0$ and $l(\epsilon_2, \epsilon_2) = 1$. Thus we choose $\epsilon_1 = (x_1, \dots, x_r)$ subject to the conditions that $x_1^2 + \cdots + x_r^2 = 1$ (i.e., $x_1 + \cdots + x_r = 1$) and $\epsilon_1 \neq (1, \dots, 1)$. The number of choices for ϵ_1 is

$$\begin{cases} q^{r-1} & \text{if } r \text{ is even,} \\ q^{r-1} - 1 & \text{if } r \text{ is odd.} \end{cases}$$

When ϵ_1 is chosen, let $V_1 = \langle \epsilon_1 \rangle^\perp$. Then $(V_1,\ l|_{V_1 \times V_1})$ is an $(r-1)$-dimensional nondegenerate symmetric bilinear space which is not alternating. Thus the number of choices for $\epsilon_2, \dots, \epsilon_r$ is $|\mathcal{G}(I_{r-1})|$. Therefore

$$|\mathcal{G}(I_r)| = |\mathcal{G}(I_{r-1})| \cdot \begin{cases} q^{r-1} & \text{if } r \text{ is even,} \\ q^{r-1} - 1 & \text{if } r \text{ is odd.} \end{cases}$$

By induction,

$$|\mathcal{G}(I_r)| = \begin{cases} q^{r^2/4} \displaystyle\prod_{i=1}^{r/2-1} (q^{2i} - 1) & \text{if } r \text{ is even,} \\ q^{(r-1)^2/4} \displaystyle\prod_{i=1}^{(r-1)/2} (q^{2i} - 1) & \text{if } r \text{ is odd.} \end{cases}$$

□

6.6.5. Witt's theorems.

In this subsection, we prove Witt's extension theorem and cancellation theorem for quadratic spaces over an arbitrary field F. The case $\operatorname{char} F \neq 2$ and the case $\operatorname{char} F = 2$ require quite different approaches.

The case $\operatorname{char} F \neq 2$.

Lemma 6.59. *Assume that* $\operatorname{char} F \neq 2$. *Let* $A_1, A_2 \in S_m(F)$ *and* $B_1, B_2 \in S_n(F)$ *be such that* $A_1 \cong A_2$ *and* $A_1 \oplus B_1 \cong A_2 \oplus B_2$. *Then* $B_1 \cong B_2$.

Note. Lemma 6.59 is false when $\operatorname{char} F = 2$. By Theorem 6.30 (ii), $[1] \oplus \left[\begin{smallmatrix} 0 & 1 \\ 1 & 0 \end{smallmatrix}\right] \cong [1] \oplus \left[\begin{smallmatrix} 1 & 0 \\ 0 & 1 \end{smallmatrix}\right]$. However, $\left[\begin{smallmatrix} 0 & 1 \\ 1 & 0 \end{smallmatrix}\right] \not\cong \left[\begin{smallmatrix} 1 & 0 \\ 0 & 1 \end{smallmatrix}\right]$.

Proof of Lemma 6.59. 1° First assume that B_1 and B_2 are nonsingular. We may assume $A_1 = A_2$. By Theorem 6.30 (i), we may assume that A_1 is diagonal. Consequently, it suffices to consider the case where A_1 is of size 1×1, i.e., $A_1 = [a]$, $a \in F$. Since $[a] \oplus B_1 \cong [a] \oplus B_2$, there exists

$$\begin{bmatrix} u & \alpha \\ \beta & Q \end{bmatrix} \in \mathrm{GL}(n+1, F),$$

where Q is of size $n \times n$, such that

$$\begin{bmatrix} u & \alpha \\ \beta & Q \end{bmatrix} \begin{bmatrix} a & 0 \\ 0 & B_1 \end{bmatrix} \begin{bmatrix} u & \beta^T \\ \alpha^T & Q^T \end{bmatrix} = \begin{bmatrix} a & 0 \\ 0 & B_2 \end{bmatrix},$$

i.e.,

$$u^2 a + \alpha B_1 \alpha^T = a,$$
$$ua\beta^T + \alpha B_1 Q^T = 0,$$

$$a\beta\beta^T + QB_1Q^T = B_2.$$

Let $P = Q + x\beta\alpha$, where $x \in F$ is to be determined. By the above equations, we have

$$\begin{aligned} PB_1P^T &= QB_1Q^T + xQB_1\alpha^T\beta^T + x\beta\alpha B_1Q^T + x^2\beta\alpha B_1\alpha^T\beta^T \\ &= B_2 - a\beta\beta^T - 2xua\beta\beta^T + x^2(a - u^2a)\beta\beta^T \\ &= B_2 + a\big[(1-u^2)x^2 - 2ux - 1\big]\beta\beta^T. \end{aligned}$$

The equation $(1-u^2)x^2 - 2ux - 1 = 0$ always has a solution $x \in F$. (If $u^2 = 1$, $x = -1/(2u)$; if $u^2 \neq 1$, $x = -1/(u \pm 1)$.) For this solution x, we have $PB_1P^T = B_2$. Since B_2 is nonsingular, $P \in \mathrm{GL}(n, F)$. Hence $B_1 \cong B_2$.

2° In general, we may assume that $B_i = B_i' \oplus 0_{n-r}$, $i = 1, 2$, where $B_i' \in S_r(F)$ is nonsingular. Since $A_1 \oplus B_1 \cong A_2 \oplus B_2$, it follows from Lemma 6.31 that $A_1 \oplus B_1' \cong A_2 \oplus B_2'$. By (i), $B_1' \cong B_2'$ and hence $B_1 \cong B_2$. □

For two quadratic spaces (V_1, f_1) and (V_2, f_2) over F, form the external direct sum $V_1 \oplus V_2$ and define $f_1 \oplus f_2 \in \mathcal{Q}(V_1 \oplus V_2)$ by $(f_1 \oplus f_2)(x, y) = f_1(x) + f_2(y)$ for $x \in V_1$ and $y \in V_2$.

Theorem 6.60 (Witt's cancellation theorem, version 1). *Assume that* $\mathrm{char}\, F \neq 2$. *Let* (V_i, f_i) *and* (W_i, g_i), $i = 1, 2$, *be quadratic spaces over* F *such that* $(V_1, f_1) \cong (V_2, f_2)$ *and* $(V_1 \oplus W_1, f_1 \oplus g_1) \cong (V_2 \oplus W_2, f_2 \oplus g_2)$. *Then* $(W_1, g_1) \cong (W_2, g_2)$.

Proof. For $i = 1, 2$, let $\mathcal{D}_i$ be a basis of V_i and $\mathcal{E}_i$ a basis of W_i. Then $\mathcal{D}_i \cup \mathcal{E}_i$ is a basis of $V_i \oplus W_i$. We have $A(l_{f_1}, \mathcal{D}_1) \cong A(l_{f_2}, \mathcal{D}_2)$ and

$$\begin{aligned} A(l_{f_1}, \mathcal{D}_1) \oplus A(l_{g_1}, \mathcal{E}_1) &= A(l_{f_1 \oplus g_1}, \mathcal{D}_1 \cup \mathcal{E}_1) \cong A(l_{f_2 \oplus g_2}, \mathcal{D}_2 \cup \mathcal{E}_2) \\ &= A(l_{f_2}, \mathcal{D}_2) \oplus A(l_{g_2}, \mathcal{E}_2). \end{aligned}$$

By Lemma 6.59, $A(l_{g_1}, \mathcal{E}_1) \cong A(l_{g_2}, \mathcal{E}_2)$, i.e., $(W_1, g_1) \cong (W_2, g_2)$. □

Theorem 6.61 (Witt's extension theorem). *Assume that* $\mathrm{char}\, F \neq 2$. *Let* (V, f) *be a nondegenerate quadratic space over* F *with two subspaces* V_1 *and* V_2 *and an isometry* $\alpha : V_1 \to V_2$. *Then there exists* $\bar{\alpha} \in \mathcal{G}(V, f)$ *such that* $\bar{\alpha}|_{V_1} = \alpha$.

Proof. 1° First assume that V_1 is nondegenerate. Then V_2 is also nondegenerate and $V = V_1 \oplus V_1^\perp = V_2 \oplus V_2^\perp$. By Witt's cancellation theorem, $(V_1^\perp, f|_{V_1^\perp}) \cong (V_2^\perp, f|_{V_2^\perp})$; let $\beta : V_1^\perp \to V_2^\perp$ be an isometry. Define $\bar{\alpha} : V \to V$ by

$$\bar{\alpha}(x + y) = \alpha(x) + \beta(y), \quad x \in V_1,\ y \in V_1^\perp.$$

Then $\bar{\alpha} \in \mathcal{G}(V, f)$ and $\bar{\alpha}|_{V_1} = \alpha$.

2° Assume that $\ker f|_{V_1} \neq \{0\}$. Let $\mathcal{E} = \{\epsilon_1, \dots, \epsilon_n\}$ be a basis of V such that $\epsilon_1, \dots, \epsilon_m$ is a basis of V_1 and $\epsilon_{r+1}, \dots, \epsilon_m$ is a basis of $\ker f|_{V_1}$. Let $A = A(l_f, \mathcal{E})$. Then

$$A = \begin{bmatrix} A_1 & 0 & A_2^T \\ 0 & 0 & * \\ A_2 & * & * \end{bmatrix},$$

where $A_1 \in S_r(F)$ is nonsingular and A_2 is of size $(n-m) \times r$. Let

$$P_1 = \begin{bmatrix} I_r & & \\ & I_{m-r} & \\ -A_2A_1^{-1} & 0 & I_{n-m} \end{bmatrix}.$$

Then

$$P_1AP_1^T = \begin{bmatrix} A_1 & 0 & 0 \\ 0 & 0 & A_3^T \\ 0 & A_3 & * \end{bmatrix},$$

where $A_3 \in \mathrm{M}_{(n-m)\times(m-r)}(F)$. Since A is nonsingular, $\operatorname{rank} A_3 = m - r$, so there exists $Q \in \mathrm{GL}(n, F)$ such that

$$QA_3 = \begin{bmatrix} I_{m-r} \\ 0 \end{bmatrix}.$$

Let $P_2 = I_m \oplus Q$. Then

$$P_2P_1AP_1^TP_2^T = \begin{bmatrix} A_1 & 0 & 0 & 0 \\ 0 & 0 & I_{m-r} & 0 \\ 0 & I_{m-r} & A_4 & * \\ 0 & 0 & * & * \end{bmatrix}.$$

Let

$$P_3 = \begin{bmatrix} I_r & & & \\ & I_{m-r} & & \\ & 2^{-1}A_4 & I_{m-r} & \\ & & & I_m \end{bmatrix}.$$

Then

$$P_3P_2P_1AP_1^TP_2^TP_3^T = \begin{bmatrix} A_1 & 0 & 0 & 0 \\ 0 & 0 & I_{m-r} & 0 \\ 0 & I_{m-r} & 0 & * \\ 0 & 0 & * & * \end{bmatrix}.$$

Let $P = P_3P_2P_1$ and

$$\begin{bmatrix} \epsilon'_1 \\ \vdots \\ \epsilon'_n \end{bmatrix} = P \begin{bmatrix} \epsilon_1 \\ \vdots \\ \epsilon_n \end{bmatrix}.$$

Then $\epsilon'_i = \epsilon_i$ for $1 \le i \le m$ and

$$\big(l_f(\epsilon'_i, \epsilon'_j)\big)_{1\le i,j\le 2m-r} = \begin{bmatrix} A_1 & 0 & 0 \\ 0 & 0 & I_{m-r} \\ 0 & I_{m-r} & 0 \end{bmatrix}.$$

With $\epsilon_1, \dots, \epsilon_m$ replaced by $\alpha(\epsilon_1), \dots, \alpha(\epsilon_m)$, the above conclusion produces a basis $\epsilon''_1, \dots, \epsilon''_n$ of V such that $\epsilon''_i = \alpha(\epsilon_i)$ for $1 \le i \le m$ and

$$\big(l_f(\epsilon''_i, \epsilon''_j)\big)_{1\le i,j\le 2m-r} = \begin{bmatrix} A_1 & 0 & 0 \\ 0 & 0 & I_{m-r} \\ 0 & I_{m-r} & 0 \end{bmatrix}.$$

Define an F-map $\alpha_1 : \langle \epsilon'_1, \dots, \epsilon'_{2m-r} \rangle \to \langle \epsilon''_1, \dots, \epsilon''_{2m-r} \rangle$ by $\alpha_1(\epsilon'_i) = \epsilon''_i$, $1 \le i \le 2m - r$. Then α_1 is an isometry and $\alpha_1|_{V_1} = \alpha$, Note that $\dim\langle \epsilon'_1, \dots, \epsilon'_{2m-r} \rangle = 2m - r > m = \dim V_1$. Using induction on $\dim V_1$, we conclude that α_1 can be extended to some $\bar{\alpha} \in \mathcal{G}(V, f)$. □

Theorem 6.62 (Witt's cancellation theorem, version 2). *Assume that* $\operatorname{char} F \ne 2$. *Let* (V, f) *be a nondegenerate quadratic space over* F *and let* V_1 *and* V_2 *be two subspaces of* V *such that* $(V_1, f|_{V_1}) \cong (V_2, f|_{V_2})$. *Then* $(V_1^\perp, f|_{V_1^\perp}) \cong (V_2^\perp, f|_{V_2^\perp})$.

Proof. Let $\alpha : V_1 \to V_2$ be an isometry. Extend α to some $\bar{\alpha} \in \mathcal{G}(V, f)$. Then $\bar{\alpha}|_{V_1^\perp} : V_1^\perp \to V_2^\perp$ is an isometry. □

The case $\operatorname{char} F = 2$.

Theorem 6.63 (Witt's extension theorem). *Assume that* $\operatorname{char} F = 2$. *Let* (V, f) *be a quadratic space over* F *with* $\ker l_f = \{0\}$, *and let* V_1 *and* V_2 *be two subspaces of* V *with an isometry* $\alpha : V_1 \to V_2$. *Then there exists* $\bar{\alpha} \in \mathcal{G}(V, f)$ *such that* $\bar{\alpha}|_{V_1} = \alpha$.

Proof. The proof is adopted from [**14**].

1° Define $\phi : V_1 \to V$, $\phi(x) = \alpha(x) - x$, $x \in V_1$, and let $K = \ker \phi$. Clearly, $\alpha|_K = \mathrm{id}$ and $K \subset V_1 \cap V_2$. We claim that

$$l_f(x, \phi(y)) = l_f(\phi(x), \alpha(y)) \quad \text{for all } x, y \in V_1. \tag{6.71}$$

In fact,

$$\begin{aligned} l_f(x,\phi(y)) &= l_f(x,\alpha(y)-y) = l_f(x,\alpha(y)) - l_f(x,y) \\ &= l_f(x,\alpha(y)) - l_f(\alpha(x),\alpha(y)) \\ &= l_f(x-\alpha(x),\alpha(y)) = l_f(\phi(x),\alpha(y)). \end{aligned}$$

It follows from (6.71) that $K \perp \phi(V_1)$.

2° Assume that $\dim K = \dim V_1 - 1$, i.e., $\dim \phi(V_1) = 1$.

Case 1. Assume that $V_1 \not\subset \phi(V_1)^\perp$. Since $K \subset V_1 \cap \phi(V_1)^\perp$ and both K and $V_1 \cap \phi(V_1)^\perp$ are hyperplanes of V_1, we have $K = V_1 \cap \phi(V_1)^\perp$. Let $x, y \in V_1$ be such that $l_f(x,\phi(y)) \neq 0$. By (6.71), $l_f(\phi(x),\alpha(y)) \neq 0$, and hence $V_2 \not\subset \phi(V_1)^\perp$. It follows that $K = V_2 \cap \phi(V_1)^\perp$. Write $\phi(V_1)^\perp = K \oplus L$ for some subspace $L \subset V$. For $i = 1, 2$, $V_i \cap L = V_i \cap \phi(V_1)^\perp \cap L = K \cap L = \{0\}$. Hence $V = V_1 \oplus L = V_2 \oplus L$. Define $\bar{\alpha} : V_1 \oplus L \to V_2 \oplus L$ by

$$\bar{\alpha}(x+y) = \alpha(x) + y, \quad x \in V_1,\ y \in L.$$

For $x \in V_1$ and $y \in L$, since $L \subset \phi(V_1)^\perp$, we have

$$\begin{aligned} f(\bar{\alpha}(x+y)) &= f(\alpha(x)+y) = f(\alpha(x)) + f(y) + l_f(\alpha(x),y) \\ &= f(x) + f(y) + l_f(x,y) = f(x+y). \end{aligned} \tag{6.72}$$

Hence $\bar{\alpha} \in \mathcal{G}(V,f)$.

Case 2. Assume that $V_1 \subset \phi(V_1)^\perp$. By (6.71), we also have $V_2 = \alpha(V_1) \subset \phi(V_1)^\perp$.

First assume that $V_1 \subsetneq \phi(V_1)^\perp$. Then $V_1 \cup V_2 \neq \phi(V_1)^\perp$ since a vector space is never a union of two proper subspaces. Choose $y_0 \in \phi(V_1)^\perp \setminus (V_1 \cup V_2)$, and define

$$\begin{aligned} \alpha_1 :\ V_1 \oplus \langle y_0 \rangle &\longrightarrow V_2 \oplus \langle y_0 \rangle \\ x + ty_0 &\longmapsto \alpha(x) + ty_0, \quad x \in V_1,\ t \in F. \end{aligned}$$

The same computation as (6.72) shows that $\alpha_1 : V_1 \oplus \langle y_0 \rangle \to V_2 \oplus \langle y_0 \rangle$ is an isometry. By induction on $\dim V_1$, we conclude that α_1 can be extended to some $\bar{\alpha} \in \mathcal{G}(V,f)$.

Now assume that $V_1 = \phi(V_1)^\perp$. Then $V_2 = \phi(V_1)^\perp$. Choose $u \in V \setminus V_1$. (Since $\phi(V_1) \neq \{0\}$, $\phi(V_1)^\perp \neq V$.) Every element of $\mathrm{Hom}_F(V_1, F)$ is of the form $l_f(v,\cdot)$ for some $v \in V$; in particular, there exists $v \in V$ such that

$$l_f(u,\alpha(x)) = l_f(v,x) \quad \text{for all } x \in V_1. \tag{6.73}$$

We claim that $v \notin V_1$. If, to the contrary, $v \in V_1$, then for all $x \in V_1$,

$$\begin{aligned} l_f(v,\alpha(x)) &= l_f(v,x) && (\text{since } v \perp (\alpha(x)-x)) \\ &= l_f(u,\alpha(x)). \end{aligned}$$

It follows that $u - v \in V_1^\perp = \phi(V_1) \subset V_1$, which is a contradiction.

Since $u \notin V_1 = \phi(V_1)^\perp$, there exists $x_0 \in V_1$ such that $l_f(u, \phi(x_0)) \neq 0$. Note that

$$(6.74) \qquad \begin{aligned} f(\phi(x_0)) &= f(\alpha(x_0) - x_0) = f(\alpha(x_0)) + f(x_0) + l_f(\alpha(x_0), x_0) \\ &= l_f(\phi(x_0), x_0) = 0. \end{aligned}$$

Let $a \in F$, which is to be determined, and define

$$\begin{array}{rccl} \alpha_1 : & V_1 \oplus \langle v \rangle & \longrightarrow & V_2 \oplus \langle u + a\phi(x_0) \rangle \\ & x + tv & \longmapsto & \alpha(x) + t(u + a\phi(x_0)), \quad x \in V_1, \ t \in F. \end{array}$$

Then for $x \in V_1$ and $t \in F$,

$$\begin{aligned} & f(\alpha_1(x + tv)) = f\big(\alpha(x) + t(u + a\phi(x_0))\big) \\ &= f(\alpha(x)) + t^2 f(u) + t l_f(\alpha(x), u) + t^2 a l_f(u, \phi(x_0)) && \text{(by (6.74))} \\ &= f(x) + t^2\big(f(u) + a l_f(u, \phi(x_0))\big) + t l_f(x, v) && \text{(by (6.73)).} \end{aligned}$$

Choose $a = (f(u) + f(v))/l_f(u, \phi(x_0))$. Then

$$f(\alpha_1(x + tv)) = f(x) + t^2 f(v) + t l_f(x, v) = f(x + tv).$$

Hence $\alpha_1 : V_1 \oplus \langle v \rangle \to V_2 \oplus \langle u + a\phi(x_0) \rangle$ is an isometry. By induction on $\dim V_1$, we conclude that α_1 can be extended to some $\bar{\alpha} \in \mathcal{G}(V, f)$.

3° We drop the assumption in 2°. Assume that $0 < \dim V_1 < \dim V$. Let U be a hyperplane of V_1. By induction on $\dim V_1$, we may assume that the isometry $\alpha|_U : U \to \alpha(U)$ has an extension $\alpha_1 \in \mathcal{G}(V, f)$. Consider the isometry $\beta = \alpha_1^{-1}\alpha : V_1 \to \alpha_1^{-1}(V_2)$, and let $\phi = \beta - \mathrm{id} : V_1 \to V$. Then $U \subset \ker\phi \subset V_1$. If $\ker\phi = V_1$, α_1 is an extension of α and we are done. If $\ker\phi = U$, by 2°, β has an extension $\beta_1 \in \mathcal{G}(V, f)$. Let $\bar{\alpha} = \alpha_1\beta_1 \in \mathcal{G}(V, f)$. Then $\bar{\alpha}|_{V_1} = \alpha$, and the proof is complete. □

Theorem 6.64 (Witt's cancellation theorem). *Assume that* $\operatorname{char} F = 2$. *Let* (V, f) *be a quadratic form over* F *with* $\ker l_f = \{0\}$, *and let* V_1 *and* V_2 *be two subspaces of* V *such that* $(V_1, f|_{V_1}) \cong (V_2, f|_{V_2})$; *then* $(V_1^\perp, f|_{V_1^\perp}) \cong (V_2^\perp, f|_{V_2^\perp})$.

Proof. Let $\alpha : V_1 \to V_2$ be an isometry. Extend α to some $\bar{\alpha} \in \mathcal{G}(V, f)$. Then $\bar{\alpha}|_{V_1^\perp} : V_1^\perp \to V_2^\perp$ is an isometry. □

6.6.6. Groups of quadratic spaces.

When $\operatorname{char} F \neq 2$, the group of a quadratic space (V, f) is the same as the group of the corresponding symmetric bilinear space (V, l_f). Therefore, we assume that $\operatorname{char} F = 2$ in this subsection.

Lemma 6.65. *Let $A \in \mathrm{M}_{2r}(F)$ be such that $A + A^T$ is nonsingular. Then*

$$(6.75)\qquad \begin{aligned} &\mathcal{G}(A \oplus 0_n + \Lambda_{2r+n}(F)) \\ &= \left\{ \begin{bmatrix} P_{11} & P_{12} \\ 0 & P_{22} \end{bmatrix} : P_{11} \in \mathcal{G}(A + \Lambda_{2r}(F)),\ P_{22} \in \mathrm{GL}(n, F) \right\} \end{aligned}$$

and

$$(6.76)\qquad \begin{aligned} &\mathcal{G}(A \oplus [1] \oplus 0_n + \Lambda_{2r+1+n}(F)) \\ &= \left\{ \begin{bmatrix} P_{11} & P_{12} \\ 0 & P_{22} \end{bmatrix} : P_{11} \in \mathcal{G}(A \oplus [1] + \Lambda_{2r+1}(F)),\ P_{22} \in \mathrm{GL}(n, F) \right\}. \end{aligned}$$

Proof. We first prove (6.75). It suffices to show that the left side of (6.75) is contained in the right side. Let

$$\begin{bmatrix} P_{11} & P_{12} \\ P_{21} & P_{22} \end{bmatrix} \in \mathcal{G}(A \oplus 0_n + \Lambda_{2r+n}(F)),$$

where P_{11} is of size $2r \times 2r$. Then

$$\begin{bmatrix} A & \\ & 0_n \end{bmatrix} \equiv \begin{bmatrix} P_{11}AP_{11}^T & P_{11}AP_{21}^T \\ P_{21}AP_{11}^T & P_{21}AP_{21}^T \end{bmatrix} \quad (\mathrm{mod}\ \Lambda_{2r+n}(F)).$$

It follows that $P_{21}AP_{11}^T = (P_{11}AP_{21}^T)^T$ and $P_{21}AP_{21}^T = (P_{21}AP_{21}^T)^T$. Therefore, $P_{21}(A + A^T)P_{11}^T = 0$ and $P_{21}(A + A^T)P_{21}^T = 0$, i.e.,

$$(6.77)\qquad P_{21}(A + A^T)[P_{11}^T\ P_{21}^T] = 0.$$

Since $\mathrm{rank}\big((A + A^T)[P_{11}^T\ P_{21}^T]\big) = 2r$, we have $P_{21} = 0$. It follows that $P_{11} \in \mathcal{G}(A + \Lambda_{2r}(F))$ and $P_{22} \in \mathrm{GL}(n, F)$.

Now we prove (6.76). Again, it suffices to show that the left side is contained in the right. Let $B = A \oplus [1]$ and let

$$\begin{bmatrix} P_{11} & P_{12} \\ P_{21} & P_{22} \end{bmatrix} \in \mathcal{G}(B \oplus 0_n + \Lambda_{2r+1+n}(F)),$$

where P_{11} is of size $(2r + 1) \times (2r + 1)$. It suffices to show that $P_{21} = 0$. We have

$$(6.78)\qquad \begin{bmatrix} B & \\ & 0_n \end{bmatrix} \equiv \begin{bmatrix} P_{11}BP_{11}^T & P_{11}BP_{21}^T \\ P_{21}BP_{11}^T & P_{21}BP_{21}^T \end{bmatrix} \quad (\mathrm{mod}\ \Lambda_{2r+1+n}(F)).$$

Same as (6.77), we have $P_{21}(B + B^T)[P_{11}^T\ P_{21}^T] = 0$. Since $\mathrm{rank}\,[P_{11}^T\ P_{21}^T] = 2r + 1$, we have $P_{21}(B + B^T) = 0$. Note that $B + B^T = \left[\begin{smallmatrix} A+A^T & \\ & 0 \end{smallmatrix}\right]$, where $A + A^T$ is nonsingular. Thus $P_{21} = [0\ u]$ for some column vector $u \in F^n$. By (6.78), all diagonal entries of $P_{21}BP_{21}^T = uu^T$ are 0, which gives $u = 0$. This completes the proof. $\square$

Lemma 6.66. *Assume that F is a perfect field. Let $A \in \mathrm{M}_{2r}(F)$ be such that $A + A^T$ is nonsingular. Then*

$$\mathcal{G}(A \oplus [1] + \Lambda_{2r+1}(F)) \cong \mathrm{Sp}(2r, F).$$

Proof. 1° We first show that

$$(6.79)\quad \begin{aligned}&\mathcal{G}(A \oplus [1] + \Lambda_{2r+1}(F))\\ &= \left\{ \begin{bmatrix} P_{11} & P_{12} \\ 0 & 1 \end{bmatrix} : P_{11} \in \mathrm{GL}(2r, F),\ P_{11}AP_{11}^T \equiv A + P_{12}P_{12}^T \pmod{\Lambda_{2r}(F)} \right\}.\end{aligned}$$

It is easy to see that the right side of (6.79) is contained in the left. Now assume

$$\begin{bmatrix} P_{11} & P_{12} \\ P_{21} & c \end{bmatrix} \in \mathcal{G}(A \oplus [1] + \Lambda_{2r+1}(F)),$$

where P_{11} is of size $2r \times 2r$. Then

$$(6.80)\quad \begin{aligned}\begin{bmatrix} A & \\ & 1 \end{bmatrix} &\equiv \begin{bmatrix} P_{11} & P_{12} \\ P_{21} & c \end{bmatrix} \begin{bmatrix} A & \\ & 1 \end{bmatrix} \begin{bmatrix} P_{11}^T & P_{21}^T \\ P_{12}^T & c \end{bmatrix} \pmod{\Lambda_{2r+1}(F)}\\ &= \begin{bmatrix} P_{11}AP_{11}^T + P_{12}P_{12}^T & P_{11}AP_{21}^T + cP_{12} \\ P_{21}AP_{11}^T + cP_{12}^T & P_{21}AP_{21}^T + c^2 \end{bmatrix}.\end{aligned}$$

From (6.80), we have $P_{21}AP_{11}^T = (P_{11}AP_{21}^T)^T$; hence $P_{21}(A + A^T)P_{11}^T = 0$. Since $P_{21}AP_{21}^T$ is of size 1×1, we also have $P_{21}(A + A^T)P_{21}^T = 0$. Since $\mathrm{rank}\big((A+A^T)[P_{11}^T\ P_{21}^T]\big) = 2r$, we have $P_{21} = 0$. By (6.80) again, $P_{11}AP_{11}^T + P_{12}P_{12}^T \equiv A \pmod{\Lambda_{2r}(F)}$, and $c^2 = 1$, i.e., $c = 1$. Thus we have proved that the left side of (6.79) is also contained in the right side.

2° In (6.79), note that the condition $P_{11}AP_{11}^T \equiv A + P_{12}P_{12}^T \pmod{\Lambda_{2r}(F)}$ implies that $P_{11} \in \mathcal{G}(A + A^T)$. For each $P_{11} \in \mathcal{G}(A + A^T)$, $A + P_{11}AP_{11}^T$ is symmetric. Since F is perfect, there is a unique $P_{12} \in \mathrm{M}_{2r\times 1}(F)$ such that $P_{12}P_{12}^T \equiv A + P_{11}AP_{11}^T \pmod{\Lambda_{2r}(F)}$. Define

$$\begin{aligned}\phi:\ \mathcal{G}(A + A^T) &\longrightarrow \mathcal{G}(A \oplus [1] + \Lambda_{2r+1}(F))\\ P_{11} &\longmapsto \begin{bmatrix} P_{11} & P_{12} \\ 0 & 1 \end{bmatrix},\end{aligned}$$

where $P_{12}P_{12}^T \equiv A + P_{11}AP_{11}^T \pmod{\Lambda_{2r}(F)}$. Then it is easy to verify that ϕ is an isomorphism; hence, $\mathcal{G}(A \oplus [1] + \Lambda_{2r+1}(F)) \cong \mathcal{G}(A + A^T) \cong \mathrm{Sp}(2r, F)$. □

Assume that q is even, and choose $a \in \mathbb{F}_q$ such that $\mathrm{Tr}_{q/2}(a) = 1$. By Corollary 6.48, the canonical forms in $\mathcal{Q}(\mathbb{F}_q^n)$ are

$$f_{n,r} = \mathtt{X}_1\mathtt{X}_2 + \mathtt{X}_3\mathtt{X}_4 + \cdots + \mathtt{X}_{2r-1}\mathtt{X}_{2r} + \mathtt{X}_{2r+1}^2, \quad 0 \le r \le (n-1)/2,$$
$$g_{n,r} = \mathtt{X}_1\mathtt{X}_2 + \mathtt{X}_3\mathtt{X}_4 + \cdots + \mathtt{X}_{2r-1}\mathtt{X}_{2r}, \quad 0 \le r \le n/2,$$
$$h_{n,r} = \mathtt{X}_1\mathtt{X}_2 + \mathtt{X}_3\mathtt{X}_4 + \cdots + \mathtt{X}_{2r-1}\mathtt{X}_{2r} + \mathtt{X}_{2r-1}^2 + a\mathtt{X}_{2r}^2, \quad 1 \le r \le n/2.$$

Lemma 6.67. *Let $f \in \mathcal{Q}(\mathbb{F}_q^n)$ be a nondegenerate quadratic form, and let $b \in \mathbb{F}_q$.*

(i) *If n is odd,*
$$|Z(f-b)| = q^{n-1}.$$

(ii) *If n is even and* $\mathrm{Arf}(f) = 0$,
$$|Z(f-b)| = \begin{cases} q^{n-1} + (q-1)q^{n/2-1} & \text{if } b = 0, \\ q^{n-1} - q^{n/2-1} & \text{if } b \neq 0. \end{cases}$$

(iii) *If n is even and* $\mathrm{Arf}(f) = 1$,
$$|Z(f-b)| = \begin{cases} q^{n-1} - (q-1)q^{n/2-1} & \text{if } b = 0, \\ q^{n-1} + q^{n/2-1} & \text{if } b \neq 0. \end{cases}$$

Proof. Using the above notation, we may assume that $f = f_{n,(n-1)/2}$ in (i), $f = g_{n,n/2}$ in (ii), and $f = h_{n,n/2}$ in (iii). Let $e(x) = (-1)^{\mathrm{Tr}_{q/2}(x)}$ for $x \in \mathbb{F}_q$.

(i) Let $n = 2r+1$. For each $(x_1, \dots, x_{2r}) \in \mathbb{F}_q^{2r}$, there is a unique $x_{2r+1} \in \mathbb{F}_q$ such that $f_{2r+1,r}(x_1, \dots, x_{2r+1}) - b = x_1x_2 + x_3x_4 + \cdots + x_{2r-1}x_{2r} + x_{2r+1}^2 - b = 0$. Hence $|Z(f_{2r+1,r} - b)| = q^{2r}$.

(ii) Let $n = 2r$. We have

$$\begin{aligned}
&|Z(g_{2r,r} - b)| \\
&= q^{-1} \sum_{y, x_1, \dots, x_{2r} \in \mathbb{F}_q} e\big(y(x_1x_2 + x_3x_4 + \cdots + x_{2r-1}x_{2r} - b)\big) \\
&= q^{2r-1} + q^{-1} \sum_{y \in \mathbb{F}_q^*,\, x_1, \dots, x_{2r} \in \mathbb{F}_q} e\big(y(x_1x_2 + x_3x_4 + \cdots + x_{2r-1}x_{2r} - b)\big) \\
&= q^{2r-1} + q^{-1} \sum_{y \in \mathbb{F}_q^*} \Big(\sum_{x_1, x_2 \in \mathbb{F}_q} e(yx_1x_2) \Big)^r e(-yb) \\
&= q^{2r-1} + q^{r-1} \sum_{y \in \mathbb{F}_q^*} e(-yb) \\
&= q^{2r-1} + q^{r-1} \cdot \begin{cases} q-1 & \text{if } b = 0, \\ -1 & \text{if } b \neq 0. \end{cases}
\end{aligned}$$

(iii) Let $n = 2r$. The computation of (ii) gives

$$|Z(h_{2r,r} - b)| = q^{2r-1} + q^{r-2} \sum_{y \in \mathbb{F}_q^*} e(-yb) \sum_{x_1, x_2 \in \mathbb{F}_q} e\big(y(x_1x_2 + x_1^2 + ax_2^2)\big).$$

From here we have

$$\begin{aligned}
&|Z(h_{2r,r} - b)| \\
&= q^{2r-1} + q^{r-2} \sum_{y \in \mathbb{F}_q^*} e(-y^2b) \sum_{x_1, x_2 \in \mathbb{F}_q} e\big(y^2(x_1x_2 + x_1^2 + ax_2^2)\big) \\
&= q^{2r-1} + q^{r-2} \sum_{y \in \mathbb{F}_q^*} e(-y^2b) \sum_{x_2 \in \mathbb{F}_q^*} e(y^2ax_2^2) \sum_{x_1 \in \mathbb{F}_q} e(y^2x_1x_2 + yx_1) \\
&= q^{2r-1} + q^{r-2} \sum_{y \in \mathbb{F}_q^*} e(-y^2b) \sum_{x_2 \in \mathbb{F}_q^*} e(y^2ax_2^2) \sum_{x_1 \in \mathbb{F}_q} e((yx_2 + 1)yx_1) \\
&= q^{2r-1} + q^{r-2} \sum_{y \in \mathbb{F}_q^*} e(-y^2b) \sum_{x_2 = y^{-1}} e(y^2ax_2^2)\, q \\
&= q^{2r-1} - q^{r-1} \sum_{y \in \mathbb{F}_q^*} e(-y^2b) \\
&= q^{2r-1} - q^{r-1} \cdot \begin{cases} q-1 & \text{if } b = 0, \\ -1 & \text{if } b \neq 0. \end{cases}
\end{aligned}$$

□

Theorem 6.68.

(i) *For* $0 \le r \le (n-1)/2$,

$$|\mathcal{G}(\mathbb{F}_q^n, f_{n,r})| = q^{(2r+1)(n-2r-1)+r^2} |\mathrm{GL}(n-2r-1, \mathbb{F}_q)| \prod_{i=1}^{r} (q^{2i} - 1).$$

(ii) *For* $0 \le r \le n/2$,

$$|\mathcal{G}(\mathbb{F}_q^n, g_{n,r})| = 2q^{r(2n-3r-1)}(q^r + 1)^{-1} |\mathrm{GL}(n-2r, \mathbb{F}_q)| \prod_{i=1}^{r} (q^{2i} - 1).$$

(iii) *For* $1 \le r \le n/2$,

$$|\mathcal{G}(\mathbb{F}_q^n, h_{n,r})| = 2q^{r(2n-3r-1)}(q^r + 1) |\mathrm{GL}(n-2r, \mathbb{F}_q)| \prod_{i=1}^{r-1} (q^{2i} - 1).$$

Proof. (i) Let

$$A = \underbrace{\begin{bmatrix} 0 & 1 \\ 0 & 0 \end{bmatrix} \oplus \cdots \oplus \begin{bmatrix} 0 & 1 \\ 0 & 0 \end{bmatrix}}_{r}.$$

Then by Lemmas 6.65 and 6.66,

$$\begin{aligned}|\mathcal{G}(\mathbb{F}_q^n, f_{n,r})| &= |\mathcal{G}(A \oplus [1] \oplus 0_{n-2r-1} + \Lambda_n(\mathbb{F}_q))| \\ &= q^{(2r+1)(n-2r-1)}|\mathrm{GL}(n-2r-1, \mathbb{F}_q)||\mathcal{G}(A \oplus [1] + \Lambda_{2r+1}(\mathbb{F}_q))| \\ &= q^{(2r+1)(n-2r-1)}|\mathrm{GL}(n-2r-1, \mathbb{F}_q)||\mathrm{Sp}(2r, \mathbb{F}_q)|.\end{aligned}$$

In the above, by Theorem 6.52,

$$|\mathrm{Sp}(2r, \mathbb{F}_q)| = |\mathcal{G}(K_{2r,r})| = q^{r^2}\prod_{i=1}^{r}(q^{2i}-1).$$

Combining the above equations produces the desired formula.

(ii) By Lemma 6.65,

$$|\mathcal{G}(\mathbb{F}_q^n, g_{n,r})| = q^{2r(n-2r)}|\mathrm{GL}(n-2r, \mathbb{F}_q)||\mathcal{G}(\mathbb{F}_q^{2r}, g_{2r,r})|,$$

so it suffices to determine $|\mathcal{G}(\mathbb{F}_q^{2r}, g_{2r,r})|$.

Call a basis $\epsilon_1, \dots, \epsilon_{2r}$ of $\mathbb{F}_q^{2r}$ *special* if $g_{2r,r}(\epsilon_1) = \cdots = g_{2r,r}(\epsilon_{2r}) = 0$ and

$$\begin{bmatrix}\epsilon_1 \\ \vdots \\ \epsilon_{2r}\end{bmatrix} K_{2r,r}[\epsilon_1^T \ \cdots \ \epsilon_{2r}^T] = K_{2r,r},$$

where $K_{2r,r}$ is defined in (6.70). (The standard basis of $\mathbb{F}_q^{2r}$ is special.) An element $\sigma \in \mathrm{Aut}_{\mathbb{F}_q}(\mathbb{F}_q^{2r})$ belongs to $\mathcal{G}(\mathbb{F}_q^{2r}, g_{2r,r})$ if and only if it maps the standard basis of $\mathbb{F}_q^{2r}$ to a special basis of $\mathbb{F}_q^{2r}$. Thus $|\mathcal{G}(\mathbb{F}_q^{2r}, g_{2r,r})|$ is the number of special bases of $\mathbb{F}_q^{2r}$. To count the number of special bases $\epsilon_1, \dots, \epsilon_{2r}$ $(r \geq 1)$, we first choose $\epsilon_1, \epsilon_2 \in \mathbb{F}_q^{2r}$ such that

$$(6.81) \qquad \begin{cases} g_{2r,r}(\epsilon_1) = g_{2r,r}(\epsilon_2) = 0, \\ \begin{bmatrix}\epsilon_1 \\ \epsilon_2\end{bmatrix} K_{2r,r}[\epsilon_1^T \ \epsilon_2^T] = \begin{bmatrix}0 & 1 \\ 1 & 0\end{bmatrix}. \end{cases}$$

The alternating form $l_{g_{2r,r}} \in \mathcal{A}(\mathbb{F}_q^{2r})$ is nondegenerate, and with respect to this alternating form, $\mathbb{F}_q^{2r} = \langle \epsilon_1, \epsilon_2 \rangle \mathbin{\bigcirc\!\!\!\!\perp} \langle \epsilon_1, \epsilon_2 \rangle^{\perp}$. The restriction $g_{2r,r}|_{\langle \epsilon_1, \epsilon_2 \rangle^{\perp}}$ is a nondegenerate quadratic form on $\langle \epsilon_1, \epsilon_2 \rangle^{\perp}$ with type 0 and Arf invariant 0. Therefore, the number of choices for $\epsilon_3, \dots, \epsilon_{2r}$ is $|\mathcal{G}(\mathbb{F}_q^{2(r-1)}, g_{2(r-1),r-1})|$. Thus,

$$(6.82) \qquad |\mathcal{G}(\mathbb{F}_q^{2r}, g_{2r,r})| = G_r\,|\mathcal{G}(\mathbb{F}_q^{2(r-1)}, g_{2(r-1),r-1})|,$$

where G_r is the number of $\epsilon_1, \epsilon_2 \in \mathbb{F}_q^{2r}$ satisfying (6.81). By Lemma 6.67, the number of $\epsilon_1 \in \mathbb{F}_q^{2r} \setminus \{0\}$ with $g_{2r,r}(\epsilon_1) = 0$ is $q^{2r-1} + (q-1)q^{r-1} - 1 = (q^r - 1)(q^{r-1} + 1)$. After ϵ_1 is chosen, we may assume that $\epsilon_1 = (1, 0, \dots, 0)$

since by Witt's extension theorem, $\mathcal{G}(\mathbb{F}_q^{2r}, g_{2r,r})$ acts transitively on $\{x \in \mathbb{F}_q^{2r} \setminus \{0\} : g_{2r,r}(x) = 0\}$. Let $\epsilon_2 = (x_1, \dots, x_n)$. Then (6.81) becomes

$$\begin{cases} x_2 = 1, \\ x_1 + x_3x_4 + x_5x_6 + \cdots + x_{2r-1}x_{2r} = 0. \end{cases}$$

The number of solutions $(x_1, \dots, x_{2r}) \in \mathbb{F}_q^{2r}$ of the above system is clearly q^{2r-2}. Hence

$$G_r = q^{2r-2}(q^r - 1)(q^{r-1} + 1). \tag{6.83}$$

By (6.82) and (6.83),

$$|\mathcal{G}(\mathbb{F}_q^{2r}, g_{2r,r})| = \prod_{i=1}^{r} q^{2i-2}(q^i - 1)(q^{i-1} + 1) = 2q^{r(r-1)}(q^r + 1)^{-1} \prod_{i=1}^{r}(q^{2i} - 1). \tag{6.84}$$

(iii) The proof of (iii) is similar to that of (ii). By Lemma 6.65,

$$|\mathcal{G}(\mathbb{F}_q^n, h_{n,r})| = q^{2r(n-2r)}|\mathrm{GL}(n - 2r, \mathbb{F}_q)||\mathcal{G}(\mathbb{F}_q^{2r}, h_{2r,r})|,$$

so it suffices to determine $|\mathcal{G}(\mathbb{F}_q^{2r}, h_{2r,r})|$.

Call a basis $\epsilon_1, \dots, \epsilon_{2r}$ of $\mathbb{F}_q^{2r}$ *special* if

$$(h_{2r,r}(\epsilon_1), \dots, h_{2r,r}(\epsilon_{2r})) = (0, \dots, 0, 1, a)$$

and

$$\begin{bmatrix} \epsilon_1 \\ \vdots \\ \epsilon_{2r} \end{bmatrix} K_{2r,r}[\epsilon_1^T \; \dots \; \epsilon_{2r}^T] = K_{2r,r}.$$

Then $|\mathcal{G}(\mathbb{F}_q^n, h_{2r,r})|$ is the number of special bases of $\mathbb{F}_q^{2r}$. To count the number of special bases $\epsilon_1, \dots, \epsilon_{2r}$, we first choose $\epsilon_{2r-1}, \epsilon_{2r} \in \mathbb{F}_q^{2r}$ such that

$$\begin{cases} h_{2r,r}(\epsilon_{2r-1}) = 1, \\ h_{2r,r}(\epsilon_{2r}) = a, \\ \begin{bmatrix} \epsilon_{2r-1} \\ \epsilon_{2r} \end{bmatrix} K_{2r,r}[\epsilon_{2r-1}^T \; \epsilon_{2r}^T] = \begin{bmatrix} 0 & 1 \\ 1 & 0 \end{bmatrix}. \end{cases} \tag{6.85}$$

With respect to the nondegenerate alternating form $l_{h_{2r,r}} \in \mathcal{A}(\mathbb{F}_q^{2r})$,

$$\mathbb{F}_q^{2r} = \langle \epsilon_{2r-1}, \epsilon_{2r} \rangle \oplus \langle \epsilon_{2r-1}, \epsilon_{2r} \rangle^{\perp}.$$

The restriction $h_{2r,r}|_{\langle \epsilon_{2r-1}, \epsilon_{2r} \rangle^{\perp}}$ is a nondegenerate quadratic form on $\langle \epsilon_{2r-1}, \epsilon_{2r} \rangle^{\perp}$ with type 0 and Arf invariant 0. Therefore, the number of choices for $\epsilon_1, \dots, \epsilon_{2r-2}$ is $|\mathcal{G}(\mathbb{F}_q^{2(r-1)}, g_{2(r-1),r-1})|$, and hence

$$|\mathcal{G}(\mathbb{F}_q^{2r}, h_{2r,r})| = H_r \, |\mathcal{G}(\mathbb{F}_q^{2(r-1)}, g_{2(r-1),r-1})|, \tag{6.86}$$

where H_r is the number of $\epsilon_{2r-1}, \epsilon_{2r} \in \mathbb{F}_q^{2r}$ satisfying (6.85). By Lemma 6.67, the number of $\epsilon_1 \in \mathbb{F}_q^{2r}$ such that $h_{2r,r}(\epsilon_{2r-1}) = 1$ is $q^{2r-1}+q^{r-1} = q^{r-1}(q^r + 1)$. After ϵ_{2r-1} is chosen, we may assume that $\epsilon_{2r-1} = (0,\dots,0,1,0)$ since by Witt's extension theorem, $\mathcal{G}(\mathbb{F}_q^{2r}, h_{2r,r})$ acts transitively on $\{x \in \mathbb{F}_q^{2r} : h_{2r,r}(x) = 1\}$. Let $\epsilon_{2r} = (x_1,\dots,x_{2r})$. Then (6.85) becomes

$$\begin{cases} x_{2r} = 1, \\ x_1x_2 + x_3x_4 + \cdots + x_{2r-3}x_{2r-2} + x_{2r-1} + x_{2r-1}^2 = 0. \end{cases}$$

Let $f = g_{2(r-1),r-1} = \mathtt{X}_1\mathtt{X}_2 + \mathtt{X}_3\mathtt{X}_4 + \cdots + \mathtt{X}_{2r-3}\mathtt{X}_{2r-2}$. By Lemma 6.67, the number of solutions $(x_1,\dots,x_{2r}) \in \mathbb{F}_q^{2r}$ of the above system is

$$\begin{aligned} &2|Z(f)| + \sum_{x_{2r-1}\in\mathbb{F}_q\setminus\mathbb{F}_2} |Z(f + x_{2r-1} + x_{2r-1}^2)| \\ &= 2(q^{2r-3} + (q-1)q^{r-2}) + (q-2)(q^{2r-3} - q^{r-2}) = q^{r-1}(q^{r-1}+1). \end{aligned}$$

Hence

$$H_r = q^{2(r-1)}(q^r+1)(q^{r-1}+1). \tag{6.87}$$

By (6.86), (6.87) and (6.84),

$$\begin{aligned} &|\mathcal{G}(\mathbb{F}_q^{2r}, h_{2r,r})| \\ &= q^{2(r-1)}(q^r+1)(q^{r-1}+1)\cdot 2q^{(r-1)(r-2)}(q^{r-1}+1)^{-1}\prod_{i=1}^{r-1}(q^{2i}-1) \\ &= 2q^{r(r-1)}(q^r+1)\prod_{i=1}^{r-1}(q^{2i}-1). \end{aligned}$$

□

Remark. For even q, $\mathcal{G}(\mathbb{F}_q^{2r}, g_{2r,r})$ and $\mathcal{G}(\mathbb{F}_q^{2r}, h_{2r,r})$ are called *orthogonal groups* of degree $2r$ over $\mathbb{F}_q$ and are denoted by $\mathrm{O}_+(2r,\mathbb{F}_q)$ and $\mathrm{O}_-(2r,\mathbb{F}_q)$, respectively. The group $\mathcal{G}(\mathbb{F}_q^{2r+1}, f_{2r+1,r})$ ($\cong \mathrm{Sp}(2r,\mathbb{F}_q)$) belongs to the family of symplectic groups.

Exercises

6.1. (i) Let $T_{\mathrm{L}}^*(n,\mathbb{F}_q)$ (*resp.*, $T_{\mathrm{U}}^*(n,\mathbb{F}_q)$) $< \mathrm{GL}(n,\mathbb{F}_q)$ be the subgroup of lower (resp., upper) triangular matrices with 1's on the diagonal. Then both $T_{\mathrm{L}}^*(n,\mathbb{F}_q)$ and $T_{\mathrm{U}}^*(n,\mathbb{F}_q)$ are Sylow p-subgroups of $\mathrm{GL}(n,\mathbb{F}_q)$, where $p = \operatorname{char}\mathbb{F}_q$. By Sylow's theorem, there exists $P \in \mathrm{GL}(n,\mathbb{F}_q)$ such that $P^{-1}T_{\mathrm{L}}^*(n,\mathbb{F}_q)P = T_{\mathrm{U}}^*(n,\mathbb{F}_q)$. Find such a matrix P.

(ii) When $n \geq 3$, prove that there does not exist $P \in \mathrm{GL}(n,\mathbb{F}_q)$ such that $P^{-1}AP = A^T$ for all $A \in T_{\mathrm{L}}^*(n,\mathbb{F}_q)$.

6.2. Let F be a field. Prove that all transvections in $\mathrm{SL}(2,F)$ are conjugate if and only if every element of F is a square.

6.3. A k-dimensional $\mathbb{F}_q$-subspace C of $\mathbb{F}_q^n$ is called an $[n,k]_q$ *linear code*. The dual of C is defined to be $C^\perp = \{x \in \mathbb{F}_q^n : xy^T = 0 \text{ for all } y \in C\}$. Prove that $\dim_{\mathbb{F}_q} C + \dim_{\mathbb{F}_q} C^\perp = n$. If C is the row space of

$$\begin{bmatrix} I_k & A \end{bmatrix} \in \mathrm{M}_{k\times n}(\mathbb{F}_q),$$

prove that $C^\perp$ is the row space of

$$\begin{bmatrix} -A^T & I_{n-k} \end{bmatrix} \in \mathrm{M}_{(n-k)\times n}(\mathbb{F}_q).$$

6.4. An element $M \in \mathrm{GL}(n,\mathbb{F}_q)$ is called a *monomial matrix* if each row and each column of M contains exactly one nonzero entry. The group of all monomial matrices in $\mathrm{GL}(n,\mathbb{F}_q)$ is denoted by $\mathcal{M}(n,\mathbb{F}_q)$. ($\mathcal{M}(n,\mathbb{F}_q)$ is the subgroup of $\mathrm{GL}(n,\mathbb{F}_q)$ generated by all permutation matrices and invertible diagonal matrices.)

(i) Prove that

$$\begin{array}{rccc} \alpha: & \mathcal{M}(n,\mathbb{F}_q) & \longrightarrow & \mathcal{M}(n,\mathbb{F}_q) \\ & M & \longmapsto & (M^{-1})^T \end{array}$$

is an isomorphism.

(ii) Let C be an $[n,k]_q$ linear code. The *monomial automorphism group* of C is defined as $\mathrm{MAut}(C) = \{M \in \mathcal{M}(n,\mathbb{F}_q) : CM = C\}$, where $CM = \{xM : x \in C\}$. Prove that

$$\mathrm{MAut}(C^\perp) = \alpha(\mathrm{MAut}(C)).$$

6.5. Enumerate the canonical forms of the elements of $\mathrm{GL}(3,\mathbb{F}_q)$ under conjugation, and use Theorem 6.14 to compute the cardinality of each corresponding conjugacy class of $\mathrm{GL}(3,\mathbb{F}_q)$.

Exercises 6.6–6.8 provide another proof of Theorem 6.14.

6.6. Let A be a finitely generated torsion module over a PID R such that all elementary divisors of A are powers of a fixed irreducible element $\rho \in R$. Let $\sigma \in \mathrm{End}_R(A)$. Then $\sigma \in \mathrm{Aut}_R(A)$ if and only if the induced map $\bar\sigma : A/\rho A \to A/\rho A$ is an isomorphism.

6.7. Let R be a PID and let A be a finitely generated torsion module over R with elementary divisors

$$\underbrace{\rho_1^1,\dots,\rho_1^1}_{\mu_1^{(1)}},\ \dots,\ \underbrace{\rho_1^{s_1},\dots,\rho_1^{s_1}}_{\mu_{s_1}^{(1)}},$$

$$\cdots$$

$$\underbrace{\rho_t^1, \ldots, \rho_t^1}_{\mu_1^{(t)}}, \ \ldots, \ \underbrace{\rho_t^{s_t}, \ldots, \rho_t^{s_t}}_{\mu_{s_t}^{(t)}},$$

where $\rho_1, \ldots, \rho_t \in R$ are distinct irreducible elements.

(i) Prove that there is a ring isomorphism

$$\operatorname{End}_R(A) \cong \prod_{k=1}^{t} \operatorname{End}_R(A_k),$$

where

$$A_k = \bigoplus_{i=1}^{s_k} B_k^{(i)}, \quad B_k^{(i)} = \big(R/(\rho_k^i)\big)^{\mu_i^{(k)}}.$$

(ii) Prove that the mapping

$$\begin{array}{rccc} \phi: & \operatorname{End}_R(A_k) & \longrightarrow & \displaystyle\prod_{1\le i,j\le s_k} \operatorname{Hom}_R(B_k^{(i)}, B_k^{(j)}) \\ & \sigma & \longmapsto & (\pi_j \circ \sigma \circ \iota_i)_{1\le i,j\le s_k} \end{array}$$

is an R-isomorphism, where $\iota_i : B_k^{(i)} \to B_k^{(1)} \oplus \cdots \oplus B_k^{(s_k)}$ is the inclusion and $\pi_j : B_k^{(1)} \oplus \cdots \oplus B_k^{(s_k)} \to B_k^{(j)}$ is the projection. Moreover, for $\sigma \in \operatorname{End}_R(A_k)$, $\sigma \in \operatorname{Aut}_R(A_k)$ if and only if $\pi_i \circ \sigma \circ \iota_i \in \operatorname{Aut}_R(B_k^{(i)})$ for all $1 \le i \le s_k$.

(iii) Prove that for $1 \le i, j \le s_k$, there are R-isomorphisms

$$\operatorname{Hom}_R\big(R/(\rho_k^i),\, R/(\rho_k^j)\big) \cong R/(\rho_k^{\min(i,j)})$$

and

$$\operatorname{Hom}_R(B_k^{(i)}, B_k^{(j)}) \cong \big(R/(\rho_k^{\min(i,j)})\big)^{\mu_i^{(k)}\mu_j^{(k)}}.$$

(iv) For $\sigma \in \operatorname{End}_R(B_k^{(i)})$, prove that $\sigma \in \operatorname{Aut}_R(B_k^{(i)})$ if and only if the induced map $\bar\sigma \in \operatorname{End}_{R/(\rho_k)}\big(B_k^{(i)}/\rho_k B_k^{(i)}\big)$ belongs to $\operatorname{Aut}_{R/(\rho_k)}\big(B_k^{(i)}/\rho_k B_k^{(i)}\big)$.

6.8. Let $A \in \mathrm{M}_n(F)$, where F is a field. The F-vector space F^n is an $F[\mathtt{X}]$-module with the scalar multiplication defined as

$$fa = f(A)a, \quad f \in F[\mathtt{X}],\ a \in F^n.$$

Denote this $F[\mathtt{X}]$-module by V_A.

(i) Prove that V_A is a finitely generated torsion module over the PID $F[\mathtt{X}]$. Moreover, the elementary divisors of V_A are precisely those of A.

(ii) Let

$$\begin{array}{rccc} \phi: & \mathrm{M}_n(F) & \longrightarrow & \mathrm{End}_F(V_A) \\ & B & \longmapsto & \phi(B) \end{array}$$

be the ring isomorphism, where

$$\begin{array}{rccc} \phi(B): & V_A & \longrightarrow & V_A \\ & a & \longmapsto & Ba. \end{array}$$

Prove that

$$\phi(\mathrm{C}_{\mathrm{M}_n(F)}(A)) = \mathrm{End}_{F[\mathtt{X}]}(V_A),$$
$$\phi(\mathrm{C}_{\mathrm{GL}(n,F)}(A)) = \mathrm{Aut}_{F[\mathtt{X}]}(V_A).$$

(iii) Assume that $F = \mathbb{F}_q$. Let the elementary divisors of A be

$$\underbrace{f_1^1, \dots, f_1^1}_{\mu_1^{(1)}}, \quad \dots, \quad \underbrace{f_1^{s_1}, \dots, f_1^{s_1}}_{\mu_{s_1}^{(1)}},$$
$$\dots$$
$$\underbrace{f_t^1, \dots, f_t^1}_{\mu_1^{(t)}}, \quad \dots, \quad \underbrace{f_t^{s_t}, \dots, f_t^{s_t}}_{\mu_{s_t}^{(t)}},$$

where $f_1, \dots, f_t \in \mathbb{F}_q[\mathtt{X}]$ are distinct monic irreducible polynomials with $\deg f_k = d_k$. Prove that

$$|\mathrm{C}_{\mathrm{GL}(n,\mathbb{F}_q)}(A)| = |\mathrm{Aut}_{\mathbb{F}_q[\mathtt{X}]}(V_A)|$$
$$= q^{\sum_{k=1}^t d_k \sum_{1\le i,j\le s_k} \min(i,j)\mu_i^{(k)}\mu_j^{(k)}} \prod_{k=1}^{t}\prod_{i=1}^{s_k}\prod_{u=1}^{\mu_i^{(k)}} (1 - q^{-d_k u}).$$

6.9. Let F be a field and $P \in \mathrm{Sp}(2r, F)$. Prove that $\det P = 1$.

6.10. Let q be even. Recall that the canonical forms in $\mathcal{Q}(\mathbb{F}_q^n)$ are

$$f_{n,r} = \mathtt{X}_1\mathtt{X}_2 + \mathtt{X}_3\mathtt{X}_4 + \cdots + \mathtt{X}_{2r-1}\mathtt{X}_{2r} + \mathtt{X}_{2r+1}^2, \quad 0 \le r \le (n-1)/2,$$
$$g_{n,r} = \mathtt{X}_1\mathtt{X}_2 + \mathtt{X}_3\mathtt{X}_4 + \cdots + \mathtt{X}_{2r-1}\mathtt{X}_{2r}, \quad 0 \le r \le n/2,$$
$$h_{n,r} = \mathtt{X}_1\mathtt{X}_2 + \mathtt{X}_3\mathtt{X}_4 + \cdots + \mathtt{X}_{2r-1}\mathtt{X}_{2r} + \mathtt{X}_{2r-1}^2 + a\mathtt{X}_{2r}^2, \quad 1 \le r \le n/2,$$

where $a \in \mathbb{F}_q$ is such that $\mathrm{Tr}_{q/2}(a) = 1$. Let $e(x) = (-1)^{\mathrm{Tr}_{q/2}(x)}$, $x \in \mathbb{F}_q$. Prove that

$$\sum_{x\in\mathbb{F}_q^n} e\big(f_{n,r}(x)\big) = 0,$$
$$\sum_{x\in\mathbb{F}_q^n} e\big(g_{n,r}(x)\big) = q^{n-r},$$

$$\sum_{x\in\mathbb{F}_q^n} e\big(h_{n,r}(x)\big) = -q^{n-r}.$$

6.11. Let $q = p^e$, where p is an odd prime, and let $f \in \mathcal{Q}(\mathbb{F}_q^n)$ be a quadratic form with rank r and discriminant $d(f) = d\,\mathbb{F}_q^{*2}$. Prove that

$$\sum_{x\in\mathbb{F}_q^n} e(f(x)) = \eta(d)g(\eta)^r q^{n-r},$$

where $e(\ ) = \zeta_p^{\mathrm{Tr}_{q/p}(\ )}$ ($\zeta_p = e^{2\pi i/p}$), η is the quadratic character of $\mathbb{F}_q$ and $g(\ )$ is the Gauss sum. Conclude that $\operatorname{rank} f$ and $d(f)$ are determined by the sum $\sum_{x\in\mathbb{F}_q^n} e(f(x))$.

6.12. Let $f \in \mathcal{Q}(F^n)$. Prove that $\operatorname{rank} f$ is the smallest integer $m \geq 0$ such that f is equivalent to some $g \in \mathcal{Q}(F^m)$.

6.13. Let $q = 2^e$, and let $H = \left[\begin{smallmatrix} 1 & 1 \\ 0 & a \end{smallmatrix}\right] \in \mathrm{M}_2(\mathbb{F}_q)$, where $a \in \mathbb{F}_q$ is such that $\mathrm{Tr}_{q/2}(a) = 1$. For $A, B \in \mathrm{M}_n(\mathbb{F}_q)$, write $A \approx B$ if there exists $P \in \mathrm{GL}(n, \mathbb{F}_q)$ such that $A \equiv PBP^T \pmod{\Lambda_n(\mathbb{F}_q)}$. Let

$$A_n = \begin{bmatrix} 0 & 1 & \cdots & 1 \\ & \ddots & \ddots & \vdots \\ & & \ddots & 1 \\ & & & 0 \end{bmatrix}, \quad B_n = \begin{bmatrix} 1 & 1 & \cdots & 1 \\ & \ddots & \ddots & \vdots \\ & & \ddots & 1 \\ & & & 1 \end{bmatrix} \in \mathrm{M}_n(\mathbb{F}_q).$$

(i) Prove the following statements using the algorithm in the proof of Theorem 6.46:

$$\begin{aligned} &A_n \approx A_2 \oplus B_{n-2}, \quad n \geq 2,\\ &B_n \approx A_2 \oplus B_2 \oplus B_{n-4}, \quad n \geq 4,\\ &B_2 \oplus B_2 \approx A_2 \oplus A_2,\\ &B_3 \approx B_2 \oplus [0],\\ &B_2 \oplus [1] \approx A_2 \oplus [1],\\ &B_2 \approx \begin{cases} A_2 & \text{if } e \text{ is even},\\ H & \text{if } e \text{ is odd}. \end{cases} \end{aligned}$$

(ii) Define $A^{\oplus k} = \underbrace{A \oplus \cdots \oplus A}_{k}$. Moreover, if $A \approx B \oplus 0$, we also write $A \approx B$. Use (i) to verify Tables 6.3 and 6.4 about the canonical forms of A_n and B_n in $\mathrm{M}_n(\mathbb{F}_q)$ under $\approx$.

6.14. Determine the canonical forms of the following quadratic forms in $\mathcal{Q}(\mathbb{F}_q^n)$. For even q, use Exercise 6.13 when applicable.

Table 6.3. Canonical forms of A_n under $\approx$

n	canonical form of A_n	
$8k$ or $8k+1$	$A_2^{\oplus 4k}$	
$8k+2$	$A_2^{\oplus(4k+1)}$	
$8k+3$	$A_2^{\oplus(4k+1)} \oplus [1]$	
$8k+4$ or $8k+5$	$A_2^{\oplus(4k+2)}$	if e is even
	$A_2^{\oplus(4k+1)} \oplus H$	if e is odd
$8k+6$	$A_2^{\oplus(4k+3)}$	if e is even
	$A_2^{\oplus(4k+2)} \oplus H$	if e is odd
$8k+7$	$A_2^{\oplus(4k+3)} \oplus [1]$	

Table 6.4. Canonical forms of B_n under $\approx$

n	canonical form of B_n	
$8k$	$A_2^{\oplus 4k}$	
$8k+1$	$A_2^{\oplus 4k} \oplus [1]$	
$8k+2$ or $8k+3$	$A_2^{\oplus(4k+1)}$	if e is even
	$A_2^{\oplus 4k} \oplus H$	if e is odd
$8k+4$	$A_2^{\oplus(4k+2)}$	if e is even
	$A_2^{\oplus(4k+1)} \oplus H$	if e is odd
$8k+5$	$A_2^{\oplus(4k+2)} \oplus [1]$	
$8k+6$ or $8k+7$	$A_2^{\oplus(4k+3)}$	

(i) $\sum_{1\le i,j\le n} \mathtt{X}_i\mathtt{X}_j$.
(ii) $\sum_{1\le i\le j\le n} \mathtt{X}_i\mathtt{X}_j$.
(iii) $\sum_{1\le i<j\le n} \mathtt{X}_i\mathtt{X}_j$.

6.15. Determine the canonical form of the quadratic form $\mathtt{X}_1\mathtt{X}_2 + \mathtt{X}_3\mathtt{X}_4 + \cdots + \mathtt{X}_{n-1}\mathtt{X}_n + \mathtt{X}_n\mathtt{X}_1 \in \mathcal{Q}(\mathbb{F}_q^n)$ $(n \ge 2)$. Consider the even and odd q's separately.

Bibliography

[1] E. Artin, *Geometric Algebra* (reprint of the 1957 original), John Wiley & Sons, New York, 1988.

[2] J. Ax, *Zeros of polynomials over finite fields*, Amer. J. Math. **86** (1964), 255–261.

[3] E. A. Bender and J. R. Goldman, *On the applications of Möbius inversion in combinatorial analysis*, Amer. Math. Monthly **82** (1975), 789–803.

[4] E. R. Berlekamp, *Algebraic Coding Theory*, McGraw-Hill, New York, 1968.

[5] B. C. Berndt and R. J. Evans, *The determination of Gauss sums*, Bull. Amer. Math. Soc. (NS) **5** (1981), 107–129.

[6] A. Cafure and G. Matera, *Improved explicit estimates on the number of solutions of equations over a finite field*, Finite Fields Appl. **12** (2006), 155–185.

[7] C. Chevalley, *Démonstration d'une hypothèse de M. Artin*, Abh. Math. Sem. Univ. Hamburg **11** (1936), 73–75.

[8] C. Chevalley, *Introduction to the Theory of Algebraic Functions of One Variable*, Mathematical Surveys, Vol. 6. American Mathematical Society, Providence, RI, 1951.

[9] P. Delsarte, J. M. Goethals, F. J. MacWilliams, *On generalized Reed-Muller codes and their relatives*, Inform. Control **16** (1970), 403–442.

[10] W. Fulton, *Algebraic Curves: An Introduction to Algebraic Geometry*, Addison-Wesley Publishing Company, Advanced Book Program, Redwood City, CA, 1989.

[11] C. F. Gauss, *Summatio quarumdam serierum singularium*, Comment. Soc. Reg. Sci. Gottingensis **1** (1811); *Werke*, vol. 2, pp. 11–45, Königl. Gesellschaft der Wissenschaften, Göttingen, 1876; *Untersuchungen über Höhere Arithmetik* (H. Maser ed.), pp. 463–495, Springer, Berlin, 1889.

[12] D. Gorenstein, *Finite Simple Groups: An Introduction to Their Classification*, Plenum Press, New York, 1982.

[13] D. Gorenstein, R. Lyons, R. Solomon. *The Classification of the Finite Simple Groups*, American Mathematical Society, Providence, RI, 1994.

[14] L. C. Grove, *Classical Groups and Geometric Algebra*, Graduate Studies in Mathematics, Vol. 39, American Mathematical Society, Providence, RI, 2002.

[15] X. Hou, *Solution to a problem of S. Payne*, Proc. Amer. Math. Soc. **132** (2004), 1–6.

[16] X. Hou, *A note on the proof of a theorem of Katz*, Finite Fields Appl. **11** (2005) 316–319.

[17] X. Hou, *Permutation polynomials over finite fields — a survey of recent advances*, Finite Fields Appl. **32** (2015), 82–119.

[18] T. W. Hungerford, *Algebra*, Springer-Verlag, New York–Berlin, 1980.

[19] A. Hurwitz, *Mathematische Werke*, Band II, Birkhäuser Verlag, Basel–Stuttgart, 1963.

[20] K. Ireland and M. Rosen, *A Classical Introduction to Modern Number Theory*, Springer-Verlag, New York, 1990.

[21] N. Jacobson, *Basic Algebra I*, Freeman, New York, 1985.

[22] N. M. Katz, *On a theorem of Ax*, Amer. J. Math. **93** (1971), 485–499.

[23] S. Lang, *Algebraic Number Theory*, Springer-Verlag, New York, 1994.

[24] S. Lang, *Algebra*, Springer-Verlag, New York, 2002.

[25] S. Lang and A. Weil, *Number of points of varieties in finite fields*, Amer. J. Math. **76** (1954), 819–827.

[26] E. Leducq, *A new proof of Delsarte, Goethals and MacWilliams theorem on minimal weight codewords of generalized Reed-Muller codes*, Finite Fields Appl. **18** (2012), 581–586.

[27] R. Lidl and H. Niederreiter, *Finite Fields*, Cambridge University Press, Cambridge, 1997.

[28] G. L. Mullen and D. Panario (eds), *Handbook of Finite Fields*, Discrete Mathematics and Its Applications, CRC Press, Boca Raton, FL, 2013.

[29] S. E. Payne, *A complete determination of translation ovoids in finite Desarguian planes*, Atti Accad. Naz. Lincei Rend. Cl. Sci. Fis. Mat Natur. (8) **51** (1971), 328–331.

[30] D. Robinson, *A Course in the Theory of Groups*, Springer, New York, 1995.

[31] J-P. Serre, *A Course in Arithmetic*, Springer, New York, 1973.

[32] W. M. Schmidt, *Equations over Finite Fields, An Elementary Approach*, Springer-Verlag, Berlin–Heidelberg–New York, 1976.

[33] C. L. Siegel, *Über das quadratische Reziprozitätsgesetz in algebraischen Zahlkörpern*, Nachr. Akad. Wiss. Göttingen Math.-Phys. Kl. II (1960), 1–16.

[34] S. A. Stepanov, *Arithmetic of Algebraic Curves*, Plenum Publishing, New York, 1994.

[35] H. Stichtenoth, *Algebraic Function Fields and Codes*, Springer, Berlin, 1993.

[36] B. L. Van der Waerden, *Algebra II*, Springer-Verlag, Berlin, 1959.

[37] D. Wan, *An elementary proof of a theorem of Katz*, Amer. J. Math. **111** (1989), 1–8.

[38] D. Wan, *A Chevalley-Warning approach to p-adic estimates of character sums*, Proc. Amer. Math. Soc. **123** (1995), 45–54.

[39] E. Warning, *Bemerkung zur vorstehenden Arbeit von Herrn Chevalley*, Abh. Math. Sem. Univ. Hamburg **11** (1936), 76–83.

[40] L. C. Washington, *Introduction to Cyclotomic Fields*, Springer-Verlag, New York, 1997.

[41] A. Weil, *Sur les Courbes Algébriques et les Variétés qui s'en Déduisent*, Hermann et Cie., Paris, 1948.

List of Notation

$\mathbb{N}$	set of natural numbers $0, 1, 2, \ldots$
$\mathbb{Z}$	set of integers
$\mathbb{Z}^+$	set of positive integers
$\mathbb{Q}$, $\mathbb{R}$, $\mathbb{C}$	fields of rational, real, and complex numbers
$\mathbb{F}_q$	finite field with q elements
$[K : F]$	degree of field extension
$\overline{F}$	algebraic closure of F
F^*	multiplicative group of field F
$R^\times$	multiplicative group of ring R
$\mathtt{t}, \mathtt{T}, \mathtt{X}, \mathtt{Y}, \mathtt{Z}$	indeterminates
$\gcd(f, g)$	greatest common divisor of f and g
$\operatorname{lcm}(f, g)$	least common multiple of f and g
$\mathrm{M}_{m\times n}(R)$	set of $m \times n$ matrices over R
$F[\mathtt{X}]$	polynomial ring
$F(\mathtt{X})$	field of rational functions
$\|X\|$	cardinality of X
$\emptyset$	empty set
$\cong$	isomorphism, equivalence
$\subset$	subset
$\subsetneq$	proper subset
$\operatorname{char} F$	characteristic of F, 1
(f)	ideal generated by f, 3
$o(\alpha)$	order of α, 5
$\langle\ \rangle$	cyclic group generated by an element, 5
$\operatorname{Aut}(K/F)$	Galois group of K over F, 5
$\operatorname{Tr}_{\mathbb{F}_{q^n}/\mathbb{F}_q}$, $\operatorname{Tr}_{q^n/q}$	trace from $\mathbb{F}_{q^n}$ to $\mathbb{F}_q$, 7

$\mathfrak{S}_n$, $\mathfrak{S}_X$	symmetric group, 143
E_{ij}	143
$A \oplus B$	block sum of matrices, 143
$\mathrm{GL}(n, F)$	general linear group, 144
$\mathrm{AGL}(n, F)$	affine linear group, 144
$\mathrm{SL}(n, F)$	special linear group, 144
$\mathrm{PGL}(n, F)$	projective general linear group, 145
$\mathrm{PSL}(n, F)$	projective special linear group, 145
$T(n, F)$	set of invertible lower triangular matrices, 146
$T^*(n, F)$	set of lower unitriangular matrices, 146
$M(f)$	companion matrix of f, 148
$\sigma \overset{A}{\sim} \tau$	160
$\boxplus$	161
N_n, J_n	163
$\lambda \vdash n$	λ is a partition of n, 168
σ_λ, $\sigma_{\lambda,t}$	168
$\mathcal{S}(V)$, $\mathcal{A}(V)$, $\mathcal{S}^\sigma(V)$	175
$\ker l$	kernel of sesquilinear form l, 175
$A(l, \mathcal{E})$	matrix of l with respect to basis $\mathcal{E}$, 175
$\mathrm{Aut}_F(V)$	automorphism group of F-vector space V, 176
$S_n(F)$, $\Lambda_n(F)$, $S_n^\sigma(F)$	176
$\mathrm{diag}\,(a_1, \ldots, a_n)$	diagonal matrix, 177
l_f	symmetric bilinear form of quadratic form f, 181
$\mathcal{Q}(V)$	set of quadratic forms on V, 181
$\ker f$	kernel of quadratic form f, 182
$\mathrm{type}\, f$	type of quadratic form f, 183
$\mathrm{Arf}(f)$	Arf invariant, 187
$\mathcal{G}(V, l)$	group of sesquilinear space (V, l), 193
$\mathcal{G}(V, f)$	group of quadratic space (V, f), 193
$\mathcal{G}(A, \sigma)$, $\mathcal{G}(A)$	group of matrix A, 193
$U \oplus W$	194
$K_{n,r}$	196
$\mathrm{Sp}(2r, F)$	symplectic group, 197
$\mathrm{U}(n, \mathbb{F}_{q^2})$	unitary group over $\mathbb{F}_{q^2}$, 198
$\mathrm{O}(2k+1, \mathbb{F}_q)$	orthogonal group over $\mathbb{F}_q$, q odd, 202
$\mathrm{O}_\pm(2k, \mathbb{F}_q)$	orthogonal group over $\mathbb{F}_q$, q odd, 202
$\mathrm{O}_\pm(2r, \mathbb{F}_q)$	orthogonal group over $\mathbb{F}_q$, q even, 215
$\mathrm{End}_R(A)$	endomorphism ring of R-module A, 216

Index

SELECTED PUBLISHED TITLES IN THIS SERIES

190 **Xiang-dong Hou,** Lectures on Finite Fields, 2018

187 **John Douglas Moore,** Introduction to Global Analysis, 2017

186 **Bjorn Poonen,** Rational Points on Varieties, 2017

185 **Douglas J. LaFountain and William W. Menasco,** Braid Foliations in Low-Dimensional Topology, 2017

184 **Harm Derksen and Jerzy Weyman,** An Introduction to Quiver Representations, 2017

183 **Timothy J. Ford,** Separable Algebras, 2017

182 **Guido Schneider and Hannes Uecker,** Nonlinear PDEs, 2017

181 **Giovanni Leoni,** A First Course in Sobolev Spaces, Second Edition, 2017

180 **Joseph J. Rotman,** Advanced Modern Algebra: Third Edition, Part 2, 2017

179 **Henri Cohen and Fredrik Strömberg,** Modular Forms, 2017

178 **Jeanne N. Clelland,** From Frenet to Cartan: The Method of Moving Frames, 2017

177 **Jacques Sauloy,** Differential Galois Theory through Riemann-Hilbert Correspondence, 2016

176 **Adam Clay and Dale Rolfsen,** Ordered Groups and Topology, 2016

175 **Thomas A. Ivey and Joseph M. Landsberg,** Cartan for Beginners: Differential Geometry via Moving Frames and Exterior Differential Systems, Second Edition, 2016

174 **Alexander Kirillov Jr.,** Quiver Representations and Quiver Varieties, 2016

173 **Lan Wen,** Differentiable Dynamical Systems, 2016

172 **Jinho Baik, Percy Deift, and Toufic Suidan,** Combinatorics and Random Matrix Theory, 2016

171 **Qing Han,** Nonlinear Elliptic Equations of the Second Order, 2016

170 **Donald Yau,** Colored Operads, 2016

169 **András Vasy,** Partial Differential Equations, 2015

168 **Michael Aizenman and Simone Warzel,** Random Operators, 2015

167 **John C. Neu,** Singular Perturbation in the Physical Sciences, 2015

166 **Alberto Torchinsky,** Problems in Real and Functional Analysis, 2015

165 **Joseph J. Rotman,** Advanced Modern Algebra: Third Edition, Part 1, 2015

164 **Terence Tao,** Expansion in Finite Simple Groups of Lie Type, 2015

163 **Gérald Tenenbaum,** Introduction to Analytic and Probabilistic Number Theory, Third Edition, 2015

162 **Firas Rassoul-Agha and Timo Seppäläinen,** A Course on Large Deviations with an Introduction to Gibbs Measures, 2015

161 **Diane Maclagan and Bernd Sturmfels,** Introduction to Tropical Geometry, 2015

160 **Marius Overholt,** A Course in Analytic Number Theory, 2014

159 **John R. Faulkner,** The Role of Nonassociative Algebra in Projective Geometry, 2014

158 **Fritz Colonius and Wolfgang Kliemann,** Dynamical Systems and Linear Algebra, 2014

157 **Gerald Teschl,** Mathematical Methods in Quantum Mechanics: With Applications to Schrödinger Operators, Second Edition, 2014

156 **Markus Haase,** Functional Analysis, 2014

155 **Emmanuel Kowalski,** An Introduction to the Representation Theory of Groups, 2014

154 **Wilhelm Schlag,** A Course in Complex Analysis and Riemann Surfaces, 2014

153 **Terence Tao,** Hilbert's Fifth Problem and Related Topics, 2014

152 **Gábor Székelyhidi,** An Introduction to Extremal Kähler Metrics, 2014

151 **Jennifer Schultens,** Introduction to 3-Manifolds, 2014

150 **Joe Diestel and Angela Spalsbury,** The Joys of Haar Measure, 2013

149 **Daniel W. Stroock,** Mathematics of Probability, 2013

148 **Luis Barreira and Yakov Pesin,** Introduction to Smooth Ergodic Theory, 2013

147 **Xingzhi Zhan,** Matrix Theory, 2013